MANAGEMENT OF URBAN BIODEGRADABLE WASTES

Collection, occupational health,
biological treatment, product quality
criteria and end user demand

Edited by
Jens Aage Hansen
Aalborg University

Published by
James & James (Science Publishers) Ltd, London
for
International Solid Waste Association
ISWA, Copenhagen

© 1996 ISWA, Bremerholm 1, 1069 Copenhagen K, Denmark

Published by James & James (Science Publishers) Ltd
Waterside House, 47 Kentish Town Road, London NW1 8NX, UK

All rights reserved. No part of this book may be reproduced in any form or by any means electronic or mechanical, including photocopying, recording or by any information storage and retrieval system without permission in writing from the copyright holder and publisher

A catalogue record for this book is available from the British Library

ISBN 1-873936 58 3

Printed in Great Britain by
Antony Rowe Ltd, Chippenham, Wiltshire

| Preface | V |
| Acknowledgement | VI |

Section 1: Existing and Emerging Legislation and Regulation

Conceptual Approach in Legislation and Regulation Concerning Organic Wastes from Cities... 1

Regulations in the Field of Biowaste Collection and Treatment in Austria... 9

Agreements Concerning the Application of Sewage Sludge and Household Compost in Farmland Areas ... 17

Section 2: Local Composting & Collection

Local Composting in Multi-Storied Housing... 25

Separate Collection of Biowaste from Households - Report on a Large-Scale Pilot Project in the City of Copenhagen ... 35

Biowaste Collection Project in the Central City of Vienna... 42

The Installation of Biobin in Salzburg, Austria ... 54

Section 3: Systems, Cases & Trends

Systems Analysis of Organic Waste... 65

Implementation of Source Separation of Municipal Solid Waste in the City of Kristiansand... 76

Trends in European Management of Urban Organic Wastes ... 82

Section 4: Occupational Health

Emission of Spores of *Aspergillus Fumigatus* from Waste Containers and Piles... 91

Occupational Bioaerosol Exposure in Collecting Household Waste... 98

Work Related Symptoms and Metal Concentration in Danish Resource Recovery Workers .. 106

Occupational Health Problems in Waste Recycling... 115

Section 5: Anaerobic Processing

Anaerobic Digestion Processes for the Energetic Valorisation of the Green Fraction of Municipal Solid Waste and the Recovery of Volatile Fatty Acids... 125

Codigestion of Manure with Organic Toxic Waste in Biogas Reactors ... 132

A Mathematical Model for Dynamic Simulation of Anaerobic Digestion of Complex Substrates: Focusing on Codigestion of Manure with Lipid Containing Substrates... 142

Innovative Pretreatment to Optimize Anaerobic Conversion of Biowaste... 153

Wet Mechanical Pretreatment of Organic Wastes - the Full Scale Experience ... 162

Anaerobic Fermentation of Biowaste at High Total Solids Content - Experiences with ATF-System.. 172

The Paques Anaerobic Digestion Process: A Feasible and Flexible Treatment for Solid Organic Waste... 181

Anaerobic Digestion of Biowaste in Full-Scale Plants in Brecht, Belgium and Salzburg, Austria by Means of the Dranco Process 193

Biological Waste Treatment through Biogas Digesters in Rural Nepal........................ 201

Section 6: Biodegradability

Delignification of Wheat Straw by Wet Oxidation Resulting in Bioconvertible Cellulose and Hemicellulose... 209

C/N Ratio Effect on Degradation of Cellulose in Composting of Food Waste and Paper...... 218

Options for a Common Treatment of Biodegradable Plastics and Biowaste.................... 228

Section 7: Aerobic & Other Processing

A Steady-State Model of the Compodan Composting Process... 237

Accelerated Composting in Tunnels 245

ATAD - an Effective Technology for Stabilisation and Disinfection of Biosolids 253

New Concepts for Biowaste Composting - Dutch Current Practice 262

Operational Aspects of Aerobic and Anaerobic Treatment of Biowaste in The Netherlands... 272

Section 8: Product Quality, Marketing & End User Demands

Reduction of Pathogenic Bacteria and Viruses by Anaerobic Digestion........................ 281

Use of Sewage Sludge and Composted Household Waste as a Fertiliser Source............... 287

Biowaste Composting - Constraints and Advantages 295

Application of Compost in Amenity Areas... 305

Organic Waste Recycling: A New Market with the Farmers.. 317

Consumer Demands 323

Section 9: Status & Opinion

Urban Biodegradable Wastes - Status and Opinion.. 331

Author Index 343

Preface

Increasing environmental concern, resistance to the 'produce, use and throw-away' pattern of life, and landfill capacity still more limited by both public opinion and physical availability are some of the reasons for seeking new ways of reducing and managing wastes.

Biological treatment of urban biodegradable wastes is one option for routing some of the organic wastes away from the landfill and back to the soil from where much of it originates as plant biomass. In this context plant biomass is a primary resource and urban biodegradables become secondary organic resources. After adequate collection, treatment and transportation these secondary organic resources can be reused and recycled. Including aspects such as legislation, regulation, health of workers, technology and biology of processes, end user demands, and product quality and marketing, this book deals with recycling and management of urban biodegradables.

The book is a result of a process that started with papers presented at the BIOWASTE '95 conference in Aalborg, Denmark in May 1995. BIOWASTE is a very comprehensive term. For practical reasons focus is, therefore, on the urban biodegradables from households, trade, light industry, gardens and parks. Sludge from wastewater treatment is not included, except where co-treated or controlled by guidelines similar to other wastes. Hazardous wastes are included only to the extent that they appear in e.g. household wastes. And agricultural wastes, even though they are by volume, weight and total flow greater the urban organics, are not included except where they are used for relevant comparisons, e.g. regarding quality criteria.

Each manuscript has been reviewed by at least two peers. Subsequently, authors have been asked to revise their manuscripts according to specific requests from the editor, based on the advice from the peers and with emphasis on presentation of concepts, results and discussion. Given the fact that most authors use English only as their second language, the intention of the editor has been to obtain a clear and meaningful text, but not necessarily a scholarly English.

Biological treatment of urban organic wastes is on the increase in Europe. Of the total capacity for biological treatment installed over the last 20 years (since 1975) as much as 80% was installed in the period 1991-1995. Impressive as this rise may be, the total capacity installed in Europe in 1995 is most likely still only around 5 million tons, equivalent to approximately 4% of the household waste production; or 10% of the organic wastes from households, readily available for biological treatment and recycling. On this background, the book deals with an issue with a high potential for improved environmental management of biodegrable wastes. And an area of increased activity in the next decade in research, development and management.

The Editor

Acknowledgement

The source of information for this book is the material presented at the BIOWASTE '95 conference May 21-24, 1995 in Aalborg.

The conference was organised jointly by ISWA, The International Solid Waste Association; DAKOFA, The Danish National Solid Waste Association, and Aalborg University. Aalborg City sponsored the conference and economic risks were guaranteed by BATES A/S, a collection equipment and systems manufacturer located in Aalborg.

A Programme Committee with members from the ISWA working Group on Biological Waste Treatment and the corresponding Work Committee of DAKOFA provided valuable assistance in selecting authors for the conference and hence this book. I served as chair of that committee.

During the conference, a number of reporters chosen from these committees assisted in extracting key issues from the presentation of papers and the following discussions.

I wish to express my gratitude to these persons and mentioned institutions. Their assistance and enthusiasm has made my work possible, interesting and challenging.

Jens Aage Hansen
Editor and chair of programme committee

Section 1

Existing and Emerging Legislation and Regulation

Conceptual Approach in Legislation and Regulation Concerning Organic Wastes from Cities

Werner Schenkel, Head of Department, Federal Environmental Agency, Bismarckplatz 1, 14191 Berlin, Germany

ABSTRACT

The composting of organic household waste has seen a considerable increase over the last years. The root of this trend is the target to recycle waste such as kitchen and garden wastes.

The author gives an overview about the present administrative framework like the German Closed Substance Cycle and Waste Management Act and the different statutory ordinances at Federal and Länder-level and the regulations on the communal levels. One of the restrictive regulations is the Soil Protection Act. The concept of this act pursues the aim of preservation the multifunctional use of soil.

A very important question is the quality of the compost. The author suggests ecolabeling and other quality standards.

KEYWORDS

Quality of compost, regulations and ordinances, centralised and decentralised composting, licensing of plants.

INTRODUCTION

In Germany, composting of organic household waste such as kitchen waste and garden green has seen a considerable increase over the last five years. Whilst prior to 1985, only 250,000 tonnes of compost per year were produced in some 12 composting plants, approx. 1.1 million tonnes of compost were produced in 99 plants in 1993. An additional 28 plants were under construction that year, and another 61 plants were approved. It is expected that an additional 1.6 million tonnes of compost per year will be produced when these plants have been implemented. The total biogenic potential is estimated at 10 million tonnes final product per year.

This trend is a result of the debate on strategies to reduce waste quantities, whereby one came to deal intensively with the fraction which accounts for 50% of the weight and 30% of the volume. If one were able to utilise that fraction of organic waste, there would be the chance decisively to reduce the amount of remaining residual waste.

Composting of biogenic household and garden waste and also the anaerobic digestion is regulated by numerous ordinances and laws. The regulations are graded according to the general administrative structure that exists in Germany: federal, state, and local. The regulations lay down recycling obligations as well as provisions concerning the reduction of pollutants, marketing and joint fertiliser application.

STATUTORY REGULATIONS GOVERNING RECYCLING

Regulations at federal level

Waste management has progressed considerably in Germany since 1986. In 1994, the Closed Substance Cycle and Waste Management Act was passed. Its provisions will not become effective until 1996. The new Act has adopted the EU's terminology. This means that those in possession of organic waste must first examine whether the waste is recyclable and, if so, actually recycle it. The holder of the waste has to perform this examination on his own responsibility. Only if the waste originates from households are the regional administrative bodies responsible, as before. The legislators' intention with this provision was to make it clear that waste management is a part of industrial undertakings and must be the responsibility of industry rather than the responsibility of the public sector. Today, the focus is more on savings in raw materials and waste prevention than on aspects such as hygiene and averting danger, which in the past were the dominant aspects.

Some basic principles and obligations on the part of producers and holders of waste and the parties responsible for waste management apply. Waste must primarily be avoided and secondarily be subjected to materials recycling or energy recovery.

A measure is considered a materials-recycling measure if, in keeping with an economic perspective and taking into account the impurities contained in the waste, its main purpose is to utilise the waste, and not to eliminate the pollutant potential.

Producers and holders of waste are obliged to recycle the waste. Where no other consequences result from the Act, waste recycling has priority over waste disposal. High-quality recycling appropriate to the type and nature of the waste concerned is to be pursued. Waste for recovery shall be separated for storage and treatment. Recycling takes place safely if, given the waste's nature, the amount of impurities it contains and the type of recycling concerned, no impairment of the public interest is expected and, in particular, if no increase of pollutant concentration occurs within the closed substance cycle. The obligation to recycle waste is to be met, to the extent this is technically possible and economically reasonable, especially when a market exists or can be created for a recovered material. Waste recycling is economically reasonable if the costs it entails are not disproportionate in comparison with the costs waste disposal would entail. Composts can be used both in agriculture and in landscape gardening, e. g. in the recultivation of devastated land pursuant to Article 7.2 of the Closed Substance Cycle and Waste Management Act.

The Federal Ministry for the Environment, Nature Conservation and Nuclear Safety, acting in agreement with the Federal Ministry of Food, Agriculture and Forestry and the Federal Ministry for Health, is authorised to mandate by statutory ordinance, with the consent of the Bundesrat, requirements for the agricultural sector to ensure proper and safe recycling.

If waste for recovery as secondary raw-material fertiliser or farm fertiliser is applied to land used for agricultural, silvicultural or horticultural purposes, then the following in particular can be mandated for such supply and application, with regard to the relevant pollutants:

- prohibitions or restrictions depending on factors such as type and nature of the soil, application site and time and natural site conditions, and
- studies of the relevant waste, farm, fertiliser or soil, measures for pretreatment of the substances, or suitable other measures. This applies to farm fertiliser to the extent that the usual quantities employed in good and proper practices are exceeded. The prerequisite for waste utilisation is the benefit to be derived from it. Waste used as fertiliser must promote plant growth, increase the yield or improve quality (Article 1 of the Fertiliser Act).

The *Länder* (federal states) prepare management plans for these respective areas, in keeping with supraregional perspectives. The waste management plans include descriptions of the following:

- the aims of waste avoidance and recovery
- the waste management facilities required to ensure domestic waste management
- authorised waste management facilities.

The description of requirements must take into account future developments expected to occur over a period of at least ten years. Waste management planning must take into account the aims and requirements of regional and *Land* planning.

Regulations at state level

In accordance with the German Constitution, the regulations enacted by the federal states serve to implement prescriptions made in federal legislation. For the subject dealt with here, the federal states have regulated details in their own waste legislation or in special ordinances. The matters regulated are, essentially, what waste is to be kept separate, how this should be done, what quality standards have to be met and how compliance with them has to be monitored. The first federal state to require the separate collection of biowaste throughout its territory was Hesse. A distinction is made between green waste, i.e. grass, hedge and tree cuttings, wreaths and flowers, and biocompost, i.e. organic waste from kitchens and storage. The currently most stringent regulation regulating quality and application quantities, calculated as loads at an annual application quantity of 10 t of dry matter per hectare, has been issued by Baden-Württemberg. These regulations will remain in effect until a compost ordinance has been issued at the federal level. Baden-Württemberg's regulation includes stringent orientative values; non-compliance with these values does not automatically lead to a ban on the sale of such composts.

Important to be mentioned in this context are the Technical Instructions on Waste from Human Settlements [3]. It lays down requirements for the quality of wastes to be subjected to pretreatment and the methods to be used for their disposal. The aim is to prevent the generation of contaminated waste disposal sites, wherever possible. At present, the required quality of residual waste destined for landfill can only be achieved if the waste is treated thermally. As alternatives, biological methods for the treatment of residual waste are being discussed. These are methods for the treatment of biogenic waste, but the product of such treatment is not fit for use because its pollutant content is too high.

Regulations at the local level

Centralised composting

Nearly 30 - 50% of waste materials are biodegradable. The successful experience with glass and paper recycling as well as the active participation of individual households has prompted administrations to separate and reutilise this fraction as well.

Early experience from the seventies showed that the composting of unsorted municipal waste cannot produce high-quality compost. High quality can only be achieved with a high-quality input. A prerequisite for this is a system for the separate collection of waste. Households have to separate biogenic material in 'biobins' or 'green bins'. The collected material is then treated in central plants. The system needs a lot of continuous information and extension. The households must be convinced about the positive effects to be gained from biological waste management.

Decentralised composting

In addition to the biobin collection system, organised collection of green waste is being practised in nearly every town to dispose of the high amounts of seasonal garden wastes. This is organised as a street collection of as a bring system, which means the wastes are taken to a central plant. Nearly every larger town organises its own composting system or hires a private company. There are nearly 800 composting plants for horticultural waste.

Another form of decentralised composting takes place at the private level. This form of composting ranks as waste avoidance activity, because the waste is not collected and treated by the community. In order to help people to practice private composting many towns make a lot of efforts to provide information and counselling to motivate people. Some of them pay subsidies to garden owners for the purchase of composting vessels.

All this can be regulated by way of local ordinances. Upon submission by the competent administrative department, the local parliament passes the respective regulation within the framework of the rules prescribed in *Land* legislation. Therefore, it is not surprising that chosen solutions are highly distinctive and differ from town to town and district to district.

Regulations governing quality

The quality of compost is characterised by its constituents, notably the nutrients and pollutants it contains.

The requirements to be met by compost constituents are determined by the uses, e.g. agriculture or landscape gardening.

A federal Soil Protection Act is being prepared in Germany. The purpose of the Act is to preserve the multifunctional uses of soil. The input of pollutants due to treated waste should not exceed the amount withdrawn by plants. Very sensitive soil can be protected that way.

The only federal regulation governing the application of organic waste is the Sewage Sludge Ordinance, which has been issued under waste legislation. It allows 5 t of sewage sludge per hectare to be applied every three years. In the case of composts, the application of such an amount has no nutritive or structural effects on soil. Composts also differ from sewage sludge in terms of pollutant content. Composts usually have a much lower content of pollutants and nutrients than does sewage sludge. The usefulness of composts depends on how they are used. When used permanently in agriculture, e. g. at 20 t dry matter/3 years, their value lies mainly in the nutrients they supply. An increase of 0.1% in organic matter in soil takes place in this way. One-time application of a large amount of compost can also be useful for improving soil structure.

The federal states have an instruction sheet which regulates the application of composts to farmland [4]. First issued several years ago, it is now being revised. It is expected to prescribe the following limits:

Table 1: Maximum-permissible concentrations of heavy metals in biocomposts used for recultivation measures(in mg/kg) (related to 30% dry organic matter in compost)

	Composts	Sewage sludge
Lead	150	900
Cadmium	1.5	10
Chromium	100	900
Copper	100	800
Nickel	50	200
Mercury	1	8
Zinc	400	2,500

Should the Federal Soil Protection Act and the regulations to be issued on the basis of it in fact be passed, they could also have an impact on the utilisation of waste. In any soil uses, the natural functions of the soil, notably its filter and storage functions, have to be safeguarded for the long term. To this end, work is being done to derive precautionary values. As this work has not yet been completed, the 50 percentiles of the background concentrations in agricultural soils are used as a makeshift. As an example, the following table shows such values for the lignite mining region in central Germany.

Table 2: 50 percentiles of background concentrations in agricultural soils (in mg/kg dry matter)

Central Germany (Saxony)	Lead	Cadmium	Chromium	Copper	Nickel	Mercury	Zinc
Sand	22	0.1	12	8	5	0.08	32
Loess	30	0.2	20	14	12	0.1	60
Gneiss	75	0.5	28	21	15	0.1	131
Argillaceous slate	50	0.3	27	28	27	0.2	125

There are some sets of criteria or quality standards whose development resulted from pragmatic considerations, i.e. the user's problems in selling their products. Two quality standards are currently being applied:

- the German ecolabel, the 'Blue Angel' [5]
- the requirements of the *Bundesqualitätsgemeinschaft Kompost* (federal association on compost quality).

In addition to these regulations, a Fertiliser Application Ordinance is currently being prepared at federal level, which will set limits, in t/ha, for the application of the fertiliser substances nitrogen and phosphorus. With respect to the limitation of nutrients, the Fertiliser Application Ordinance's impact on the competing sectors sewage sludge, manure and compost application will largely be neutral. The limitation of application quantities means that, in future, a larger area of land will be needed to accommodate the same amount of material. The competition between organic fertilisers will become more severe. This trend will be accelerated by the set aside of farmland that is being promoted under the EU's agricultural policy and by the trend to reduce the use of additional amounts of fertilisers in ecologically oriented farming.

The consequences of this limitation will make it necessary to step up counselling and promotion activities to prevent compost sales from slumping. The market needed to accommodate the compost must be created. This is an extremely difficult and ambitious task if compost quantities are to be increased at the same time.

The focus should be on the needs of the consumers, and not those of the producers. Food shoppers expect cereal products to be healthy, hygienic, tasty, fresh and produced authentically. Compost is not authentic; it is, among other things, subject to the suspicion of contamination.

LICENSING OF COMPOSTING PLANTS

As of 1993, the licensing of composting plants has been subject to the provisions of the Federal Emissions Control Act. They have superseded the provisions of the Waste Avoidance and Waste Management Act and of the ordinances issued on the basis of it.

The way composting plants are licensed depends on the annual treatment capacity. A differentiation is made between plants with a treatment capacity of

< 0.75 t/h
≥ 0.75 t/h - 10 t/h
≥ 10 t/h.

Steps had to be taken to reduce pollution in the form of noise, odour and pathogens, etc., inside and in the vicinity of the plants as well as pollution caused by water seeping through windrows.

The majority of plants are open air windrow systems. Some of them are roofed over. Most of these plants are so-called decentralised plants, which work on a low productivity level. The main point of criticism concerns the fact that only centralised plants with high input capacities can meet high environmental standards. This includes the requirement that the systems must be fully enclosed. Because of odour, dust and possibly disease-causing pathogens, emission control is becoming more and more indispensable. This will force the use of enclosed systems.

CONCLUSIONS

Composting of wastes like kitchen wastes and green wastes is increasing. The composting plays an important role in the recycling strategy in Germany. It exist a very detailed framework of regulation for licensing, quality standards and soil protection.

REFERENCES

1. Kreislaufwirtschafts- und Abfallgesetz (KreW-/AbfG). Gesetz zur Vermeidung, Bewertung und Beseitigung von Abfällen (Act for promoting closed substance cycles and ensuring environmentally compatible waste disposal - Waste Avoidance, Recycling and Disposal Act of 27 September 1994.) BGBl. Part 1, No. 66, 6 October 1994.
2. Composting Decree of the Federal State Baden-Württemberg of 30 June 1994 (reference no.: 48-8981.31/264). To be obtained from: Umweltministerium Baden-Württemberg, Postfach 10 34 39, 70029 Stuttgart.
3. Dritte Allgemeine Verwaltungsvorschriftr zum Abfallgesetz (TA Siedlungsabfall). Technische Anleitung zur Verwertung, Behandlung und sonstigen Entsorgung von Siedlungsabfällen. (Third General Administrative Regulation on the Waste Avoidance and Waste Management Act (Technical Instructions on Waste from Human Settlements) of 14 May 1993. Technical Instructions on the Recycling, Treatment and Other Management of Wastes from Human Settlements. BAnz., p. 4967 and Supplement).
4. Merkblatt M 10: Qualitätskriterien und Anwendungsempfehlungen für Kompost. (Instruction sheet M 10: Quality criteria and recommendations for the application of compost). Status: 15 February 1995. LAGA-Heft No. 21, Erich Schmidt Verlag GmbH & Co., Genthiner Str. 30 g, 10785 Berlin.

5. Bodenverbesserungsmittel/Bodenhilfsstoffe aus Kompost (soil amelioration agents/soil auxiliaries made of compost), RAL-UZ 45, July 1994 version, effective since 1 January 1995.
6. Kompost, Gütesicherung (compost quality assurance). RAL-GZ 21, status: 2/1992. To be obtained from: RAL Deutsches Institut für Gütesicherung und Kennzeichnung e. V., Siegburger Str. 39, 53757 Sankt Augustin. Tel.: (02241) 160 50; fax: (02241) 160511.
7. Gesetz zum Schutz vor schädlichen Umweltauswirkungen durch Luftverunrei- nigungen, Geräusche, Erschütterungen und ähnliche Vorgänge (Bundes-Immissionsschutzgesetz - BImSchG) (Act on the prevention of harmful effects on the environment caused by air pollution, noise, vibration and similar phenomena - Federal Immission Control Act) of 15 April 1974, as amended and promulgated on 14 May 1990 (BGBl. Part I, p. 880), last amended on 22 April 1993 (BGBl. Part I, p. 466).
8. Gesetz über die Vermeidung und Entsorgung von Abfällen (Abfallgesetz - AbfG) (Waste Management and Waste Avoidance Act) of 27 August 1986 (BGBl. Part I, p. 1410, corr. p. 1501), last amended on 22 April 1993 (BGBl. Part I, p. 466).

Regulations in the Field of Biowaste Collection and Treatment in Austria

Franz Mochty, Bundesministerium für Umwelt, Stubenbastei 5, 1010 Wien, Austria

ABSTRACT

Austria has introduced an ordinance on the nation-wide separate collection of biowaste. Due to the complex situation arising from the state organisation and respective competencies of federal and provincial authorities on one hand and due to the necessity for the various ministries involved to amend existing ordinances or create new ones in the field of environmental protection, the situation of composting even composting of separately collected biowaste is not satisfactory.

In order to guarantee the recycling of biowaste in the future an additional study on the quality of compost actually produced is carried out by the Ministry for the Environment in co-operation with the Ministry for Agriculture and Forestry and the Ministry for Health. Based on the results of this study the content of the legislative measures (pollutants to be controlled, limit values, control mechanism, labelling requirements for biowaste compost, etc.) will be specified.

The paper describes in detail the reasons for that situation to occur and the new approach to tackle the problem.

KEYWORDS

Biowaste, separate collection, soil improver, use of compost, selling of compost, Fertiliser Act, competencies, legislative measurements

COMPETENT AUTHORITIES

The division of the legislative competencies between the federal and the provincial authorities is characteristic to the federation principle embodied in the Austrian Constitution. Environmental protection represents a so-called 'crossover matter' which means that a number of regulatory aspects are assigned to different competencies of the Republic and of the provinces.

With regard to the collection, treatment and use of biowaste legislative and executive competencies are regulated in the following way:

Collection of biowaste

In general the provinces are responsible for the collection of non hazardous waste. However, if there is a need for an uniform regulation on a national level, the federal authorities are enabled to enact also regulations in this field. In order to ensure an Austrian wide collection with a high quality, the Federal Ministry for the Environment has enacted an ordinance on the separate collection of biowaste [1].

Treatment of biowaste

The provinces are responsible for the treatment of non hazardous waste. For the establishment and the operation of composting as well as biogas plants several regulations on national and provincial level apply (e.g. Waste Management Acts [2-3], Water Act [4], Industrial Act [5])

Application to agricultural land

The Federal Constitutional Law on Comprehensive Environmental Protection embodies the provinces to enact regulations for the application of compost to agricultural land. Most of the provinces have enacted Soil Protection Acts [6,7] which contain already general principles for the application of compost. Quality requirements and maximum quantities of sewage sludge and partly of municipal solid waste compost as well as limit values for the agricultural soil to which these materials may be applied are laid down by the way of ordinances. Examples of limit values of the heavy metal content in sewage sludge applicable to agricultural land according to the ordinances of two provinces are shown in table 1.

Detailed requirements for the use of composted biowaste are still in elaboration. These considerations are based on the Austrian standard S 2200 'Quality Requirements for Biowaste- Compost' [12]. In this standard a high quality class (class I) and a standard quality class (class II) of compost is defined (see table 2). The limit values for heavy metals in sewage sludge and the guide values for compost are not to be compared directly. On the one hand the guide values for compost are related to 30% ignition loss and the references for the limit values for sewage sludge differ in the regulations of the provinces (values are related to 'dry matter' or to 'inorganic dry matter', that means 0% ignition loss). On the other hand there are different regulations in the provinces limiting the maximum amount of sewage sludge applicable to agricultural land. Due to the lower concentrations of heavy metals in biowaste-compost the recommended amount of compost within the Austrian standard S 2200 is higher than these limit values.

In addition more detailed guidelines (depending on the field of application) for the proper use of biowaste-compost are finalised within the next months (Austrian standard S 2202 'Guidelines for Application of Compost'). Both Austrian standards are not legally binding for the user of compost.

Although the soil protection is regulated by the provinces, the Republic is responsible for the legislative and executive competence of the water protection. According to the Water Act [4] the amount of nitrogen and therefore also the maximum amount of compost applied to agricultural area is restricted. The above mentioned standards have already taken into account this limitation.

Use of compost in the production of fertilisers, soil improvers or growing medias

The republic is responsible for the legislative competence concerning the selling of fertilisers, soil improvers or growing medias. An Fertiliser Act [8] and an ordinance on fertilisers [10] were enacted by the Ministry for Agriculture and Forestry in 1994.

Table 1: Examples for limit values of the heavy metal content in sewage sludge applicable to agricultural land

Heavy metals in mg/kg dry matter; organic pollutants	Upper Austria related to dry matter	Lower Austria Class III [1] related to inorganic dry matter	Lower Austria Class II related to inorganic dry matter
Pb	400	400	100
Cd	5	8	2
Cr	400	500	50
Co		100	10
Cu	400	500	300
Ni	80	100	25
Hg	7	8	2
Zn	1,600	2,000	1,500
PCDD/PCDF	100 ng TE/kg dry matter		
PCB	each component 0.2 mg/kg dry matter		
AOX	500 mg/kg dry matter		

[1] The use of sewage sludge of this class is forbidden after a transition period of 10 years

As long as compost of separately collected biowaste is regarded as waste, the waste management regulations apply also for the selling of such biowaste compost.

COLLECTION AND TREATMENT OF BIOWASTE

As of January 1, 1995 the Austrian ordinance on the separate collection and treatment of certain organic waste is in force [1]. This ordinance requires the setting up of a nation-wide collection and recycling system. It has a twofold objective: On the one hand it aims at reducing the amount of household waste and the negative effects on the environment in case of bio-degradation of landfilled household waste; on the other hand it intends to improve the material cycles by enhancing the production of high quality compost [9].

Moreover, the ordinance specifies the kind of organic matter that has to be either collected separately or composted by the waste owner himself. The separate collection comprises:

- organic waste from the garden;
- kitchen waste, except meat residues which may only be collected if the treatment facility is suited for this kind of waste;
- plant residues from agriculture and the processing of agricultural products;

- paper used for the collection of biowaste is not prohibited.

It is of great importance to restrict the materials to be collected in order to guarantee a high quality of the end product of the composting process. Several studies have shown good results (e.g. [9]).

At the federal level the treatment procedure (aerobic or anaerobic treatment) is not regulated. The ordinance, therefore, can be understood as a framework model for more detailed ordinances to be worked out by the provinces in accordance with their local and regional circumstances. Several provinces have already enacted the obligation on the separate collection and treatment of biowaste within their territory, before the ordinance was set into force at federal level. In Fig. 1 the development of the separate collection of kitchen waste and non bulky organic waste from gardens via biowaste containers during the last seven years is shown (see [11]). In addition to the 29 kg per inhabitant of biowaste arising from this collection, 13 kg per inhabitant of bulky organic waste from gardens were collected for composting in 1994. In that year the total amount of source separated biowaste from households, which was brought to composting plants, was 335.000 tons. Not included in this sum is the amount of organic matter which is composted in the household area by the waste owner himself (according to the Austrian Waste Management Act this material is not regarded as waste).

The installation of treatment facilities is regulated by a number of rules at federal and provincial levels, such as the Water Act, the Industrial Act, the Waste Management Acts of the Provinces, Nature Protection Acts of the Provinces, etc.).

THE DISTRIBUTION, LABELLING AND USE OF COMPOST

In 1994 the Austrian Act on Fertilisers [8] was revised by the Ministry for Agriculture and Forestry. This Act regulates only the selling (including import and transport) of fertilisers (including organic fertilisers), soil improvers and growing media. The ordinance on fertilisers [10] precises the general requirements in this subject, defines types of fertilisers and lays down the quality which has to be achieved and the labelling of the products.

According to the competencies described above the use of these products is covered by different regulations such as Water Act and Soil Protection Acts.

Compost regarded as waste

Not within the scope of the Fertiliser Act falls the selling of waste like sewage sludge, composted sewage sludge and municipal solid waste compost (compost of household waste). According to the opinion of the Ministry for Agriculture and Forestry the term 'municipal solid waste compost' comprises also compost of separately collected biowaste.

The use of these materials is regulated primarily by the Soil Protection Acts of the provinces and ordinances based on these regulations. However, not all of the provinces have already enacted special regulations in this field (see 'application to the agriculture' above).

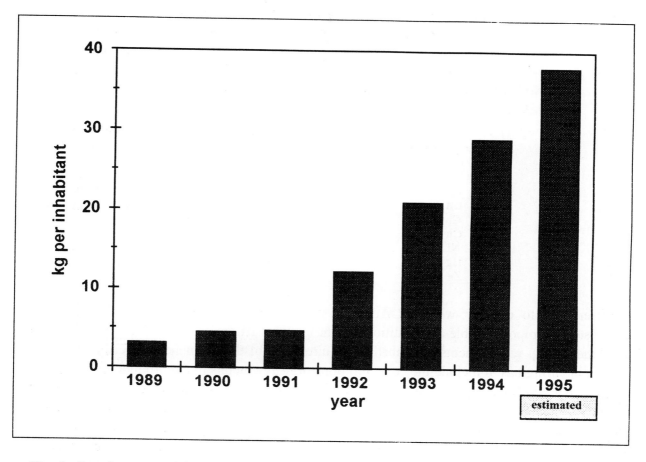

Fig. 1: Development of the collection of kitchen waste and non bulky garden waste from households related to the Austrian population

Compost used for the production of fertilisers

Based on the worries of the representatives of the Ministry for Agriculture and Forestry on the possible bad quality of compost, the selling of fertilisers containing compost of separately collected biowaste has been prohibited by the Fertiliser Act. Representatives of all of the Austrian Provincial Governments and the Minister for the Environment requested that the use of biowaste-compost (and of uncontaminated sewage sludge) for this purpose should not be banned categorically but there should be defined quality requirements.

As a result of this discussion, an escape clause was introduced into the Fertiliser Act enabling the Ministry for Agriculture and Forestry to concede an exception for uncontaminated compost of natural origin (and of uncontaminated sewage sludge) by way of an ordinance. In the case that this ordinance is being finalised and put into force, compost of defined quality and origin may be used for the production of organic fertilisers but not as part of a growing media or as soil improver. The compost will have to meet the requirements established in this ordinance **and** the quality requirements for compost laid down in the ordinance on fertilisers. According to the ordinance on fertilisers, a compost used for the production of organic

Table 2: Guide values for heavy metals in compost according to the Austrian standards 2200: 'quality criteria for compost of biowaste' and to the ordinance on fertilisers

Heavy metals mg/kg dry matter related to 30% of ignition loss	Austrian standard S 2200		Ordinance on Fertilisers
	Class I	Class II	
Pb	70	150	150
Cd	0,7	1	1
Cr	70	70	100
Cu	70	100	100
Ni	42	60	60
Hg	0,7	1	1
Zn	210	400	300

fertilisers has to comply with the Austrian standard S 2200 'Quality Requirements for Biowaste - Compost' (table 2). Additionally the product (the organic fertiliser) will have to fulfil the quality requirements and labelling requirements of the ordinance on fertilisers (table 2). The labelling has to include the concentrations of heavy metals in the product and contain a recommendation for the proper use of the product. If the amount of heavy metals derived from the concentration as well as the recommended amount per area exceeds the very stringent limit values (table 3), the selling of this compost as organic fertiliser is prohibited. Yet, according to the opinion of the Austrian Ministry for Environment, this concept is unreasonable and is likely to interfere with the competence's of the provinces that encompass the protection of the soil including the use of compost and sewage sludge.

Table 3: Ordinance on fertilisers - tolerable amount of heavy metals in compost sold as organic fertiliser, derived from the concentration of heavy metals and the recommended amount of compost per area

Heavy metals g/ha (=10.000m^2) in a period of 2 years	Limit values for a transition period of 10 years		Limit values for a transition period of 10 years	
	arable land	grassland vegetable and fruit production	arable land	grassland vegetable and fruit production
Pb	1250	625	625	315
Cd	20	10	10	5
Cr	1250	625	625	315
Cu	1250	625	625[1]	315[1]
Ni	750	375	375	190
Hg	20	10	10	5
Zn	5000	2500	2500[1]	1125[1]

[1] exceptions possible

Waste management system

Following the interpretation of the Ministry for Agriculture and Forestry the selling of biowaste-compost as soil improver shall be covered by waste management regulations. It is intended to elaborate a uniform regulation on the national basis to ensure that only compost of high quality is produced. The aim of such an regulation is to improve the material cycles on the one hand and on the other hand to prevent the selling of contaminated compost. This requires first of all to introduce the legal basis within the Federal Waste Management Act.

FURTHER MEASURES

Against this background in Austria the following measures will be taken in the near future: A study on the quality of compost is carried out in co-operation with the Ministry for Agriculture and Forestry and the Ministry for Health to investigate the quality of the collected biowaste and the produced compost depending especially on the origin of the biowaste (e.g. towns, villages), on the type of composting plant and on the size of the plant.

Based on the results of this study an ordinance on the production and selling of biowaste compost shall be elaborated. This ordinance shall regulate quality and labelling requirements and control mechanism.

The implementation of the mentioned objectives does not only depend on legislative measurements but also on the acceptance of the population concerned. Therefore information campaigns, which shall inform about the actual quality of the produced biowaste composts have high priority.

CONCLUSION

In order to guarantee the recycling of biowaste in the future, a solution of the unsatisfactory legal situation of composting of separately collected biowaste has to be found as soon as possible.

To achieve this aim it is necessary to reach an agreement of the involved authorities in two major points:

- the introduction of the legal basis for such a regulation in the Waste Management Act
- the detailed regulations in the planned ordinance:
    - parameters to be limited (e.g. is there a need for a dioxin limit value in compost?) taking into account the costs for the chemical analyses
    - frequency of controlling of the product, of the producer
    - recommendation for the maximum amount of compost to be applied on agricultural land (what differences to the ordinance on fertilisers are acceptable?)

To establish nation wide uniform requirements for the selling of compost as waste is a first step to solve the problem but the situation is still not satisfactory.

Therefore the Ministry for the Environment and the authorities of the provinces shall work together in convincing the Ministry for Agriculture and Forestry to change the regulations in the Fertiliser Act and the ordinance on fertilisers to allow the marketing of high quality compost not only as fertiliser but also as soil improver and as part of a growing media under realistic conditions.

On the European level the Ministry for the Environment should improve the co-operation with countries which are interested in high quality compost to introduce quality requirements in European standards (standards by CEN/TC 223, 'soil improver and growing media').

ACKNOWLEDGEMENTS

The author wishes to thank Evelyn Wolfslehner for assistance in legal aspects.

REFERENCES

1. Ordinance on the separate collection of biowaste, BGBl. 1992/68, BGBl. 1994/456 available at: Oesterreichische Staatsdruckerei, Rennweg 12 a, A-1030 Wien, Austria
2. Waste Management Act of Lower Austria, LGBl. 1992/8240 Information available at: Verbindungsstelle der Bundeslaender, Schenkenstrasse 4, A-1014 Wien, Austria
3. Waste Management Act of Upper Austria, LGBl. 1991/28, 1993/13, 1993/24 Information available at: Verbindungsstelle der Bundeslaender, Schenkenstrasse 4, A-1014 Wien, Austria
4. Water Act, BGBl. 1959/215 i.d.F. 1993/185 available at: Oesterreichische Staatsdruckerei, Rennweg 12 a, A-1030 Wien, Austria
5. Industrial Act, BGBl. 1994/194, 1994/314 available at: Oesterreichische Staatsdruckerei, Rennweg 12 a, A-1030 Wien, Austria
6. Soil Protection Act of Lower Austria, LGBl. 1988/6160 Information available at: Verbindungsstelle der Bundeslaender, Schenkenstrasse 4, A-1014 Wien, Austria
7. Soil Protection Act of Upper Austria, LGBl. 1991/115 Information available at: Verbindungsstelle der Bundeslaender, Schenkenstrasse 4, A-1014 Wien, Austria
8. Fertilizer Act, BGBl. 1994/513 available at: Oesterreichische Staatsdruckerei, Rennweg 12 a, A-1030 Wien, Austria
9. Amlinger, F. (1993). Biotonne Wien, Verlag Anton Schroll & Co, Wien.
10. Ordinance on fertilizers, BGBl. 1994/1007 available at: Oesterreichische Staatsdruckerei, Rennweg 12 a, A-1030 Wien, Austria
11. Raninger, B. (1995). Sammlung und Verwertung kommunaler biogener Abfälle in Oesterreich, published by: Bundesministerium für Umwelt, Stubenbastei 5, A-1010 Wien, Austria, ISBN 3-901271-29-5
12. OENORM S 2200, (1993). Guetekriterien fuer Komposte aus biogenen Abfaellen, Oesterreich. Normungsinstitut, Heinestrasse 38, Postfach 130, A-1021 Wien, Austria

Agreements Concerning the Application of Sewage Sludge and Household Compost in Farmland Areas

Senior adviser Leif Knudsen, the National Department of Plant Production, The Danish Agricultural Advisory Centre, Udkaersvej 15, DK-8200 Aarhus N

ABSTRACT

Following amendment of Danish legislation on the application of sewage sludge in farmland areas in 1989, negotiations were initiated between the agricultural organisations and the National Association of Local Authorities in Denmark on the formulation of a special contract to be used in connection with agreements between the municipalities and the farmers. A standard contract of this nature was agreed in 1992, and this contract was subsequently recommended both by the agricultural organisations and the National Association of Local Authorities in Denmark. The recommendation of this standard contract was an important manifestation on the part of the agricultural organisations which underlined the need for greater reliability and consistent rules in relation to the use of sludge on farmland. In 1994 a similar contract was agreed to the use of compost from household waste on farmland.

During the period from 1993 to 1995, the public debate on the application of sludge was revived both within the agricultural community itself, between the agricultural community and the authorities and in the general public. The consequences of this debate were that certain firms introduced restrictions in connection with their purchases of plant products grown on sludge fertilised fields, which in turn led to a situation where the agricultural advisory services did not feel able to recommend sludge fertilisation to the farmers.

New amended legislation on sludge is expected for the summer of 1995. It is important that this legislation re-establish confidence in the application of sludge in farmland areas. It is equally important that the farmers, the consumers and the food industry receives a guarantee that the application of sludge is appropriate and does not cause any significant agricultural, health or environmental problems. It is thus also essential that this new legislation is followed-up by an information campaign.

KEYWORDS

Household waste compost, sewage sludge, farmland, attitude, agricultural community; contracts.

FARMERS AS 'SLUDGE AND COMPOST USERS'

The interest that farmers will have in receiving sludge or compost will of course depend on their evaluation of the advantages and disadvantages involved in using sludge or compost, both in the short and the long run. Seen in comparison to the use of many other products, the decision to use sludge or compost will also be characterised by attitudes. The farmers' evaluation of short and long-term advantages and disadvantages will include the following considerations:

- will the cultivation value of the soil be preserved in the long run?
- will there be any restrictions in the choice of crops if sludge or compost is used?

- will there be any disease transmission risk involved in fertilising fields with sludge, and can products cultivated on sludge or compost fertilised fields or field sludge heaps transmit diseases to humans, livestock, wild animals, etc.
- will there be any problems with odours, spreading, etc.
- aesthetic considerations in connection with field sludge heaps, sludge or compost fertilised areas, etc.
- the financial gains involved in receiving sludge or compost

It is relatively easy to evaluate some of the above-mentioned items. The short-term financial gains involved can be calculated on the basis of nutrient value and the calculated savings that can be achieved through reduced purchase of commercial fertiliser or animal manure. Odour and aesthetic considerations will have to depend on previous experience. Restrictions in the choice of crops will appear from legislative regulations concerning the application of sludge, and an insight into the reservations that might exist in the market against products grown on sludge or compost fertilised fields can be gained through sales contracts, etc.

To the farmer it is most essential whether the agricultural quality of the soil can be preserved in the long run. The farmer must include the risk in his considerations that restrictions might be introduced for land fertilised with sludge or compost in previous years. The farmer will base his evaluation of this question on various types of information. As this will be a subjective evaluation, the result of the farmer's considerations will be strongly influenced by the opinions he hears from various parties. To some extent the farmer is bound to rely on the advice he obtains from his usual sources, such as his plant production adviser, but as this is a relatively abstract question, his decision will also be strongly influenced by the general attitude the public, the media, etc.

Another very important factor, on which the farmer can base his evaluation of risks involved in using sludge or compost, are the restrictions in the choice of crops are laid down in the shape of legislative measures or demands made by the firms that buy, process and sell plant products. If a large number of such restrictions exists, the farmer will be bound to consider the application of sludge or compost problematic, because 'some people' think that the use of sludge or compost is dangerous.

THE VOLUNTARY AGREEMENT CONCERNING SLUDGE - THE STANDARD CONTRACT

In connection with a new Government order [1], which was prepared on the basis of the EU directive [2], and as a consequence of ever increasing quantities of waste products, there was a debate in the late '80s both within the agricultural community itself and in the public at large about the application of sewage sludge on arable farmland.

Primarily, this debate centred on the heavy metal content of the sludge or compost. The health risks were not discussed quite as much and the same holds true about the risk of organic impurities contained in sludge or compost. The health risks had been evaluated during preparation and adoption of the above-mentioned Government order, and the order laid down requirements for different types of sludge or compost based on the degree of deactivation. Thus requirements were laid down that untreated sludge had to be injected directly into the soil on application, and that sludge or compost could not be used in connection with consumable crops[1]. In addition, the order resulted in the adoption of limit values applying to the contents and inputs of heavy metals - that are considerably lower than those defined in the EU regulations[4].

In order to ensure that the interests of the individual farmer would be duly considered when agreeing to receive sludge from the municipalities, the agricultural organisations initiated the negotiation of a standard contract with the National Association of Local Authorities in Denmark, which was to be used in connection with agreements concerning the delivery and application of sludge or compost on arable farmland. This was in the municipalities interest because it greatly increased the likelihood of selling the sludge or compost to the agricultural sector.

In general, the Danish agricultural organisations have a great influence on the way the agricultural community will react to certain questions. The agricultural organisations run a well-developed network of advisory services with approx. 120 local advisory offices that have plant production specialists, who can advise the farmers on all possible matters to do with plant production. The farmers rely heavily on these advisers for the preparation of field and fertiliser programmes. And thus the information and attitudes that are imparted to the farmers via the agricultural advisory services have a great influence on the farmers' willingness to use sludge. In addition to the local advisory services, a national advisory centre has been established, which provides the local advisory services with specialised advice within specific subject areas and which co-ordinates local advisory and research activities. Another important task of this central national advisory centre is to support the agricultural organisations in their negotiations with e.g. the public authorities. The Danish Agricultural Advisory Centre was thus also involved in the negotiations between the agricultural organisations and the National Association of Local Authorities in Denmark concerning the preparation of a standard contract about sludge, i.e. they assisted the agricultural organisations with specialist advice on both technical and legal matters.

The most significant obstacle that had to be overcome in the formulation of the contract was the question concerning the future agricultural quality of the soil. The agricultural organisations wanted a guarantee that compensations would in future be payable to farmers who as a consequence of the legal application of sludge would incur income losses on account of future legislation that might impose restrictions on the farmer with regard to his choice of crops or the like in relation to areas where sludge had previously been applied. The municipalities were not able to give this guarantee, and for a long period the negotiations were deadlocked over this question. The difficulties were solved by introducing conditions applying to certain subject areas into the standard contract that were more restrictive than the rules that had been laid down on the same subjects in the Government Order. This contributed to the decision made by the agricultural organisations in 1992 to not demand a guarantee for compensation for future income losses. An agreement was thus reached on the contract, and it was recommended both by the agricultural organisations and the National Association of Local Authorities in Denmark.

The contract consisted of 5 pages plus a 5-page annex. The contract lays down the quality requirements that have to be fulfilled with regard to heavy metal contents. It is specially noted that the requirements applying to the cadmium contents are more stringent than those laid down in the Government order. The limit values applying to the cadmium content which were agreed in the standard contract are identical to the ones applying to commercial fertilisers. In concrete terms this means a limit value of 150 mg of cadmium per kg total phosphorus until 1 July 1995 and subsequently 110 mg per kg total phosphorus - in comparison the limit values fixed in the Government order were 320 mg per kg total phosphorus and 200 mg per kg total phosphorus, respectively. With regard to the limit values applying to the remaining heavy metals, identical values were fixed in the standard contract and in the Government Order.

In addition, the contract states the nutrient contents to be expected in the sludge with the intervals of variation allowed. The annex to the contract is used to record the results of the three most recent sludge or compost analyses, and more detailed conditions concerning the frequency with which sludge or compost analyses have to be carried out are laid down here.

Moreover, the standard contract also includes details on the sludge quantities, delivery terms, etc. agreed on. Finally, it includes conditions on the correct way to solve conflicts.

Since its adoption, the standard contract has been extensively used as a contract formula, and as a standard on which other contracts about the delivery and application of sludge have been based. The most important feature of the standard contract is that by recommending it the agricultural organisations express their opinion that the application of sludge and compost to arable farmland as a reasonable procedure in which the individual farmer can safely engage.

DEVELOPMENTS 1993 TO 1995

After a calm period between 1990 and 1993, the debate on the sludge application issue was re-opened at the end of 1993 and has continued ever since. The reason for this renewed debate can be found in several circumstances.

As of 1994 more restrictive rules were introduced for application intervals and animal manure storage capacities. The introduction of these new rules meant that the application of animal manure became subject to stricter rules than the application of sludge. With a few exceptions, liquid animal manure or slurry can thus only be applied during the spring, and a storage capacity sufficient to accommodate the animal manure produced over a 9 months period is required. No such requirements are made with regard to sludge - there is no fixed time of application nor any requirements as to storage capacity. Within the agricultural community many farmers now resent the idea of using sludge, as it is impossible to understand why the municipalities are not subject to the same rules as the farmers - especially as it is these same municipalities that are supposed to control the farmers.

In addition, numerous examples could be mentioned to illustrate the fact that sludge was not handled in accordance with existing legislation. There were e.g. many non-covered field sludge heaps, the distancing requirements applying to field sludge heaps were not fulfilled, no control measures were used to check the sludge quantities actually applied, etc. - all of which contributed to destroy confidence in sludge or compost application

There were also difficulties with sludge imports to Denmark from Germany. These imports could transmit livestock diseases that no longer exist in Denmark. The application of sludge was discussed in a televised debate in the spring of 1994. The programme gave outsiders the impression that the storage and application of sludge on arable farmland involved hygienic problems. For instance, the programme cited an example of a child (not documented) that had fallen ill after playing on a field sludge heap. This television programme had a significant effect on public attitudes towards sludge - both in the agricultural community and among consumers.

In the spring of 1994 several firms reacted to this debate and introduced restrictions against the use of sludge or compost into their contracts with farmers. For example the Danish association of commercial flour mills introduced a ban on the application of sludge on areas used for the cultivation of bread wheat. This was a very serious restriction as farmers, who are able to use sludge, are also often the ones who sell bread wheat. In addition the firm Scanola, which processes and sells rape seed oil, announced that no sludge could be used in fields sown with the best grade of oil seed rape. The frozen food products industry had had reservations about the use of sludge in vegetable fields all along.

The agricultural organisations contacted the firms in question with the purpose of obtaining more detailed explanations for these reservations. The firms were, however, only able to point out the current uncertainty among consumers about the application of sludge.

These new reservations expressed by the above-mentioned firms about purchasing products grown on sludge fertilised fields made the agricultural advisory services very wary of recommending the application of sludge. The same was true for the farmers.

As a consequence of the debate, both the agricultural organisations and the food industry demanded new legislation on the application of sludge that could serve to re-establish confidence in it. In the summer of 1994, the Danish National Agency of Environmental Protection thus initiated the preparation of a new Government Order which is expected to come into operation during the summer of 1995.

In March 1995, the new Government Order plus the entire uncertain situation with regard to the application of sludge resulted in the agricultural organisations giving notice to terminate the agreement regarding the standard contract. In addition, they also gave notice of a boycott against receiving sludge, which will take place as from the summer of 1995, provided that the new Government Order is not able to restore confidence in the application of sludge.

CONCLUSION

The way the debate developed during the period from 1989 to 1995 tells us something about how sensitive the attitudes regarding application of sludge on arable farmland are. In the beginning, the agricultural organisations and the agricultural advisory services were favourably inclined to use sludge or compost on farmland areas, and this attitude did not get mixed up with other agricultural policy or environmental questions.

On account of several different circumstances the attitudes to the use of sludge for agricultural purposes changed during the period from 1993 to 1995. The most important reasons were the incoherent rules applying to the handling of animal manure on the one hand and to sludge on the other - together with too many examples of the rules on the handling of sludge being broken and a lack of control on the part of the municipalities. As a consequence of the public debate several firms introduced restrictions in connection with their purchases of crops grown on sludge or compost fertilised fields. In turn these restrictions resulted in the agricultural advisory services becoming more reserved about recommending the application of sludge to farmers.

In order to re-establish confidence in the application of sludge it is important that the new Government Order on sludge and compost expected to take effect in the summer of 1995 convinces the food industry, the farmers and especially the consumers that application of sludge or compost to farm land does not entail any kind of risks. Experiences gathered during the years from 1993 to 1995 also serve to show that attitudes towards the application of sludge can easily change. It is thus necessary that both farmers and consumers are well-informed on the implications of using sludge so that attitudes may be formed on the basis of knowledge rather than 'myths' and 'rumours'. The new Government Order should therefore also be followed-up by an information campaign.

REFERENCES

1. The Danish National Agency of Environmental Protection, 1989: 'Bekendtgørelse om anvendelse af slam, spildevand og kompost m.v. til jordbrugsformål' (Government order on the application of sludge, sewage and compost, etc. in agriculture). Danish Ministry of the Enviroment's order no. 736 of 26 October 1989.
2. Council Directive of 12 June, 1986 (86/278/ECC) on the protection of the enviroment, and in particular the soil, when sewage sludge is used in agriculture. No. L181/6 Official Journal of the European Communities.
3. The Danish National Agency of Environmental Protection, 1992: 'Bekendtgørelse om ændring af bekendtgørelse om anvendelse af slam, spildevand og kompost til jordbrugsformål' (Government order on the amendment of the Government order on the application of sludge, sewage and compost in agriculture). The Danish Ministry of the Environment's order no. 145 of the 25 February 1992.
4. Steenberg, E.C.T., 1994: 'Jordbrugsmæssig anvendelse af organisk affald. Regler, mængder og behandlingsmetoder. Seminar om spredning af smitstoffer fra husdyrgødning og organisk affald' (Agricultural application of organic wastes. Rules, quantities and treatments. Seminar on the transmission of infectious diseases via animal manure and organic wastes). The Danish Agricultural Advisory Centre, Aarhus, 1994.
5. Anon., 1992: 'Kontrakt om levering af kommunalt spildevandsslam' (Contract on the delivery of municipal sewage sludge). The National Association of Local Authorities in Denmark, Copenhagen, 1992.
6. Anon., 1994: 'Kontrakt om levering af kompost/afgasset biomasse' (Contract on the delivery of compost, degassed), Not published.

Section 2

Local Composting & Collection

Local Composting in Multi-Storied Housing

Ulrik Reeh, Danish Forest and Landscape Research Institute, Hørsholm Kongevej 11, 2970 Hørsholm, Denmark

ABSTRACT

Local composting in multi-storied housing, i. e. composting of source separated biodegradable waste and use of the compost on the adjacent amenity areas, has proved to be a possible alternative to central treatment. The advantages are found in a saving of energy and costs for transportation and marketing of the compost whereas the weakness lies in lack of experience in establishing the procedures necessary to control and ensure continuous, satisfying, sorting efficiency and process conditions. The quality of the system operation is very much dependent on the existence of a responsible person who is able to communicate the functioning of the system to the dwellers and implement and run the system, so that the incoming waste is treated in accordance with environmental and hygienic standards.

KEYWORDS

Local composting, local cycles, sustainable housing, waste minimisation, urban ecology.

INTRODUCTION

Since 1986 a number of experiments on composting of source separated organic household waste and garden and park waste has been carried out in Danish municipalities, cf. [1]. The experiments include intermunicipal organisations, cf. [2] as well as home composting in single-family houses, cf. [3]. Lately, home composting has become quite popular either as a supplement to collection and central treatment or as the only solution offered to the citizens. Some municipalities are doubtful of the number of households that will be willing and able to compost their kitchen and garden waste themselves. Only a minority of municipalities are including multi-storied housing in the calculation of the potential waste reduction due to a home composting system. The consequence is that home composting is not regarded as relevant in municipalities with many housing blocks, and that housing blocks are precluded or ignored in municipalities with a majority of garden households. Obviously, there is a need for the development of composting systems for multi-storied housing. Such systems must also avoid creating a demand for a collection system and central treatment facilities.

The overall objective of the project was to demonstrate how local composting of the organic household and garden waste and subsequent use in the adjacent amenity areas could substitute the use of fertilisers and commercial soil improvers in the local dwelling area.

More specifically, the project should cover subjects such as:

- the amount of recycled waste and compost that is generated in local composting systems,
- the quality of the compost produced with respect to heavy metals and plant nutrients and implications on necessary land areas for local application,

- process conditions with specific focus on temperature and implications on hygienic standards.

Furthermore, the project should focus on the organisational, practical and economical circumstances concerning local composting and also give an impression of the dwellers' opinion on a new way of dealing with their own waste.

The results from the project are described in detail in a Danish report with an English summary [4, in print].

METHODS

Experiments with different systems for internal collection and local composting have been carried out in 9 multi-storied housings situated in 6 municipalities in the period 1991-92. The housings were council owned except the 'Tolderlundsvej 23', which consists of flats under a multi-ownership scheme, i. e. the house is owned and run by the dwellers themselves. In addition, this house also differed from the rest by having only 7 apartments, being built in 1897, and having the smallest amenity areas which are placed close to the centre of Odense, one of the largest cities in Denmark. The rest of the experimental sites consisted of 48-222 households, of which two are only minor parts of housings of 728 and 836 apartments, cf. table 4. They are all built after the Second World War, mostly surrounded with plenty of green areas from 12 to 84m^2 per dweller.

Two different collecting principles have been implemented: 1) direct delivery of waste to the compost boxes and 2) collection in containers and further internal transport to common compost facilities for the whole housing. The fraction suitable for local composting was generally defined as waste of vegetable origin according to Danish legislation, with an exception allowed at one housing where animal waste was included. As a consequence of this extension of the compostable fraction, the waste was stored in insulated and metal-armed closed boxes, thereby seeking a high level of hygienics and preventing rats and such from being attracted.

A few other restrictions were imposed by the municipal authorities. Some of them did it indirectly by choosing the collecting equipment placed at the dwellers' disposal. However, in most cases the systems were designed in co-operation with the person in charge and the author who worked as a consultant. In some of the housings the dwellers could choose between different systems of waste collection in the kitchen. Apart from that active choice, the dwellers did not participate in the planning process except for spontaneous comments given currently and used in trimming the systems. Further, the dwellers in some of the housings were asked whether they would accept a closing of the ordinary waste chutes, which they would not. In other housings chutes have never been established or have been closed at an earlier occasion without asking the dwellers directly.

At the end of the experimental period which lasted 1.5 years, a questionnaire was distributed to the households. The respondents who did not answer within the deadline were visited at their home.

RESULTS AND DISCUSSION

Communication with the dwellers

A good time to give an oral introduction of the new waste system was found to be when the collection system for the kitchen was offered and installed. The study showed that it is not sufficient to base an implementation solely on written information. In addition, the oral communication gave the experiment a positive start even amongst critical dwellers by letting them have some influence on the part of the system that influences on their daily routine. The communication of information has to be persistently repeated, until source separation and other relevant behaviour have become part of the society culture. The process might take years or even a whole generation.

Amount of waste

The amounts of household waste sorted in the different housings are shown in table 1.

Table 1: Amount of waste sorted and delivered for composting

Housing association buildings	Observed potential[a]	Amount of waste sorted out (kg/person/week)	Sorting efficiency[a](%)	Registration period (weeks)
AKB, Etagehusene	-	0.28	-	1
AKB, Vænget	-	-	-	-
VA, Grønningen	-	-	-	-
Lejerbo, afd. 52	1.70	0.42[a]	29	1
Lejerbo, afd. 47	1.77	0.89/0.87/0.74/0.74[a]	50	4*1
B.A.B., Grantoften	1.27	0.65/0.51/0.45/0.57/0.53[a]	45	3*8+11+1[a]
Tolderlundsvej 23	1.82	1.22/0.93/0.89/1.33[a]	80	8+15+8+1[a]
BB, Hinderuphave	1.77	0.79/0.93[a]	52	64+1[a]
S.A.B., Solgården	-	1.41/0.76	-	47 (8 pers.)+ 85
Average	1.67	-	51	-

a) Related to analysis of waste from one week

The average waste potential found in the five housings analysed is very close to the potential found in a comprehensive Danish survey [5], 1.67 and 1.64 kg per person per week, respectively. The amounts of compostable waste that is separated correctly, vary from 0.28-1.33 kg per person per week, most commonly around 0.8 kg or approx. 50% of the waste potential. The highest amounts were achieved in the smallest block of flats characterised by an informal but efficient communication among the dwellers. The smallest amounts were generally collected in the biggest block of flats, e.g. 'Etagehusene' characterised by an ineffective staff organisation and weak communication channels from the staff to the dwellers and between the dwellers. The difficulties were enhanced by language problems and a generally low motivation level of the dwellers to follow directions from the administration.

24 kg per 100 m² were found in the two areas where the amount of garden waste was estimated, which is in good accordance with the range from 10-30 kg/m². In two housings, the amount of available garden waste was not sufficient to act as bulking agent and thereby improve the structure of the kitchen waste. In one case, 'Tolderlundsvej 23', this was due to a permanent lack of green areas and was managed by letting the municipality deliver shredded branches from their recycling sites. In the other case, Grønningen, the problem was temporary as it was caused by the fact that the green areas were newly established. The problem could be managed by getting wood material from the well established neighbour housings.

Compost quality

The purity of the waste delivered for composting was very good and in most cases free of visible impurities such as plastic and glass. A few exceptions were primarily related to use of plastic bags in the collecting system which were not emptied at delivery. In another case inexpediently placed containers were used unintentionally by ignorant pedestrians passing by.

Concentrations of the heavy metals are summarised in table 2.

Table 2: Heavy metal concentrations and ranges in compost from the different housings compared with other data (unit: mg/kg ts)

Heavy metal	Compost from local composting	Compost from central composting [7]	Standards in Denmark [8]	Mean soil in Denmark [9]
Cd	0.5 (<0.05-1.6)	0.05-4.5	0.8[1]	0.1
Pb	31 (10-88)	3-210	120/60[2]	20
Ni	8.5 (5-11)	-	30	3

1) Working from 95.07.01
2) 60 mg/kg is the restrictive limit for use in private gardens

In all of the housings, Pb and Ni were below the newly tightened limits of Danish legislation, with one exception for Pb. Cd was mostly below or close to the Danish standards, though in two cases analyses showed that in the double sample one value was close to the limit and the other exceeded it by 100%. The Danish legislation is one of the most restrictive in Europe.

Application of the compost

The dwellers' demand for compost was overwhelming especially in springtime, and the first compost offered turned out to be an important milestone in the introduction of the system as a whole. In most cases, kitchen gardens were established in the amenity areas around the housings, thereby enhancing the local cycle of organic resources. Other purposes of use were dwellers' gardens placed elsewhere, flower boxes, pot plants and common green spaces.

The weight reduction during the composting process could only be precisely calculated at one site, where a 60% weight loss was found in good accordance with [1]. Danish legislative restrictions for application of compost, cf. [8], determine a minimum green area of 4 m² per person with the maximum dose of 40 kg P per ha as the limiting factor, cf. table 3.

Further results from the nine experimental sites are summarised in table 4 illustrating dwellers' opinions, process temperature, time consumption and operating costs.

Table 3: Calculation of soil area necessary for environmentally sound application of the nutrient content in the waste produced yearly by one person

43 kg of vegetable waste per person per year
Results in 17 kg compost after 60% weight loss
Total solids, TS: 36%
P in TS: 0.26% or 0.0161 kg P
N in TS: 1.4% or 0.086 kg
The Danish legislation: 40 kg P/ha or 250 kg N/ha per year
Area need per person: 4 m² with P as the limiting factor

Table 4: A summary of some selected quantitative results concerning dwellers' opinion, process temperature, staff time consumption and operating costs.

Housing association, buildings.	No. of households[0]	Sorting out compostable waste (%)[1]		Local-composting is a good idea (%)[2]		Normal temp. range °C[3]	Time consumption in (man hour per 100 households / year)[4]	Operating costs incl. write off +interest per year DKK[5]
		yes	no	yes	no			
AKB, Etagehusene	144 (728)	70	22	60	8	22-35	410	460
AKB, Vænget[6]	48 (122)	33	57	-	-	-	0	20
VA, Grønningen	148	64	21	78	6	33-60	36	200
Lejerbo, afd. 52	80	66	17	67	4	frozen-38	34	55
Lejerbo, afd. 47	144	68	18	67	7	15-72	96	130
B.A.B., Grantoften	88 (836)	62	21	68	3	10-40	89	195
Tolder-lundsvej 23	7	100	0	86	0	5-57	125[7]	180
BB, Hinderuphave	222	77	11	76	5	22-64	57	105
S.A.B., Solgården	96	73	12	67	1	5-30	90	130
Average	-	69	18	73	5	13-50	104	164

0) The total number of households in the housings is given in brackets if different from the number of households involved.
1) The difference up to 100% belongs to the answer 'now and then'.
2) The difference up to 100% belongs to the answer 'don't know'.
3) In several cases much higher temperatures have been recorded.
4) The value 0 refers to Vænget, where the dwellers are supposed to handle the compost themselves.

5) Covers writing off and interest of collection and composting equipment, salaries and other well defined operating costs.
6) Only 16 out of the 48 households had accepted and thereby access to a compost box.
7) The value 125 hours/year covers work done by some of the dwellers voluntarily.

Dwellers' opinion

Answers on two out of eight questions are presented here, representing the general picture of the 76% of the participating households, that answered the questionnaire. Approx. 70% describe themselves as positive and actively engaged in sorting the vegetable fraction and delivering it to the composting system. Approx. 15% are contributing now and then and the remaining 15% are ignorant of or negative towards the system. Comparing those statements with the limited amount of organic waste sorted, it seems as if the dwellers have a more positive picture of their waste sorting behaviour than the actual amount of waste shows.

The generally very positive opinion to local-composting, where only 5% think it is a bad idea compared to transportation and central treatment, does not mean that everyone will or can play an active part in running the system. Thus it was only 6% who answered positively on the question whether they wanted to be responsible for the composting activities. Several answers commented on this apparent inconsistency, by stating that they appreciated the concept of local-composting although they were not able to take part in the related physical activities. On the other hand, 6% of a housing counting more than approx. 30 households are enough to run the system.

90% of those who claimed that they sorted out the vegetable fraction for composting amongst others chose the argument: 'to contribute to a better environment', while 39% chose the argument, that they 'wanted some of the available compost'.

Process temperature

The temperatures obtained during the composting process vary a great deal between the different systems tested. The variation is primarily dependent on the way the composting process is managed, secondly on insulation characteristics of the composting entity either as a result of insulation of the boxes used or of the batch size handled at once. Filling up a compost box immediately instead of doing it during a longer period thus gives a faster heat development (within hours or a few days), cf. fig 1. Other observations show that in addition the temperature distribution becomes more homogenous.

The design of the system or more precisely the degree of insulation influences the temperature. The insulation determines which level of temperature may be reached and maintained and the influence of changing climatic conditions during the year. Thus, not insulated boxes containing 250 l made for one family normally cannot keep the temperature above 10°C during the winter and is therefore not able to support the decomposition processes at a satisfactory speed. A bigger volume, e.g. 2 m^3, has its own insulating effect which is enough to bring the temperature above 50°C for some days and (within the mesophilic temperature range) even for months if sufficient amounts of structure improving materials are incorporated. Although [10] and [11] recommend mixing with between 25% and 33% sawdust, a reasonable temperature has been reached with as little as 15% of shredded twigs and branches ('Hinderuphave').

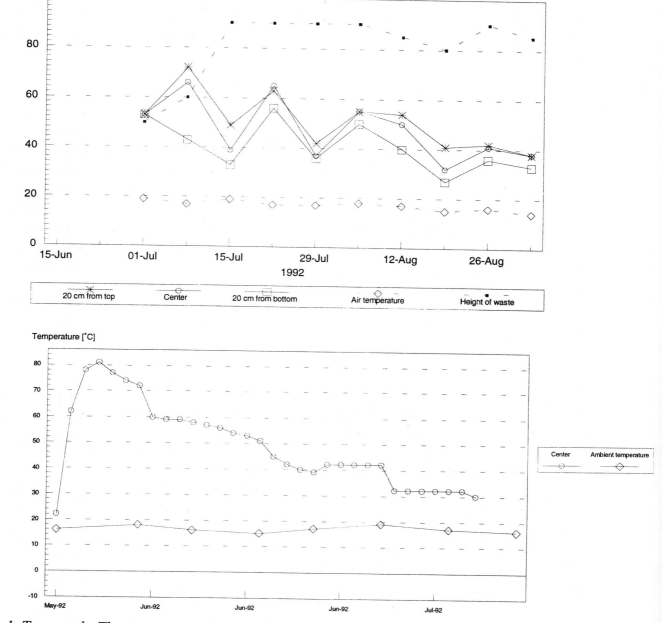

Fig. 1. Top graph: The temperature in an insulated compost box filled up by the dwellers during a month between June 15th and July 15th. Bottom graph: The temperature in the same kind of box filled up at once by the staff

A critical prerequisite for maintaining the aerobic and thermophilic process during longer periods is the ability and the willingness of the responsible persons to learn how to compost by mixing the kitchen waste which is wet and rich in nitrogen and poor in structure with dry carbon and structure rich shredded branches.

Insufficient temperature development resulted in a prolonged decomposition period, lack of hygienisation for possible pathogens and weed seeds, but most noticeable periodically explosive growth of banana flies *Drosophila spp.* which caused a lot of complaints from the dwellers.

Time consumption and economy

Time consumption is based on estimates given by the involved workers and varies from 0-410 hours per 100 households per year. The highest value is related to a very rigid organisation where every time budget is a result of a negotiation between the manager and the workers.

The costs of local composting systems as organised in this project is mainly affected either by time consumption of the staff or by investments in technical equipment to save man hours. The typical costs are within an interval from 100-200 DKK. per household per year. According to Danish municipalities their savings will exceed that level if they can avoid the establishment of a collection system and a central biological waste treatment plant. The length of the period will be dependent on whether the discount is based on an organisation where everybody has to pay to keep a central system running or whether it is financed only by those who are not willing to take care of their own organic residues from the household and green areas.

CONCLUSION

The amount of waste that is sorted for composting can vary a great deal. In the participating housings typically 50% of the potential of the vegetable fraction, 1.7 kg per person per week, is delivered to the composting system.

On the whole the compost quality is good, i. e. free of impurities and with a heavy metal content below the rather strict Danish standards and not much above the background level in Danish agricultural soils. The compost content of plant nutrients and the restrictions imposed by the Danish environmental legislation imply the need for a green area of 4 m^2 per person related to the housing.

Thus in most cases, except for older buildings near the centre of bigger cities, local composting can be a possible alternative or supplement to traditional collection and treatment on a central composting plant. To prevent a nutrient overdose in the longer run establishment of local food production in kitchen gardens must be promoted.

At least one responsible, reliable, and engaged key person or fiery soul, either employed or dweller must be found, to establish and maintain the system as a prerequisite for a successful implementation and continuation. An important part of this maintenance consists of an ongoing dialogue with the dwellers teaching them how to cope with the system and adapting the system so that users will not have any reason to criticise the functioning.

There are arguments both for and against the inclusion of waste of animal origin in the fraction to be composted locally.

The advantages lie in:

- raising the degree of recycling
- facilitating the reduction of the collection frequency for the residual waste fraction by reduced potential smell problems
- improving the economy as a consequence of reduced collection frequency

The disadvantages to be considered are:

- increased hygienic risks and consequently precautions needed
- higher demands to control the composting process
- increased investments in more advanced equipment

The more sophisticated composting equipment needed (than tested in this project) will result in a decrease in time consumption by handling the compost which will add to the lessening of the depreciation costs. Further experiments with such equipment currently being developed will set the optimal technological level for the future.

ACKNOWLEDGEMENT

All of the participating housings and the related housing associations are greatly acknowledged. Also the author wishes to thank the municipalities of Albertslund, Ballerup, Hvidovre, Odense and Slagelse and the Board of Waste and Recycling affiliated the Danish Agency of Environmental Protection who have all participated in the funding of the project. Thanks to Finn Jørgensen M.Sc. and Morten Carlsbæk M.Sc. for assistance with computer treatment of some of the collected data.

REFERENCES

1. Reeh, Ulrik & Dybdahl Jensen, Arne (1990): Det grønne affaldssystem i Høng (The Green Waste System in Høng). Miljøprojekt nr. 142, Miljøstyrelsen, 87 pp.
2. Hirsbak, Stig; Petersen, Claus; Jørgensen, Lars Vedel; Hauge, Per; Nielsen, Lars Krogsgård (1990): Det grønne affaldssystem i AFAV (The Green Waste System in AFAV). Miljøprojekt nr. 141, Miljøstyrelsen, 104 pp.
3. Plan-Energi s/i (1990): Hjemmekompostering (Local Composting). Miljøprojekt nr. 135, Miljøstyrelsen, 148 pp.
4. Reeh, Ulrik (1995): Lokal-kompostering i etageejendomme (Local Composting in Multi-storied Housing), Arbejdsrapport nr. 69, 1995, ISBN 87-7810-501-3, Miljøstyrelsen, 121 pp.
5. Nissen, Bodil; Hansen, Gert; Høeg, Gert; Nielsen Arne; Pommer, Kirsten 1994: Dagrenovation i private husholdninger (Refuse from private Households). Miljøprojekt nr. 264, 1994. Miljøstyrelsen. 165 pp.
6. Bucht, E., Hogland, W., Persson, B., Pettersson, O., Thyselius, L. (1991): Biologisk avfall på drift? (Biodegradable Waste under Changing Conditions) Stad & Land nr. 94, 1991. MO VIUM/Insti tut för landskapsplanering, Sveriges lantbruksuniversitet.

7. Mortensen, Bente & Reeh, Ulrik (1994): Klassifikation af kompostprodukter (Classification of Compost Products). Arbejdsrapport nr. 3, 1994. Miljøstyrelsen, 131 pp.
8. Miljøstyrelsens bekendtgørelse nr. 736 af 26. oktober 1989 om anvendelse af slam, spildevand, og kompost m.v. til jordbrugsformål (Use of Sludge, Waste Water, Compost etc. for Agricultural Purposes).
9. Jørgensen, Sven Erik (1993): Removal of heavy metals from compost and soil by ecotechnological methods. Ecological Engineering, 2 (1993) 89-100.
10. Persson, K. & Sveriges Lantbruksuniversitet, (1993): Roterende behållere, lokal kompostering i flerfamiliehus (Rotating Containers, Local Composting in Non-Detached Houses. Rapport 4229, Naturvårdsverket,
11. Nilsson, P., Persson, P.-E., Schroeder, H., Fergedal, S., (1993): Utvärdering av källsortering och lokal hemkompostering (Evaluation of Source Seperated Waste and Local Composting). En fallstudie i HSB brf Tusenskönan, Västerås och Västra Orminge, Nacka. FoU nr 84. Stiftelsen Reforsk.

Separate Collection of Biowaste from Households - Report on a Large-Scale Pilot Project in the City of Copenhagen

Per Nilsson, Renholdningsselskabet af 1898, Kraftværksvej 25, 2300 Copenhagen, Denmark

ABSTRACT

In the city of Copenhagen a large-scale pilot project for collection of source-separated organic waste from households was launched in 1991. It is the only collection scheme in Denmark aimed at collecting food waste from households in typical inner-city areas of a large town. The project comprises 9000 flats in mainly old multi-storey buildings with small kitchens and often difficult access from the street to collection points in the backyards. Also 1000 single-family houses are included in the project. It is a top-priority to obtain a high quality soil-conditioner (compost) as the end-product from the planned biogas-production. Hence the sorting criteria are very strict: Only food waste - no wet paper etc. Also, occupational health - both in the ergonomical and the hygienic sense - has a high priority. Three different kinds of kitchen receptacle have been tested, and for collection both 30 litre paper bags, 125 litre paper sacks and 240 litre two-wheels plastic containers have been tried out. Based on a comprehensive report on the numerous tests and evaluations performed during the year 1992 - 1993 the following topics are dealt with in this paper:
- Technical equipment, including collection vehicles
- Percentage of food waste in household waste
- Amount of food waste actually source-separated
- Quality of food waste collected
- Hygiene
- Occupational health
- Motivation and attitude of the public
- Economy

KEYWORDS

Food waste, source separation, hygiene, occupational health, receptacles for biowaste, collection vehicles.

INTRODUCTION

The main objective of the project has been to assess the feasibility of introducing separate collection and treatment of biowaste as a supplement to incineration, which has been the disposal method for ordinary unsorted household waste in Copenhagen since 1971.

Some practical experience in separate collection of biowaste from single-family houses had been gained in smaller Danish cities, but very little collection of biowaste from multi-storey buildings had taken place before the Copenhagen pilot project started in 1991.

Typically, conditions in blocks of flats differ radically from those in single-family houses:

- The percentage of food waste in household waste is smaller

- Kitchen space is more restricted
- Motivating people is more difficult

By the time the project was planned, many cities had a fresh memory of the inherent problem in composting mixed household waste: a contaminated end-product that could not be marketed. Therefore, very strict sorting criteria were considered absolutely necessary, and in principle only food waste was allowed in the bio-system.

Furthermore, the first indications that the exposure of operatives to microorganisms in degrading bio-waste could lead to serious health problems had been seen. Consequently, the collection scheme should be designed to meet any possible demand from labour inspection authorities, and of course also the user's aesthetic and hygienic demands.

Kitchen receptacles
A high quality of compost could best be obtained by using a plastic bucket in the kitchen to be emptied into a container in the yard. Alternatively, a degradable paper bag could be used in the kitchen. It was predictable that the bucket scheme would be unpopular because of the need to carry the empty bucket up the stairs again and also that some people would find the paper bag unhygienic. Therefore, a 20 micron polyethylene bag was also tested in a large number of households.

Yard receptacles
Also in the yards, different types of receptacle were tested for the storing of biowaste from the flats and transport to the collection vehicle: 190 or 220 litre two-wheels containers, 110 litre bins without wheels and 125 litre two-ply paper bags. Very soon it became obvious that the containers needed a one-way lining of sheet polyethylene to maintain a reasonable level of hygiene. In single-family houses, a 30 litre paper bag proved suitable for the biowaste.

Collection vehicles
Experience from other cities had shown that the exposure of operatives to microorganisms is most critical while loading the waste onto the collection vehicle. Hence, a modified collection vehicle with a dust-extraction system in the hopper-part, and a vehicle with a high-lifting bin-elevator were also tested.

RESULTS

The amounts of biowaste collected from 10,000 households equipped with different types of receptacle in kitchen and yard are shown in table 1.

Table 1: Biowaste collected

Blocks of flats Kitchen/yard	Kgs/person/year	Kgs/household/year
Paper bag/container 'Urban'	28.95	55.49
Paper bag/container	23.62	36.15
Paper bag/125 litre paper sack	15.87	22.31
Plastic bag/container	36.67	57.95
Plastic bag/125 litre paper sack	29.52	44.68
Bucket/container	19.12	28.87
Average	26.53	42.31
Single-family houses	**Kgs/person/year**	**Kgs/household/year**
Paper bag/30 litre paper sack	49.04	118.04
Plastic bag/30 litre paper sack	55.82	126.30
Average	52.46	122.34

The collection efficiency is of course not only a function of the collection equipment used, but also influenced by other factors, such as social conditions etc. For instance the high amount of biowaste collected in the 'Urban' can be attributed to a particular interest in the project from this well-organised building society.

However, careful analysis of these results in combination with results of the opinion poll also carried out confirm, that the plastic bag as a kitchen receptacle is best accepted and also yields the best collection results.

Sorting efficiency

In order to assess the efficiency of the collection scheme, representative samples of waste from containers for the non-biodegradable part of the waste have been handsorted carefully. By this method the amount of biowaste not sorted out according to instructions and the sorting efficiency (biowaste correctly sorted out in percent of total yield of biowaste) have been established. The results are given in table 2 below.

Table 2: Sorting efficiency

Area	Biowaste fraction in % of household waste	Sorting efficiency [percent]
Urbanplanen	29	48
Spaniensgade	37	66
Geislersgade 12	21	9
Geislersgade 16	38	63
Geislersgade 22	29	44
Geislersgade Average	28	41

Sorting Efficiency and Biowaste potential

The total amount of biowaste, separated or not separated from other household waste, in multi-storey households ranges from 1,00 kg/week/person to 1,80 kgs/week/person, averaging 1.08 kgs/week/person. The average sorting efficiency is calculated at 47%, with a considerable variation (Table 3).

Table 3: Sorting efficiency related to type of receptacle

Test area kitchen/yard	Biowaste Kgs. per week	Persons number	Sorting efficiency %
Paper bag/container 'Urban'	1,707	3,103	51
Paper bag/container	528	1,128	43
Paper bag/125 litre paper sack	141	450	29
Plastic bag/container	1,742	2,574	63
Plastic bag/125 litre paper sack	1,436	2,546	52
Bucket/container	1,755	4,479	36
Total	7,309	17,280	47

This table confirms that the best sorting efficiency is obtained where plastic bags are used as kitchen receptacles. In single-family houses the corresponding figures were:

- Total amount of biowaste: 1,32 kgs/week/person
- Sorting efficiency: 85%.

Based on these results, the total potential of biowaste in the cities of Copenhagen and Frederiksberg - 553,000 inhabitants - can be calculated at 31,500 tonnes per year.

The total amount of household waste - including some commercial waste - collected from the same two cities is approximately 175,000 tonnes.

Based on the demonstrated sorting efficiency a possible collection scheme covering all households in the two cities could be expected to yield approximately 15,000 tonnes of biowaste annually, corresponding to approximately 9% of the total amount of household waste collected.

Quality of collected biowaste

By meticulous sorting of nearly 1,500 kgs of collected biowaste, it has been established that items not meeting the sorting criteria totalled only 1,7% of the biowaste.

Similar examination of biowaste from single-family houses revealed so few sorting errors, that this biowaste can be considered as absolutely 'clean'.

HYGIENE AND HEALTH

All collection of biowaste has been carried out on a weekly basis - In the warm season - from May to September - the waste is often very malodorous and full of fungi, ants and maggots. As it was not considered realistic to wash, or expect the users to wash the containers - at least not in multi-storey houses - it soon became clear that using containers for biowaste would necessitate the use of a one-way lining in the containers. However - apart from the expenses - the replacement of this lining also represents a certain risk that the operative be exposed to microorganisms, toxins etc.

The use of plastic bags in the kitchen secures a high degree of hygiene, whereas paper bags in the kitchen tend to get soaked and to cause hygienic problems, especially in combination with paper bags as yard receptacle.

The bucket as kitchen receptacle is only acceptable in combination with a lined container in the yard.

The exposure of operatives to microorganisms etc. was measured by portable instruments for different combinations of kitchen receptacle and yard receptacle. The results did not show marked differences in the level of exposure.

The high specific weight of biowaste - approximately 400 kgs/cubic metre - makes the handling of both containers and bags a heavy job. Actually the handling of a full 240 litre or even a 190 litre two-wheels container must be considered unacceptable from an ergonomic point of view. The 110 litre bin used in houses with stairways has to be carried by two persons, as it may weigh up to 50 kgs.

The microbiologic exposure during the loading of waste onto the collection vehicle varies considerably, depending on the type of vehicle. Rear-loading compaction vehicles with a dust-extraction system are better than standard rear-loaders. But the best results are obtained by using a vehicle with a high-lifting bin-elevator, where the unloading of containers takes place far from the breathing zone of the operatives.

An overall evaluation of ergonomic, microbiological and general hygienic conditions has led to the conclusion that the optimum collection scheme for biowaste from blocks of flats would be the combination of plastic bags in the kitchen and paper sacks in the yard, preferably the new Bates Combi System now being introduced in Denmark.

MOTIVATING THE CITIZENS

Right from the beginning the new collection scheme has been followed up by a comprehensive and intensive publicity campaign aimed at:

- explaining the benefits of biologic waste treatment
- ensuring correct sorting and packing of biowaste
- informing about achieved results.

A complete communication plan was set up, including quarterly distribution of a newsletter to all households, a direct-line telephone service, regular press-releases, TV-spots etc. and special arrangements for caretakers, tenant's representatives. In addition to this, the kitchen bag holders and the kitchen bags were supplied free of charge.

It was expected, that this very costly programme would have resulted in an increase in collection efficiency from the initial level at approximately 50% to about 75%, but, as already shown, the average amount of biowaste actually separated from the household waste has not exceeded 50% of the total yield of biowaste from Copenhagen blocks of flats.

Consumer analyses conducted by telephone as well as by printed questionnaires showed that 98% of the potential users actually knew about the biowaste collection scheme, and 88% said they had been using the biowaste containers at least in some periods.

Citizens indicating they never or seldom use the system mostly gave the following reasons for this:

- I only have very little biowaste
- It's too much trouble
- It smells, is unhygienic

Asked about the kitchen receptacles, a clear majority of users prefer plastic for paper bags, which is also clearly reflected in the better sorting results from households supplied with plastic bags.

ECONOMY

Calculating the added costs of a two-stream source separation system like the Copenhagen scheme is rather complicated. On the one hand, separate collection of biowaste is very costly, on the other hand, a saving - but a relatively small one - will result from the reduction in the amount of mixed waste to the collected and incinerated.

Without entering into more details concerning economy, I will conclude that collecting a tonne of biowaste according to the Copenhagen system described, costs about twice the price of collecting a tonne of unsorted household waste in the present collection system.

This would result in an increase in the annual cost of collection per household in flats from 536 DKK to approximately 675 DKK - or approximately 26% -(figures not including cost of waste treatment).

CONCLUSIONS

- It is possible to obtain a perfect, clean organic waste fraction that will yield a first class, uncontaminated soil conditioner.
- The average total amount of food waste from flats is 1,08 kgs/person/week.

- In spite of the intensive information and motivation campaign, only 47% of the potential amount of food waste is actually source-separated in the households.
- The exposure of operatives to microorganisms, toxic agents etc. can be kept within acceptable limits.
- For both aesthetic, hygienic and health reasons, it is important to use one-way packaging for storage and collection: Plastic bags in the kitchen, paper sacks in the yard.
- A large-scale collection from all households in the cities of Copenhagen and Frederiksberg (553,000 inhabitants) is estimated to yield only 15,000 tonnes of food waste annually. The corresponding increase in collection costs will be approximately 26%.

By the spring of 1995 there is still no political decision taken as to whether separate collection of food waste shall be introduced as a permanent collection scheme covering all households in the cities of Copenhagen and Frederiksberg.

REFERENCES

1. Miljøprojekt nr. 220 - 1993: Indsamling af madaffald fra husstande i København. Miljøministeriet - Miljøstyrelsen (Collection of Food Waste from Households in Copenhagen Ministry of the Environment - Copenhagen 1993)

Biowaste Collection Project in the Central City of Vienna

Walter Hauer, Technisches Büro Hauer, Hauptplatz 22, A-2100 Korneuburg, Austria

ABSTRACT

The main objective of the project was to evaluate whether a placement of biowaste containers in the densely populated city districts of Vienna is successful and accepted by the public. A further objective is to gather experiences from the model project and make use of them for a possible large-scale introduction of the biowaste collection system in the central city of Vienna.

KEYWORDS

Biowaste collection, biowaste containers, organic wastes, waste stream, composting, impurities, densely populated city districts.

INTRODUCTION AND PROJECT OBJECTIVES

In the city of Vienna, compostable wastes are collected in specific containers which are set up in various places round the town. In the inner city districts such biowaste containers are exclusively found at collection sites for recyclables, but are more densely placed in the other districts. Nearly 29,500 biowaste containers are available all over Vienna and in 1994 alone they served to separate about 65,000 tons of compostable wastes from the waste stream, which is more than 40 kg per resident per year.

A proper strategy to increase collection quantities is to improve convenience for the citizen by placing collection containers next to his or her home. Maximum convenience is achieved when each residential building receives its own collection container. It is assumed that the majority of the population will reject biowaste collection at their homes with the arguments that it produces odour, vermin, and other nuisances.

The project's aim was to evaluate whether a placement of biowaste containers in every house is successful and accepted by the public.

An expansion of the collection of organic wastes - especially in the central city - will naturally affect the quality of the biowaste material when it is processed; the compost will then have entirely different characteristics than the one that is currently being derived from organic material collected in the city's greener suburbs. Another objective of the project must therefore be to evaluate the composition of the collected biowaste.

In another project phase, the biowaste-derived compost is analysed by experts of the Vienna Municipal Department 48 in order to determine how the finished compost can finally be applied.

A further objective is to gather experiences from the model project and make use of them for a possible large-scale introduction of the biowaste collection system in the central city of Vienna. On the basis of these experiences, it would be easier to find out which steps are required before and during collection to minimise impurities in the biowaste material, to maintain or achieve public support, and to separate a maximum share of organic wastes from the waste stream while achieving a compost of at least adequate quality.

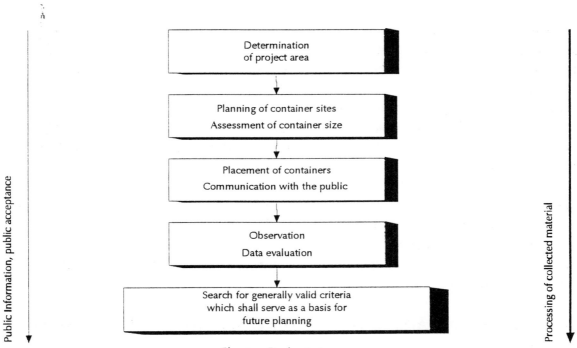

Fig. 1: Project steps

DIMENSIONING THE PROJECT AREA

Number of residents and households

The number of residents and households are a vital prerequisite for dimensioning the size of the project area. The test area had to be large enough to make sure that the experiences gathered from the project can also be applied for organising biowaste collection in the other densely populated city districts. It was assumed that this requirement could be fulfilled if a representative number of 5,000 persons were incorporated in the project. According to a public census in 1991, the community living in the project area comprises 6,471 residents. It could therefore be expected that the results obtained would have the necessary accuracy.

Approximately one million residents live in the densely populated central area of Vienna. In the green suburbs of the city, providing a home to 600,000 residents, containers for the disposal of organic wastes are already available.

Technical requirements: composting

Another dimension necessary for determining the project area was provided by the technologies which are currently available for processing the collected biowaste quantities. In order to achieve that the composting process can take place under conditions comparable to those in a large composting facility, a minimum collection quantity of about 3,000 kg per week is required.

Technical requirements: collection vehicles

The collection vehicle applied in the test area had a permissible payload of 5,900 kg. The collection staff and the vehicle were only available one day per week. To keep the costs low, one single truck load was considered sufficient for each collection tour, which means that it was not allowed to empty the vehicle before the tour was finished.

The collected biowaste was transported to the local waste treatment facility where it was composted by means of a static-pile system. The digestion process was accompanied by the following activities:

- the composting material was inspected at regular intervals to control impurities; undesirable substances were sorted out by hand before the digestion process started
- photographic documentation
- the heavy metal content of the finished compost was analysed by the staff of the Municipal Department 48

CRITERIA FOR CONTAINER SITES AND DETERMINATION OF CONTAINER VOLUME

Selection of number and size of containers on the basis of a waste stream analysis

The collection quantity to be expected in one collection tour was estimated by referring to the results of the so-called 'Wiener Systemmüllanalyse 93/94'. According to this analysis, which provides figures on the non-recoverable waste stream produced in Vienna in 1993 and 1994, there is a further collection potential of 90 kg per resident per year. Additional to that potential 40 kg have been separately collected in 1994.

To determine the number and size of the biowaste containers needed for each site, an average biowaste quantity of 3 litres per resident per week was assumed.

As a new waste stream analysis is currently being elaborated for Vienna, all further planning must be based on the potential derived from this analysis.

Planning of container sites

The containers set up in the test area had three different sizes: 120-l containers were to cover a biowaste quantity of 17,760 litres or 42% of the total container volume, 240-l containers were set up to tackle 11,280 litres of biowaste, corresponding to 27% of the container volume, and 770-l containers - making up 31% of the container volume - were to swallow a quantity of 13,090 litres.

In other words: 55% of the 6,473 residents use 120-l containers, 30% put their biowaste into 240-l containers, and 15% utilise 770-l containers for biowaste disposal.

COMMUNICATION ACTIVITIES

The people living in the test area were informed of the project before and during the placement of biowaste containers:

1. A personal letter from the head of the Municipal Department 48 was sent to each household in the test area to announce the placement of biowaste containers in the various residential buildings. The letter was accompanied by an information leaflet containing instructions on how to separate the wastes that belong in the biowaste bin from those which do not.
2. In the week in which the biowaste containers were set up, the respective letter and information leaflet issued by the Municipal Department 48 were pinned to the notice-board hanging in the corridor of the flats.
3. In each residential building, the information leaflet sent out by the Municipal Department 48 was attached to the door that leads to the biowaste container site in German, Serbo-Croatian and Turkish language. Also a sign was added saying 'Biowaste collected on Saturdays - do not lock door! For information please phone the Municipal Department 48´s waste hotline 55 16 61'.
4. Clearly visible stickers were attached to the containers to identify them as biowaste containers.
5. Another sticker was attached to each biowaste container which indicated that this was a test program for biowaste collection.
6. The biowaste containers were outfitted in the classical colours of biowaste collection: green bin with brown lid.
7. A public event was organised together with a Viennese high school (AHS Rainergasse) on October 16, 1993.

8. On the occasion of a public event in one of Vienna's districts at the end of September 1993, an information stand (a so-called 'city bus') was installed to provide the citizens with details on the biowaste collection program.
9. Waste hotline

DATA COLLECTION AND TEST RESULTS

Collection quantities

Biowaste quantities in the test area remained largely constant throughout the project. The specific collection quantity was varying between 22 and 35 kg per resident per year.

As shown in figure 2 the quantity varied in the first weeks and became relatively constant about three months after the beginning of biowaste collection.

On a 38-week average, it amounted to slightly more than 27 kg per resident per annum, corresponding to 0.54 kg per resident and week, or one kg per household and week.

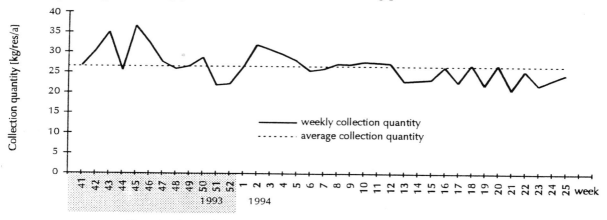

Fig. 2: Specific biowaste collection quantities in the test area

When considering a potential of 100 kg per resident per year in the urban areas of Vienna[1], one can assume that the biowaste quantity separated from the waste stream is about 27%.

With an intensivation of advertising the amount of separate collected biowaste could be increased but the test has been done under realistic conditions. That means a small financial advertising budget has to be sufficient for connecting more than one million people to the biowaste collection. So no extra PR-work or advertising have been done.

On-site monitoring: the results

During the entire project period of seven months, the biowaste containers and the sites at which they had been set up were continuously observed for impurities. Contamination of biowaste with materials belonging to other waste fractions was recorded for each site. The monitoring was aimed at determining the number of sites where the organic matter used to be unpolluted, slightly polluted, heavily polluted, or not useable for composting at all.

It was particularly important to assess those sites where the collected biowaste was frequently or always mingled with a lot of waste from other fractions.

Regular on-site monitoring of the collected biowaste was intended to provide objective indicators for the quality of the collected material. The goal was to find out before the placement of containers in which households collection does not work, and develop solutions which would afford only modest means to provide biowaste material of a certain minimum quality over a longer period.

The biowaste was inspected in order to find out regularities and work up a link between the purity of the collected material and the specific characteristics of the collection site. On-site monitoring was also intended to bring forward any problems with public acceptance.

The data derived from monitoring were linked to statistical demographic data through the household address. These links served to discover certain dependencies and obtain details on how to minimise impurities in the collected material and increase collection quantities - always on condition that public acceptance was not impaired.

Container inspection immediately after emptying

Sticky residues
52 containers were inspected as to their cleanliness after having been emptied. It turned out that

- 16 containers were clean or hardly dirty
- 36 containers had sticky residues on their inside
- 11 containers had rather persistent residues on their inside.

Odour
Just as with sticky residues, three groups were set up to classify the odour criterion:

- There was hardly any smell in 28 containers
- Odour was normal in 30 containers
- Heavy odour emission was found in 5 containers

Despite the uncommonly mild weather in January, with temperatures ranging around +8°C, there were only few cases where odour was regarded as a nuisance.

Evaluation of container contents

Collected material wrapped in paper
Each time the biowaste material was inspected and data were recorded, also the share of paper-wrapped biomaterial inside the containers was estimated. The weighted average share of paper-wrapped material in all inspected containers was 30%.

Purity of the collected material

When the biowaste containers were inspected, the pollutant content of the container-collected biowaste was assessed by means of a scale. As the project proceeded, the share of containers with heavily polluted or unusable material did not change much.

According to the inspections, 6% of all containers were found to include heavily contaminated material and 2% of the biowaste containers were even used to dispose of ordinary garbage.

In larger containers, which are naturally used by more people, it is more difficult to avoid impurities. Smaller containers are, however, much more affected by impurities. Containers whose content had to be discarded were exclusively found among the 120-l containers.

Most of the impurities that are found among container-collected biowastes are plastic bags which are handed out at food retail stores. Such bags are used in the household to collect organic kitchen waste and are finally thrown into the biowaste bins along with their contents. Spoiled food is often not even unwrapped before dumping it into the biowaste containers. It still needs to be analysed in the laboratory whether wastes not intended for disposal in biowaste bins may actually contain substances which, on being set free, become aggressive during the digestion process and finally spoil the compost quality.

Relationship between biowaste impurities and collection performance per resident

Is it true that container sites with a smaller collection performance per resident have a higher pollution rate? If yes, this would provide us with a perfect criterion for sorting out waste that comes from specific addresses. More than 90% of the containers showed little or no impurities at all. Most of the undesirable substances found in biowaste containers were plastic bags in which the residents use to put their organic wastes before throwing them into the containers.

Table 1: Share of biowaste containers depending on biowaste impurities and collected amount per resident

	Collection performance above average [% of containers]	collection performance below average [% of containers]	average project area [% of containers]
collection material clean	60	63	62
slightly polluted	34	29	30
heavily polluted	4	6	6
unuseable	2	2	2
total	100	100	100

At sites where the collection performance per resident is below the average, 63% of the biowaste bins contain unpolluted material. Sites with a collection performance above average have a 60-percent share, which shows that there is no significant difference.

Assessment of collection containers

Odour emission

During inspection, the collection containers were opened and odour emanating from the collected material or from sticky residues was determined. The strength of odour was marked on a scale with three grades.

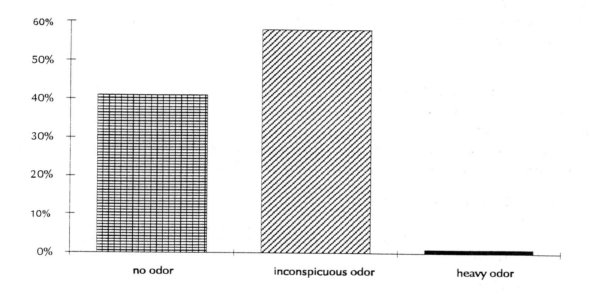

Fig. 3: Odour emission from biowaste containers

41% of the containers showed no or only little odour emission, while 58% of all biowaste containers emitted moderate odour.

Pestilent odour was experienced in merely 1% of all containers. Even during the hot summer month of June (with temperatures rising to 35°C), there was no striking odour increase. Odour is mostly considered a nuisance in collection containers that have just been emptied: when the organic residues that have remained inside the container for several weeks are suddenly exposed and not immediately covered with fresh material, or when foul liquids come to the foreground and disseminate their stench. A comparison among various container sizes reveals that small containers emit less odour than larger ones.

Maggots

Table 2: Biowaste infestation with maggots

	not affected [% of containers]	slightly affected [% of containers]	heavily affected [% of containers]
120-litre containers	99	1	0
240-litre containers	99	1	0
770-litre containers	100	0	0
Total	99	1	0

The rare occurrence of maggots is in part due to the fact that the recommendation not to put meat leftovers in the biowaste is obviously observed by the public. Meat leftovers should also be excluded from biowaste collection in the future.

Insects

Insects to be found were only some flies and ants, which prefer solitary outdoor containers.

Wasps were only found in two biowaste containers. This small amount cannot be expressed in percentage.

The occurrence of wasps in biowaste containers, as it sometimes happens in the greenbelt districts, does not seem to be a problem in the central city.

Table 3: Occurrence of insects

	none [% of containers]	some insects [% of containers]	a lot of insects [% of containers]
120-litre containers	97	3	0
240-litre containers	98	2	0
770-litre containers	98	2	0
Total	97	3	0

A B C analysis of biowaste container sites based on the filling quantity of containers

During the project, samples were taken to assess the material quantity with which the biowaste containers at the sites were filled. The gathered data revealed that biowaste collection at about 35 sites (which is slightly more than 20% of all sites) is enough to achieve approximately 50% of the entire collection quantity.

A further objective was to find out already at the planning stage in which houses a satisfactory collection result is to be expected.

FACTORIAL ANALYSIS

Based on the availability of site-specific data of the buildings, it was possible to assess the impact of objective factors on the collection results in the test area by comparing household categories with regard to certain parameters.

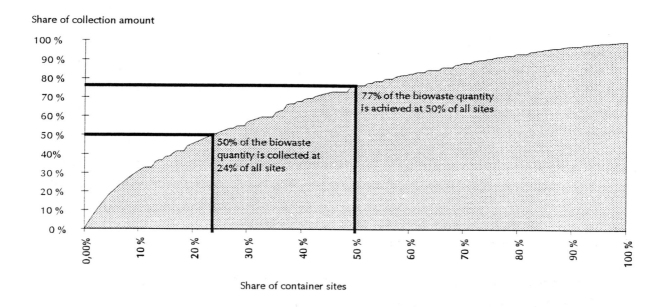

Fig. 4: Correlation between collection quantity and number of container sites

The collection performance of the residents (litre/res.) was chosen as a criterion to divide the households in two groups. To facilitate assignment to either of the groups, the average collection quantity per resident throughout the test area - corresponding to 1.76 litres per resident and week - served as a reference.

Group 1:   57 houses with above-average collection performance of residents
Group 2:   132 houses with below-average collection performance of residents

Based on such a factorial analysis, the development of collection quantities per resident may allow to make out the following trends:

- The younger the residents, the poorer the collection result per person.
- The smaller the households, the better the collection result per person.
- The smaller the number of residents in a household community, the better the collection result per person.
- The smaller the surface area of the building, the better the collection result per person.
- The larger the ground area of the premise, the higher the collection quantity per person.
- The smaller the share of foreign residents, the higher the collection quantity per person.
- The older the house, the worse the collection result per person.

What must be pointed out is that these are no binding rules which could help assign individual buildings and addresses to specific groups and form categories. The various factors have not the same prominence in all households and may be superimposed in some cases.

They can, however, serve as a useful guideline for later planning when it must be decided for each specific case whether a waste collection program shall be launched or not. Later on, in the course of container site planning, an integrated assessment of each household community is indispensable.

The planning of biowaste collection programs on the basis of specific data on buildings and residents as well as the inspection of household communities by experts make it possible to select the buildings in the planning area of the central city of Vienna by the following means

a) *a positive approach* involves *inclusion* of only those *high in quantity*
b) *a negative approach* involves *exclusion* of those *low in quality*

FULL-SCALE BIOWASTE COLLECTION IN THE CENTRAL CITY OF VIENNA ON THE BASIS OF PROJECT EXPERIENCES

Biowaste containers which are set up at people's homes contain much less non-organic matter than containers placed at public sites. Besides, impurities found in the collected biowaste material can be decisively reduced by sieving.

The fact that the people living in the densely populated districts greatly support biowaste collection in containers is mainly due to the City of Vienna's recommendation not to place meat leftovers in the bin for biowaste. Such a step will be helpful in keeping odour and ugly maggots out and must be recommended whenever it comes to a full-scale introduction of the biowaste collection program in the inner city districts. The weekly emptying of containers is just as important for avoiding odour and maggots.

The collection quantity to be expected in the central area of Vienna ranges between 25 and 30 kg per resident per year. In other words, between 25,000 and 30,000 tons of collected material can be achieved in this area. When containers are set up at selected sites, the collection quantity decreases in accordance with the density of container sites.

It must be considered that during the digestion process the organic matter is already reduced by 80%, so that only 20% of the collected matter are actually turned into compost. Correspondingly, the amount of compost produced and put on the market is 5,000 to 6,000 tons per year, while the landfill quantity that is saved by composting actually corresponds to an expected collection quantity of 25,000 to 30,000 tons.

Setting up containers in the densely populated city districts requires a careful disposition of sites and container sizes.

Further activities will include a thorough evaluation of the heavy metal contents and their sources. Investigation in this field has not yet been concluded.

CONCLUSIONS

Broad acceptance
Assumptions that there is general public opposition to the collection of biowaste in containers have proved to be wrong.

No odour emission, no insects
Even considering that biowaste containers are used over longer periods and are exposed to high temperatures in summer, neither extensive odour nor insects were observed.

Large-scale participation, broad acceptance
Public participation (83%) in the biowaste collection project has rendered satisfactory results. 86% of those questioned said they would be glad if biowaste collection were to continue after the project.

Quality of the collected material
Undesirable substances found in the material were about 2% by mass (5% by volume). Sieving would further reduce the content to less than 1%.

Dimensioning of biowaste containers and site selection
It will be possible in the future to determine the container volume that is needed for each site on the basis of planning parameters which have been derived from observation during the project and statistical analysis.

a) positive approach
On the basis of the experiences made in the project, it could become possible to optimise the collection result and the quality of the collected material through a positive selection of households. For example, 50% of the total collection quantity would be achieved by involving only 25% of all houses in the collection program.

b) negative approach
Although it is planned to set up biowaste bins at every household community, it would be possible already at the planning stage to exclude those houses from collection where the collected material is most likely to be heavily contaminated. It may be expected that if 10% of the houses were excluded from the biowaste collection program, impurities could be reduced by half.

REFERENCES

1. Scharff, C., Vogel, G.: Wiener Systemmüll- und Altstoffanalysen 1990/91, Wien 1991, S. 83

The Installation of Biobin in Salzburg, Austria

Winfrid Herbst, Abfallwirtschaftsamt der Stadt Salzburg, Siezenheimer Straße 20, A-5020 Salzburg, Austria

ABSTRACT

Studies have shown that the household refuse collected in the city of Salzburg contains approximately 30% of biological materials. On 1st July 1994 it became law in the whole of Austria that this fraction be collected and processed separately from the remaining refuse. For this reason, a total of 14,000 biobins were delivered to properties within the city of Salzburg between November 1993 and June 1994.

The introduction of separated collection was preceded by an information campaign with radio spots, large-scale posters and advertisements. Every household was visited, given advice and provided with informational material about correct separation. Competitions were run in schools and kindergartens and meetings were held in the individual parts of the city.

At the present time approximately 25% of the previous household quantity is collected as biological refuse and processed. 78% of the inhabitants of Salzburg use the biobin and the collected biowaste is contaminated to the extent of about 1%.

KEYWORDS

Biowaste, introduction biobin, separated collection, public relations, quality.

INTRODUCTION

Salzburg is the capital of the identically named province in the west of Austria and has approximately 145,000 inhabitants. The city is economically well developed, and has a university (14,000 students) and a thriving tourist industry (1.65 million overnight stays in 1994). During 1993, inhabitants and guests produced 40,000 tons of household waste.

Legal situation

In Austria, the political competence for administering the laws relating to waste is divided between the federal and the provincial government. The federal government is responsible for hazardous waste and also for the overall Austrian waste policy (Abfallwirtschaftsgesetz, BGBl. No 325/1990) [1]. The so-called Organic Waste Decree (Bioabfallverordnung, BGBl No. 68/1992) lays down which items of organic waste must be subjected to a separated collection.

The following materials must be separately collected: a) natural organic waste from house gardens and parks, in particular grass, leaves, flowers; b) solid vegetable waste from the preparation of food; c) vegetable waste from the industrial preparation of food; d) paper which was used for packaging food.

The provincial governments are responsible for organising the biowaste collection and may extend the list of separately collected biowaste. The Salzburg Biowaste Decree

('Getrennte Abfuhr und Behandlung biogener Abfälle', Bioabfallverordnung, IGBl. No. 37/1992) [2] lays down the conditions for the biowaste collection. Because the decision had been made to build a plant for anaerobic treatment of biowaste, the list was extended to include separate collection of food residues from household and catering. Both decrees came into force on 1st July 1994 although the federal decree first came into force on 1st January 1995.

Political aims

The biowaste decree has three objectives:

1. To lay down regulations for the municipal collection and processing of biowaste;
2. To list all the types of biowaste which must be separated from household refuse;
3. To reduce to a minimum the amount of organic material that is deposited (or incinerated).

REALISATION

The situation

Since 1976, the municipal waste of Salzburg has been treated in a combined household refuse and sewage sludge composting plant. The original plan to produce marketable compost for farming and similar activities did not succeed, but this method of processing the waste is still used as a pre-conditioning step before deposition. A study from 1988 (Engineering Society Prof. Tabasaran, Stuttgart) [3] recommended that biowaste be collected and processed separately, in order to produce composted material suitable for agricultural and other uses. Other studies were also carried out:

- Municipal waste was analysed to determine the percentage of biogenic material in household refuse.
- The market demand for biocompost was investigated.
- Because the plant is situated near two villages to the north of Salzburg, methods for treating all types of biowaste were compared in the search for a process which did not generate new unpleasant smells.

These studies produced the following results:

- The municipal waste from Salzburg contains approximately 29.8% biogenic material (fig. 1) amounting to 12,300 tons per year [4].
- The best possible method for treating biowaste is the DRANCO process (dry anaerobic composting), which can be used for all kinds of biological waste (except of course for carcasses) [5]. A capacity of approximately 18,000 tons per year was planned with a total cost of ATS 140 million. The plant was put into operation in December 1993. In the plant,

organic waste is converted into compost in two steps: 1. The closed system anaerobic phase which produces biogas and has a residence time of approximately three weeks at 55 °C. 2. The subsequent aerobic phase during which the material is rotted together with chopped up branches, twigs etc. The material becomes hygienically acceptable during this phase. The plant was ordered by 'Salzburg Abfallbeseitigung (SAB)', which is responsible for processing the communal waste from Salzburg (city) and two neighbouring districts, for a total population of approximately 400,000 people.

- It is possible to sell several thousand tons of good quality compost per year [6].

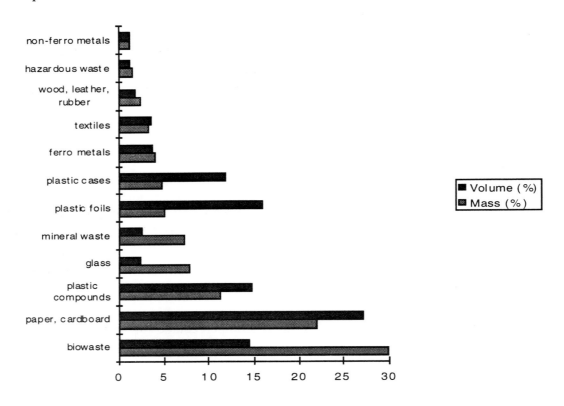

Fig. 1. Composition of household waste 1988/19890, Scharff, Ch., Vogel, G. [4].

Organisation of the biowaste collection

After consultations with the Prof. Vogel advisory institute and the Austrian Ecology Institute, the following scheme was selected for the installation of the biobin:

- Biowaste is the more unpleasant part of household refuse, being both unsightly and odorous. It is important to improve its 'image'. It is environmentally important to collect biowaste separately and thus to close the natural cycle. People must be given the possibility of collecting the right things in the right way so as to avoid negative consequences.

- As much biowaste as possible should be collected, including the remains of meals and the faeces of domestic animals. This means every household has to have its own biobin. Apart from the convenience, this also makes monitoring easier. Every biobin owner is responsible for correct collection and keeping his biobin clean. There is no communal biobin cleaning service.
- Individual composting is not only tolerated but is regarded as a desirable alternative to the biobin. For hygienic reasons it is very important that this is done properly, especially in densely populated areas [7]. Registration as an individual composter, (or de-registration) should involve as little bureaucracy as possible.
- The types of biowaste containers are restricted to bins with a volume of 120 or 240 litres. The volume limitation is a result of the high density of biowaste. A completely filled 240 litre bin used by a gastronomic undertaking can weigh as much as 200 kg. After examining several possible designs, simple, two-wheeled bins were selected, coloured mint green. A rim inside the lid leads condensation water back into the bin when the lid is opened.
- The biowaste is collected by vehicles constructed on the rotating principle (14 cubic metres, maximum permitted load 8 tons), equipped with a ventilation system and liquid-tight towards the liquid biowaste phase.
- It was decided to empty the bin weekly throughout the entire year in spite of the difference in quantity between winter and the other seasons. 800 bins have to be emptied per day using four rounds, each performed by a crew of three. The bins were installed between November 1993 and June 1994 in ten stages, corresponding to ten urban districts.
- The cost of using the biobin is included in the general refuse tax and is not charged separately. This avoids the temptation to save money (in the wrong place) by refusing the biobin. A collection service for bulky garden waste is available to every biobin user. Every biobin user receives a biowaste allowance amounting to one third of the weekly volume of his residual waste bin(s). Those requiring more (e.g. hotels or people with large gardens) must pay for the extra volume. People composting individually receive a rebate amounting to 15% of the price of their residual waste bin but they cannot use the collection service for bulky garden waste.

Preparing for installation

The introductory phase started in July 1992 with mapping of all the sites for the refuse bins, finding out whether space is available for the installation, whether a biobin is actually needed (not every installation needs one) etc. In an old village with narrow streets and narrow houses it is often extremely difficult to find suitable sites for installing waste bins.
This mapping procedure revealed that a minimum of 13,000 and a maximum of 17,000 biobins would be needed. The exact number would first be known after every household had been visited by our team of instructors. 15,000 SULO biobins were ordered and delivered to the ten districts in ten instalments. This was the moment when the first stage of the public relations

programme was started. 'What use is the separate collection? What use is individual composting? Why is biowaste different from normal refuse?'

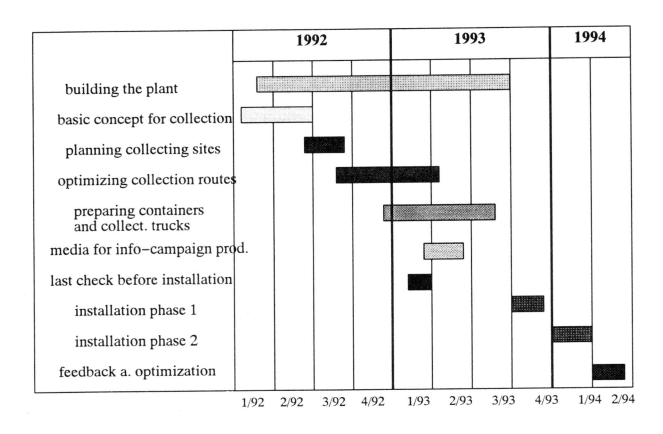

Fig. 2. Time table of planning and installation the biobin in Salzburg. Ocenasek, Ch., Scharff, Ch. Planungsarbeiten zur Biotonne [8].

Installation

The crucial phase began in November 1993. Every household received a letter from the urban government containing all the information about the biobin. The motto was: 'bio-logically' and gave answer to such questions:

- What is the sense of the biowaste collection?
- How to collect and what to collect?
- When will the biobin be installed in my district?
- Where can I get advice?

November 11th: A letter announcing the visit of our instructors and the imminent start of the biowaste collection was sent to all households.

November 22nd to 24th: 65 specially informed students started their visits to a) take informational material to the households about what should be separated (sample list), what is the best place for the biobin, how should the collecting be carried out (as dry as possible, biowaste wrapped in a lot of paper, biobin not misused for other refuse), b) to hand over a small 5 l bucket for collection in the kitchen and c) to ask whether the person visited wants to use the biobin or compost individually.

November 24th - 27th: The biobins were installed in accordance with the list of biobin users.

December 1st: The first biowaste collecting round started.

December 16th: First meeting with the biobin users to give them the chance for discussion and for airing any complaints.

Public relations

In this phase of the introduction the inhabitants of Salzburg were informed in various ways, by television and radio spots, exhibitions, special information for teachers, schools and kindergartens, folders, painting competitions for children in collaboration with the biggest regional newspaper (Salzburger Nachrichten). The winning picture (from 5000 entries) was printed on a bag which was distributed to all the pupils. A winning puzzle about the biobin and what should be placed in it was published in another newspaper (Kronenzeitung, 10,000 people taking part). A bus was painted with subjects from the bio-campaign and a large number of articles appeared in Salzburg's newspapers. The inhabitants of the city reacted very positively to the public relations effort.

Calculation of the economic success

The tax for the municipal refuse for a family of five persons is currently (1995) ATS 1638. This price includes the following benefits:

1. emptying the 120 l residual waste bin once a week
2. emptying the 120 l biobin once a week
3. using the municipal waste paper collection service
4. using the recycling centre
5. using the hazardous waste collection service
6. using the separate bulky waste collection service (whenever required)
7. using the separate bulky biowaste collection service (whenever required)
8. providing the two bins

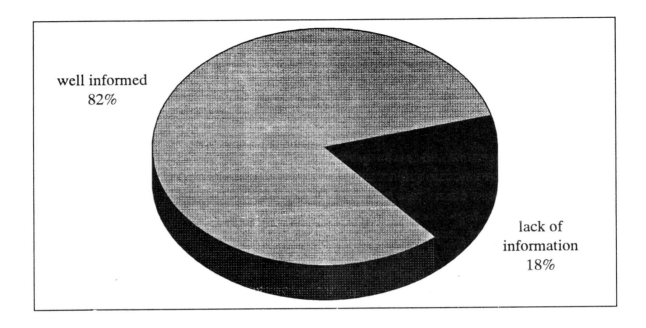

Fig. 3. How people feel informed about separate collection of biowaste.

(The cost of processing biowaste (1995) = ATS 1000,- per ton and for processing household refuse = ATS 1177,- per ton.) Although the tariffs have been raised by approximately 16.5% this year, the costs of installing the biobin and of the separated collection and treatment would have only required a rise of 5%. However it was also necessary to compensate for loss of revenue due to the reduced quantity of residual refuse, as a result of the separation.

The quality of the compost is also very good (first quality grade). The only problem is the degree of nitrogen enrichment. This must be countered by mixing the biowaste compost with compost from the bulky biowaste collection in the ration of 1:2. A give-away scheme will start shortly. Everybody will be offered biocompost to demonstrate the closing of the biomaterial cycle.

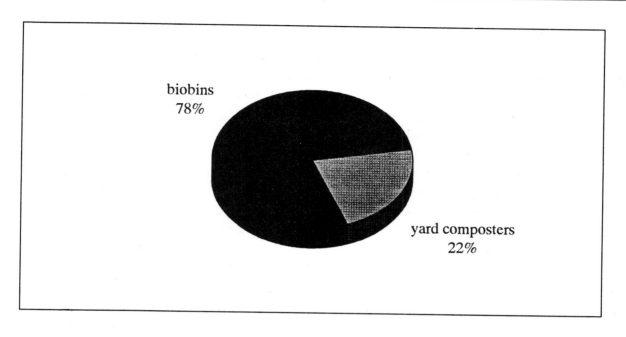

Fig. 4. Percentage of users of biobins and yard composters.

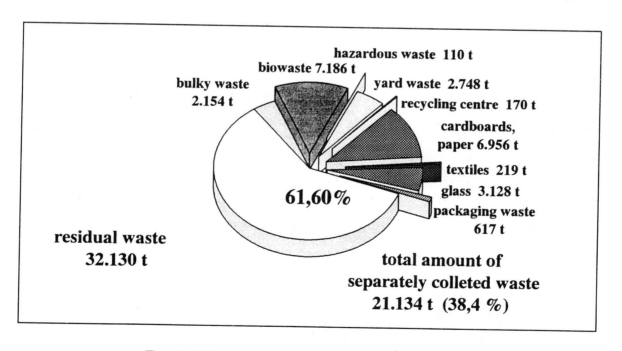

Fig. 5. Splitting of the total waste 1994 in Salzburg (town).

CONCLUSION

The separate collection and processing of biogenic refuse is an important objective of the recycling industry. The measure of success is the quality of the compost generated by the processing and also the proportion of biogenic refuse which remains in the residual waste.

The crucial factor for success is motivation of the general public. The requirement for this is a transparent system which enables the entire path travelled by the biowaste, from collection to finished product, to be followed and understood.

After extensive preparations (establishment of need, selection of systems for collection, transport and processing), the project of introducing the biobin in the city of Salzburg ran from November 1993 to June 1994.

The information campaign which immediately preceded and accompanied the installation of the biobins, attempted to communicate the political sense of the separation and the necessary know-how for doing it, as well as to transfer a sense of responsibility for correct collection. It is a fact that unpleasant side effects of biowaste collection such as the appearance of maggots, can only be avoided through careful collection.

The results indicate that the biowaste campaign was a great success (collected quantity 65 kg / inhabitant and year; impurity level approximately 1%). Although the summer of 1994 was the hottest in the history of Salzburg, acceptance of the biobin by the general public remained intact. This was confirmed by an opinion poll in December 1994 [9]. The separate collection of biogenic refuse became part of the everyday behaviour of the population.

REFERENCES

1. Doralt, W. (ed.) 1992. Abfallwirtschaftsgesetz mit Verordnungen. 2. Auflage, Verlag Orac, Wien, Austria.
2. Landesgesetzblatt für das Land Salzburg, 8.Stück, Jahrgang 1992, Nr. 37 Verordnung der Salzburger Landesregierung - Getrennte Abfuhr und Behandlung biogener Abfälle (Boioabfallverordnung), Salzburg, Austria.
3. Amt der Salzburger Landesregierung, Abt. 16, Referat für Umweltschutz (1990), Abfallwirtschaftsplan des Landes Salzburg, Salzburg, Austria.
4. Scharff, Ch., Vogel, G. (1990). Projekt Biotonne Salzburg, Teil 1. Planungsgrundlagen und Ergebnisse der Salzburger Müllanalysen 1988 - 1990, Endbericht. Ein Forschungsprojekt im Auftrag der Salzburger Landesregierung. Wien, Austria.
5. Lutz, W. (1990), Studie über den Einsatz des Systems Biotonne und die Verwertung biogener Abfälle im Bundesland Salzburg. Konzept der Biomüllbehandlung und -verwertung; Band 1. Verfahrenstechnische Grundlagen und Anlagensysteme; Band 2., Wien, Austria.
6. Scharff, Ch., Vogel, G. (1990). Projekt Biotonne Salzburg, Teil 2. Der Markt für Kompost aus Biotonne - Sammelmaterial und Verhaltensbereitschaften der Gartenbesitzer. Ein Forschungsprojekt im Auftrag der Salzburger Landesregierung. Wien, Austria.
7. Amt der Salzburger Landesregierung, Abteilung 9 - Gesundheitswesen und Anstaltenverwaltung (1994). Hygienerichtlinie für die Eigenkompostierung biogener Abfälle. Salzburg, Austria.
8. Ocenansek, Ch., Scharff, Ch. (1992). Planungsarbeiten zur Biotonne, Zwischenbericht. Wien, Austria.
9. Institut für Grundlagenforschung (1994). Einstellung der Salzburger Bevölkerung zur Entsorgung und Verwertung von Bioabfällen. Repräsentative Haushaltsbefragung in der Stadt Salzburg, Flachgau und Tennengau. Salzburg, Austria.

Section 3

Systems, Cases & Trends

Systems Analysis of Organic Waste

T. Nybrant, H. Jönsson, B. Frostell*, J.-O. Sundqvist, L. Thyselius, M. Dalemo, K. Mingarini* and U. Sonesson
* Corresponding authors: Royal Institute of Technology, Dept of Environmental Technology and Work Science, Osquars backe 7, S-100 44 Stockholm, Sweden

ABSTRACT

A systems analysis of various alternatives to collection, transport, treatment and disposal of organic waste has been undertaken. The types of organic waste investigated were toilet waste and food waste from households, food waste from restaurants and grocery stores, food industry wastes, park and garden waste. We have not included agricultural wastes. To facilitate the work, a computer-based simulation model - ORWARE, ORganic WAste REsearch model - was developed. The model describes the flow of material and energy through a city, as characterised by a 43 parameter vector. The model has been used to simulate various future scenarios for organic waste management in Uppsala. Our aim is to analyse the scenarios from an environmental, economic, and sociological point of view. This paper considers only some resource related and environmental aspects. For further analysis we refer to our final report. [6]. A preliminary result suggest that an increased source separation of organic waste, followed by biological treatment and recycling of residual products to farm land, is a good alternative to conventional systems, according to energy consumption, global warming potential and resource conservation. For the size of city that Uppsala represents - 170,000 inhabitants - anaerobic digestion seems to be a good choice among biological treatments. It is important to lower the degree of contamination of the waste, especially the levels of heavy metals and persistent organic substances in the waste. An important aspect to consider is that in scenarios allowing a high recycling ratio of nutrients the transport work increases. This aspect needs further investigation. We have approached this problem in our final report [6].

KEY WORDS

Organic waste, system analysis, simulation model, nutrient recirculation, energy, heavy metals, global warming potential

INTRODUCTION

A problem in discussing organic waste management in the industrialised world is that, organic wastes since long ago are collected and treated by a number of different means. Organic waste is a type of material whose amount is not possible to reduce to any significant extent. Very little effort has hitherto been devoted to discuss organic waste management from a true system stand-point. Typically, sanitary engineers discuss the possibilities to improve water supply and sewage treatment, while solid waste engineers discuss collection and treatment of solid waste. Therefore it is rather difficult to get a concerted picture of shortcomings of the present system and possibilities for the future.

Sushil [1] has performed a system analysis on waste with emphasis on macroeconomic aspects. This study considers all wastes as equal. A conclusion from this study is that there is a need for developing a database and information system on waste management. Basri and Stentiford [2] investigated expert systems on solid waste management, and thus neglected waste water. Their study is focused on the computer implementation, not on the results. They have constructed a system for localising and constructing landfills. Huang et al [3] investigated solid waste management with grey dynamic programming, and also neglected the waste water as an interesting and necessary waste. Hoffman et al. [4] have performed a screening Life Cycle Assessment on biological treatment of organic waste, but also they have omitted the waste water. Their energy comparison included coal fired power plants for marginal energy, which was very unfavourable for biological treatments, which they also pointed out. Sundberg [5] has developed an optimisation model for waste handling systems. They do not consider waste water either, and have a smaller vector for flow characterisation. Their main conclusion is that an increased source separation and composting of household waste is beneficial for the environment and economy under certain conditions.

The basic hypothesis behind the project System Analysis of Organic Waste is that existing handling systems for organic waste are sub-optimal with respect to environmental effects and efficient use of natural resources. The existing organic waste management systems in Sweden today includes centralised sewage treatment, incineration or landfilling of municipal solid waste and various solutions for industrial organic waste. In order to test the hypothesis, we decided to take a system approach to the problem, including gaseous, liquid and solid material flows. To start with therefore, we developed a conceptual model comprising different sources of organic waste, different modes of transportation, different options for treatment and possibilities for recycle or disposal (Figure 1). Early in the planning phase, we also decided to build a computer-based model from the conceptual one, in order to facilitate the simulation of different alternatives to organic waste management. We also decided to include an economic, an ecological and a sociologic evaluation of the results obtained with the model.

THE MODEL

Conceptual approach and modelling

The ORWARE model describes the entire organic waste handling system in a Swedish model city. The city of Uppsala was chosen as a basis for the model, but it still represents the situation for a large number of cities in Sweden and abroad. The model has been built in a modular mode, facilitating the inclusion or exclusion of various parts. The types of waste included in this study is easily biodegradable organic waste produced in a given area, e.g. a municipality.

In the model, each incoming quantity of a substance, e.g. carbon can be followed through the system as identified in the vector. It will finally end up either as a product, a residual product spread on farm land, waste stored in a landfill or as an uncontrolled emission. Energy is included in the model, both consumption and production. An example of energy production is methane from the anaerobic digester. Energy consumption are for example diesel oil used for transports and electricity for pumping waste water.

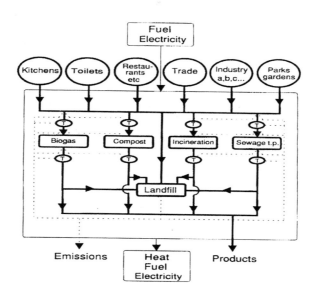

Fig. 1: The conceptual model for organic waste management

Figure 2 shows the lay-out of the conceptual model in the language MATLAB/SIMULINK which is used for implementation of the model on the computer. Below this principal sheet, there exist a large number of more detailed sheets expressing the fate of the 43 element vector through the system modelled. The sewage treatment plant e.g. contains a number of different steps corresponding to the actual design of the sewage treatment plant in Uppsala (cf. Figure 3). The way the model is implemented on the computer allows further modelling, e.g. dynamic modelling of the processes in the sewage treatment plant.

Material flows characterised in the analysis

The 43 parameters selected for the vector, table 1, may be divided into three different categories, namely parameters that are selected primarily for environmental impact, those which have impact on a process and those which have a value. In some cases it may be discussed whether a parameter should belong to one or another category. The parameter flows are expressed in kg/year. Examples are:

- BOD, COD, VOC, CH_4, CO, CO_2 anthropogenic, chlorinated volatile compounds, phenols, PCB, dioxins, NOx, N_2O, NO_3^-, NH_3, Cl, sulphur, SO_2, Pb, Cd, Hg, Cu, Cr, Ni, Zn have **an environmental significance**.

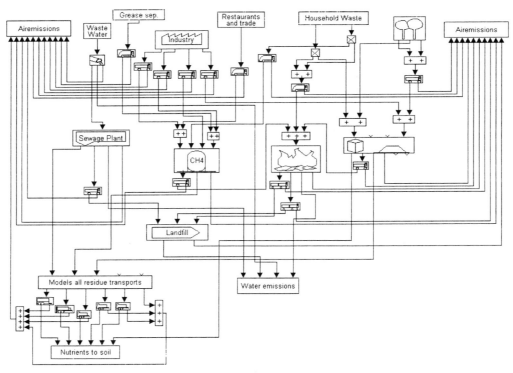

Fig. 2: Computer image of the ORWARE model.

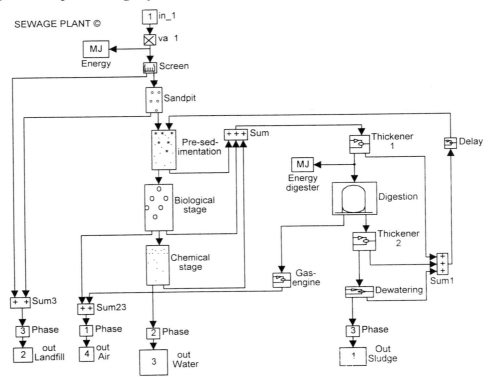

Fig. 3: Diagram showing the ORWARE sewage treatment plant sub model.

- Total carbon, slowly degradable carbohydrates, cellulose, fast degradable carbohydrates, fat, protein, VS, TS, moisture, N, P etc. have an **impact on a process**. They are also directly or indirectly important environmentally.
- P, K, CH_4 represent parameters that have a **value**.

Table 1: The vector used in the model ORWARE

No	Substance	No	Substance	No	Substance
1	C-tot	16	CO	31	Cl-tot
2	C-ch stable*	17	Phenols	32	K
3	C-ch biodegradable*	18	PCB	33	Ca
4	C-fat	19	Dioxins	34	Pb
5	C-protein	20	O-tot	35	Cd
6	BOD	21	H-tot	36	Hg
7	VS	22	H_2O	37	Cu
8	DM	23	N-tot	38	Cr
9	CO_2-f◇	24	NH_3/NH_4^+-N	39	Ni
10	CO_2-b¤	25	NOx-N	40	Zn
11	CH4	26	NO_3^--N	41	C-ch semistable*
12	VOC	27	N_2O-N	42	particles or susp. solids
13	CHX✠	28	S-tot	43	COD
14	AOX	29	SO_2-S		
15	PAH	30	P-tot		

✱ 'ch' means carbohydrate, ◇ 'f' means of fossil origin, ¤ 'b' means of biological origin
✠ 'CHX' means halogenated hydrocarbons and CFC's.

System boundaries

Our definition of organic waste is rather pragmatic and includes waste that is suitable for biological treatment. The types of organic waste discussed in the system analysis are:

- the biologically degradable fraction in household waste
- the biologically degradable fraction in industrial waste
- the biologically degradable fraction in trade waste
- municipal sewage
- park and garden waste

We have not included agricultural wastes, since this systems analysis considered an urban area. The geographical limits are in principle the community border. In the specific case of Uppsala, it is the area covered by the municipal sewer system.

Output from the model

At the moment, the output from the model is expressed in the form of six vectors:

- Total gaseous emissions

- Total liquid emissions
- Nutrients to soil
- Fodder (for example waste from the big slaughterhouse in Uppsala is transported to a fodder producing industry)
- Materials that are still in the landfill after 100 years, divided into potential gaseous and liquid emissions

SCENARIOS INVESTIGATED

The model has been used to simulate various scenarios for future organic waste management in the city of Uppsala. In different scenarios, the way the waste is collected, the degree of contamination, the mode of transportation, the treatment methods used and the way the residual products are finally disposed of have been varied. In this way it has been possible to calculate influence of different scenarios on the use of natural resources and on type and quantity of various emissions. At the moment, the scenarios tested so far are being evaluated from an economical, an ecological and from social point of view. Se our final report [6].

We have formulated 12 scenarios, but in this paper we present four selected scenarios. Consult our forthcoming report for all scenarios [6]. The **reference scenario** is the situation in the city of Uppsala in 1993, denoted 'Ref 93a'. In ref 93a all waste water is treated in a sewage plant and most solid organic waste is incinerated. Half of the sewage sludge is put in a landfill and the other half is used on arable land.

Besides the reference, we have selected three future scenarios. For these scenarios we have calculated resulting material flows in the ORWARE model and compared them to the reference scenario.

The scenarios selected are:

- **Scenario 2a**, denoted 'Sc 2a'. This scenario is supposed to illustrate a near future. It is anticipated that the contamination level of the waste has decreased to half that of today for heavy metals and persistent organic compounds. Almost all waste, except waste water, is treated with anaerobic digestion and the residues is recycled to farm land.
- **Scenario 2b**, denoted 'Sc 2b'. Similar to scenario 2a, except that the solid organic waste is composted in a rotary drum reactor.
- **Scenario 3c**, denoted 'Sc 3c'. In this scenario we have allowed more thorough changes in the collection, treatment and disposal system for organic waste. It thus gives a more futuristic picture of society. Here we have a water conserving system for the toilet waste, that enables a treatment and recycling of all residues to farm land. The waste is anticipated to be as clean as in scenario 2. The toilet waste except urine, and the solid organic waste are transported with a vacuum system to anaerobic digestion. The urine is source separated

and spread untreated on arable land. This reduces the amount of nitrogen and water to the anaerobic digester.

SELECTED PRELIMINARY RESULTS

For practical reasons, it has not been possible to analyse and compare all resulting material flows from the scenarios investigated. What we have accomplished so far, is the reduction of the simulation results to some 'key numbers'. In these figures, it is not the absolute value that is interesting, but the comparison between scenarios.

The emission profile for global warming potential is positive for biological treatment of organic waste. These figures include also emissions from use of energy, i.e. diesel oil. The high level of green house gases in reference scenario 93 a is due to the landfill, where 50% of the produced sewage sludge is being degraded. This process gives rise to emissions of methane. The landfill facility has a system for gas collection, but we estimated that approximately 45% of the methane escapes.

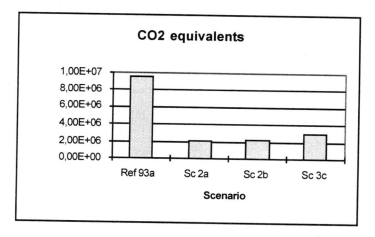

Fig. 4: *Global warming potential, expressed in CO2 equivalents.*

In Figure 5 below, we can see the load of some metals on arable land for different scenarios. The model calculates amount of cadmium in the nutrients, and the area available for spreading. This figure shows average annual input, the atmospheric precipitation of metals is not included, neither the plant uptake. The load of cadmium and mercury is significantly reduced due to the reduction of contamination in the incoming waste. There is not an reduction to 50%, because the amount of material that is put on arable land is greater, and thus the total amount of metals arriving at the soil is larger.

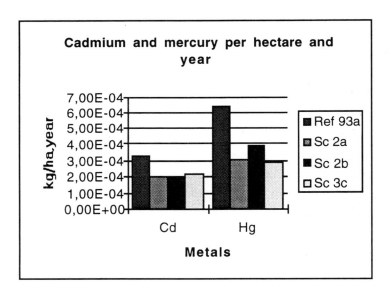

Fig. 5: Metals spread on arable land, kg per hectare and year for selected scenarios

An important aspect of the handling systems for organic waste is how much of the plant nutrients removed from the agricultural system in food is returned to arable land after it has passed human society. We have expressed this as a nutrient recycling ratio. This is the ratio between the amount of nutrient recycled to the agricultural system and the total amount of the nutrient in the waste. The nutrient recycling ratio is shown in Figure 6.

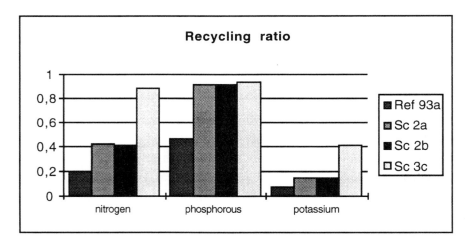

Fig. 6: The nutrient recycling ratio for selected scenarios

Another key number is the fuel energy ratio (Figure 7), where we express the efficiency of vehicle fuel use in the system. This is a quotient between produced biogas energy and used fossil fuel. This is an important figure because a lot of the environmental problems are connected with the use of energy, especially fossil fuels.

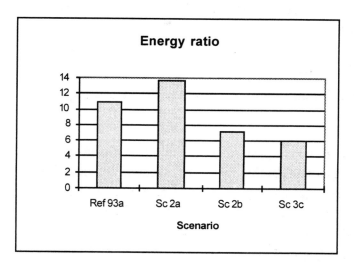

Fig. 7: Fuel energy ratio, produced MJ biogas divided per MJ used oil

DISCUSSION

The ORWARE model is an attempt to take a system approach to material- and energy flows associated with organic waste management. Thus it emphasises the consequences of different alternatives with respect to the following aspects:

- resulting emissions, where are the emissions produced and in what quantities
- energy use, what type of energy is used and in what quantities

We believe that these aspects will be increasingly important as a basis for decisions. In this respect, the results may hopefully be regarded as an indication of where more research and development is important.

Natural resources consumption is not explicitly accounted for in the ORWARE analysis. This is since ORWARE does not deal with waste minimisation as such, but with means of minimising material and energy flows once waste is generated. In a broader analysis, however, natural resource consumption should be integrated as part of our results.

Use of urine without pre-treatment is acceptable in a climate like the Swedish, where it is humid, and the precipitation exceeds the evaporation. Thus, the risk for salt accumulation is small, since salts are leached out from the soil. In a dry climate it is not advisable to add urine to soil, because the risk for salt accumulation is large.

In its present form, the ORWARE model only represents a preliminary approach. We have used literature data, but in some situations, there is a lack of data in many places, data necessary for the establishment of the desired material and energy balances. In these cases, we have often used a method of qualified estimates. With the model, however, it is possible to

PRELIMINARY CONCLUSIONS

From the numeric simulations with the ORWARE model, some preliminary conclusions may drawn:

- It is very important to take a system approach when discussing potential improvements in organic waste management. In this case, we included both liquid and solid organic waste in a system analysis. Just discussing solid organic waste would have resulted in large misjudgements, since 73% of the waste nitrogen 80% of the waste phosphorous in the city of Uppsala is disposed of into the sewage system.
- Currently, society is contaminated with various heavy metals and anthropogenic substances such as persistent organic compounds. To a considerable extent, these problems are hidden in our landfills. By introducing a more recycling society today, we would increase the present contamination. A general clean-up of products and processes used for goods and service production will therefore be necessary when developing an increased recycling of organic waste residual products. The amount of contamination of the waste management system - rather materials handling system - determines to what degree it will be possible to create eco-cycles for organic waste in the future.
- In order to make it possible to recycle a major part of the plant macro nutrients, we have to catch the nitrogen and potassium before it reaches the sewage plant (scenario 3c). It is not enough to spread all sewage sludge on farm land. The simulations also indicate that a rotary drum reactor compost system is fair at recycling plant nutrients but they are not very energy efficient (2b). The biogas scenario, 2a, seems to be a good alternative for the near future, fair at recycling, low level of global warming potential and with a good energy efficiency.
- Organic waste is rich in energy; the amount of energy that can be extracted from the organic waste exceeds what is consumed by the organic waste handling system. The major part of the recoverable energy is in the form of gas produced in the anaerobic digestion process and in the landfill. We should mention that the low fuel energy ratio in scenario 3c depends on the large amount of water that has to be transported out to the fields due to the fact that urine is used as nutrient. With a better logistic planning this transport can possibly be reduced, and thus give this scenario a better energy ratio.
- With the scenarios investigated, the transports will increase with the introduction of more recycling. With fossil fuelled engines, this causes increased natural resource consumption and increased uncontrolled emissions. It is therefore important to evaluate a scenario in which the transports are fuelled with biogas.

- In scenarios where organic waste is landfilled, the landfill process is the major contributor of greenhouse gases, even though the landfill has a system for collecting landfill gas.

REFERENCES

1. Sushil, (1993). Application of physical system theory and goal programming to modelling and analysis of waste management in national planning, *Int. J. Systems Science,* Vol. 24, No. 5, pp. 957-984.
2. Basri, H.B. and Stentiford E.I., (1995). Expert systems in solid waste management. *Waste Management and Research*, No. 13, pp. 67-89.
3. Huang G.H. et al, (1994). Grey dynamic programming for waste management planning under uncertainty. *J. Urban Plann. Dev.*, vol. 120, No. 3, pp. 132-156.
4. Hoffmann L. et al, (1995). Methodological aspects of life cycle screening of biological treatment of source separated household waste. *First International Symposium - Biological Waste Management 'A Wasted Chance'*, University of Essen, Technical University of Hamburg-Harburg, April 1995, S 10, pp. 1-8
5. Sundberg, J. (1993). Generic Modelling of Integrated Material Flows and Energy Systems. *Ph.D. Thesis*. Chalmers University of Technology, Gothenburg, Sweden.
6. Nybrant T., Jönsson H., Frostell B., Sundqvist J.-O., Thyselius L., Dalemo M., Mingarini K. & Sonesson U. (1995). System Analysis of Organic Waste - A Case Study. *AFR report*, in press.

Implementation of Source Separation of Municipal Solid Waste in the City of Kristiansand

Knut H. Kristensen, City engineer, Tollbodgata22, Postutak, 4604 Kristiansand, Norway

ABSTRACT

The object of this presentation is to demonstrate the new system of source separation of solid municipal waste adopted by the city of Kristiansand (67.000 inhabitants), with particular reference to the handling of biological waste, and the process of implementing the new system. The paper also includes a description of the differentiated price system adopted, and a preliminary evaluation of how well the system functions.

INTRODUCTION

In the spring of 1993, the Norwegian ministry of the environment invited cities and local authorities in Norway to participate in a national demonstration project for full scale source separation of municipal waste. The essence of the project was to demonstrate implementation of full scale source separation of MSW including biological waste. Four cities and four rural districts were selected. Kristiansand was successful, together with the cities of Oslo (Groruddalen), Moss and Kristiansund. The rural districts of Vik in Sogn, Hitra, Lervik in Troms and the district served by Innherred Renovasjon were also chosen. The participating authorities were given substantial grants to cover the capital investments involved. One of the conditions for this grant was that we completed the introduction of the biological waste collection system in 1994.

The system prior to 1994

In 1990 Kristiansand introduced a two-bin system for source separation of cardboard and paper. The second bin (green), was collected every fourth week by a separate team. The grey bin was collected every week.

We also had a system where hazardous household waste could be brought to one of two petrol stations for collection and safe destruction.

We had several stations where empty glass bottles can be delivered for material reuse, and used clothes can be given to the Salvation Army.

Existing landfills

We have two existing controlled landfills which are to be closed and rehabilitated no later than 01.01.1996. In co-operation with three neighbouring smaller local authorities, we are therefore establishing a new controlled landfill. The government has strongly indicated that permission to use the new landfill will depend on the implementation of source separation of biological waste. Though our choice was somewhat limited, we, and most of our politicians, were very much in favour of a better and more sustainable waste disposal system.

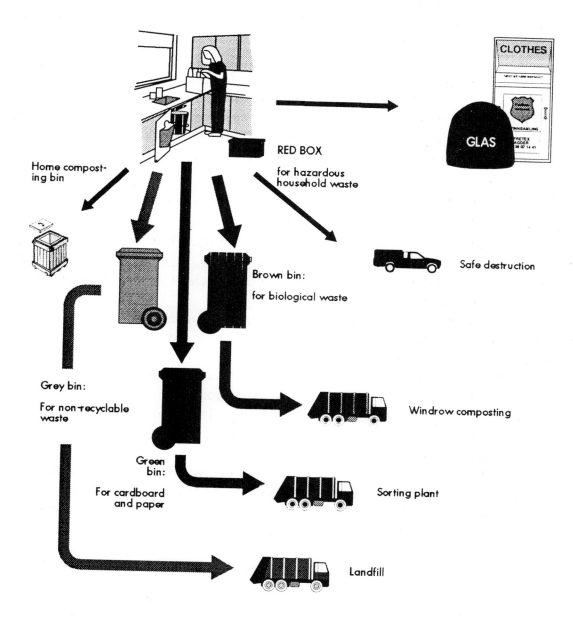

Fig. 1: The new system of household waste collection

Every household has a three bin system consisting of:

- **Green bin** for paper and cardboard, collected every 4th week for sorting and recycling

- **Brown bin** for biological waste for central composting collected on alternate weeks, or a **home composting bin** for biological waste
- **Grey bin** for waste to be landfilled, collected on alternate weeks
- **Small red plastic container** for household hazardous waste (solvents, oil, herbicides etc.) collected twice a year (the 30 litre container is to be distributed in June 1995)

In addition every household has a small **brown bucket** to keep in the kitchen cupboard. With every kitchen bucket we distribute 150 strong **paper bags** for hygienic disposal of the food waste. We believe it essential to have a good, hygienic system if we are to succeed in changing the kitchen habits of every inhabitant of Kristiansand. The paper bag also improves the working environment for the refuse collectors, and the cleanliness of the brown bin.

OTHER SERVICES

Source separated household waste can also be brought to:

Several centrally placed 'bottle banks' where glass bottles for material reuse, and clothes for reuse (Salvation army) can be deposited free of charge.

Three stations (petrol stations) where hazardous household waste (red container) can be safely delivered free of charge.

A large recycling centre for material reuse, where free of charge, metals, building waste, cardboard, green waste from gardens, glass, clothes, etc., are sorted into separate containers. The second station will be opened autumn 1995.

One repair shop where used/repairable household goods (furniture etc.) can be given for repair and resold through the Salvation Army. The repair shop provides work for the unemployed, and some income to the Salvation Army. The repair shop reduces the waste deposited on the landfill, it provides cheap furniture etc. for young people and others, and has a 'feel good factor' for the givers.

For parents of babies and children below the age of two, we have recently introduced a beginners pack with cloth nappies. For NOK 250.- they are offered a set of nappies which in the shop would cost around NOK 800.-. The hope is that this will be an incentive to substantially reduce the use of disposable paper nappies.

FLEXIBILITY

The new system is very flexible. We can use the green bin to collect waste fractions to be sorted and recycled. In late April this year we introduced a system where empty rinsed and closely packed milk and juice cartons, packed in a tied plastic carrier bag, can be placed in this bin. These are sorted together with the cardboard and paper at the sorting plant and sold for material reuse.

The 'bottle bank stations' and the recycling centres can easily be adapted to receive other waste fractions for material reuse.

CHARGES AND SHARED BINS

We have also introduced substantial system changes in the charges to the individual household/customers. We now allow shared bins between neighbouring properties (reducing the number of bins), and the charge reflects the actual cost incurred by the individual customer. In the pricing system, all the 'free services' (bottle bank, hazardous household waste, recycling centres etc.) are added on to the cost of the grey bin. This is a strong economic incentive to reduce waste to be deposited. The table below shows the actual charges:

Table 1: Price in NOK/year (ex. VAT) for different bin types and volumes

Type of bin	120 litre	240 litre	660 litre	1100 litre
Brown bin for biological waste. Collected every 2nd week**	280*	600*	---	---
Green bin for paper/cardboard. Collected every 4th week	90	140	430	695
Grey bin for waste to be deposited. Collected every 2nd week	475	880	2410	4080

* An additional annual charge of NOK. 80.- for every set of bucket/paper bags is added.
** Those choosing home composting as an alternative to the brown bin get an additional annual charge of NOK. 100.- to cover the cost of this system. The people wanting to compost their own biological waste at home, have to enter into a written contract with the municipality. They have to buy their own insulated composting bin, or make it themselves. They do however save the cost of having the brown bin collected. We run free courses in home composting. The courses are very popular. We have now about 1900 households home composting.

COLLECTING AND COMPOSTING THE BIOLOGICAL WASTE

We chose the wheeled bin, and the use of paper bags partly to improve the hygienic standard in the households, and partly to improve the hygienic standard for the refuse collectors. The experience so far is good. There are no problems with liquid seepage, or smell from the collection lorries. These are of course early days, and the summer will be the real test. We have a contingency plan prepared. If the summer should prove to bring unacceptable smells from the brown bins, we will increase collection frequency to once a week.

The biological waste is composted in windrows. We have several years experience in composting sewage sludge in this manner. To cope with the increased amount we invested in a turning machine, a sieving machine and a shredder. Apart from some initial problems with the new turning machine the system functions well. A new mechanical composting plant is to be constructed on the site of the new landfill.

PUBLIC PERCEPTION AND PARTICIPATION

The new system was necessary both to comply with modern waste management in a sustainable way, and to comply with expected government regulations for our new landfill. The political process leading to the decision in August 1994 was time consuming but almost unanimous. The little opposition we had was only concerned with the extra cost to the consumer.

The general public at this deciding stage was informed through articles in the local press, and were surprisingly acquiescent. When we sent an information brochure to every household, describing the system in detail, informing them about when the new bins would be delivered and asking them whether they would wish to have a brown bin or a home composting bin, all hell broke loose. We became the centre of local news interest with daily interviews in local press and radio. We became the target for all the customers who disagreed with the necessity for source separation of biological waste. The media storm lasted about four weeks. It was an interesting and taxing period where the whole of the local population seemed to participate.

The media attention gave ample opportunity to get information about the system and the environmental reasons why it is necessary across. In the beginning many people seemed to be against the brown bin, and disliked having to sort kitchen waste. But the tide turned as most people after a period seemed to grasp the sound reasons for the revised system. During this time we also got some letters of support to the newspapers from 'green' individuals and organisations. It is probably fair to say that the majority of people were sceptical but willing to try. They recognise the environmental arguments, but are concerned about the summer, when they fear problems with smell and insects. At present it is our impression that most people are loyal to the system and find it no problem.

The distribution of the brown bins was mostly carried out in November 1994. Due to the individual option of home composting or a brown bin, and the option of shared bins, the logistics became complicated. This was further complicated by reducing the collection of the grey bin from once a week to once a fortnight, which meant that quite a few of our customers had to change their grey bin from 120 litres to 240 litres. Every day for weeks before, during and after the change we had long queues of customers wanting to change their subscription, taxing both the office and the people distributing the bins. We also had much unrest with the contractors carrying out the collection. They feared the change would lead to more work, and wanted to re negotiate the contracts. In fact the change in total workload was marginal. We will have a review of the contracts after six months.

FOOD WASTE FROM LARGE KITCHENS AND FOOD PROCESSING PLANTS

From food processing plants, restaurants, hotels, large institutional kitchens, food shops and other places producing more than 50 litres of food waste each week, there is a separate compulsory waste food collection. The waste food is delivered to a local pig breeder for sterilisation and use as pigfeed. The food is placed in bins, which are collected once a week or more frequently. The customer gets a new clean bin at collection time.

INDUSTRY AND COMMERCE

The next stage of the source separation project is to work with local industry and commerce. In January 1995 we introduced differentiated prices at the landfill. The real aim is for industry and commerce to source separate, and cost seems to be the most effective tool to motivate them. In addition we will have a local government adviser to offer free advice. This work is at an early stage, but has been well received.

RESULTS

In February we carried out a preliminary investigation, to see how well people sort their waste. The results with regard to purity in the green and brown bin was very encouraging. Both contained less than 1% impurities, mostly plastic. The grey bin was, however, not quite so uplifting. It contained on average 16% cardboard/paper, and 24% biological waste.

Preliminary results indicate that the three bin system recovers 52% of the waste (25% paper and cardboard, 27% biological), while 48% are landfilled. The total system for household waste collection recovers 60% of the waste. The final results are likely to improve as more people becomes loyal to the new system.

The investigation also indicated that it is in established residential areas where sorting is poorest. Our impression from introducing the system is that it is in the established generation that we find the strongest opposition to source separation of household waste. Young people are clearly on average better motivated.

In the autumn we will carry out market research to find out exactly how people feel about the new system. By then we will also have more reliable figures for the actual amount of source separated waste collected. At this stage it is a clear impression that most people have accepted the new system and approve.

Trends in European Management of Urban Organic Wastes

Wojciech Rogalski, Municipal Administration of the City of Vienna - Waste Management, Einsiedlergasse 2, A-1050 Vienna

ABSTRACT

In many European countries, the treatment of natural organic wastes (biowastes) has become one of the most important tasks of modern waste management. The volume of biowastes, which is continuing to increase almost everywhere, as well as the notion that natural organic wastes are valuable recyclables, not refuse to be dumped, have triggered a dynamic development in this sector.

The ISWA working group "Biological Treatment of Waste' has for some years been studying the current state of the art in some European countries as well as future developments in this field. The book "Status and Trends for Biological Treatment of Organic Wastes in Europe'[1] was presented as a result of this work.

In the present contribution, experimental values of waste management and technical data collected in the countries presented in the above mentioned book, i.e. Austria, Denmark, Finland (Helsinki Metropolitan Area Council), France, Italy, Netherlands, Norway, Spain and partly Germany, are summarised and diagrammatically represented, with additional information supplied by Germany and Switzerland.

KEYWORDS

Natural organic wastes (biowastes), biological waste treatment, treatment plants, quantity and quality of natural organic wastes, composition of wastes, compost, quality of compost, application of compost, individual composting, separate collection.

INTRODUCTION

Due to the different political, organisational, demographic and climatic conditions of the countries studied, some of the required parameters cannot be compared directly. However, by applying the above mentioned frame conditions, it was possible to study the following aspects:

- legal basis of biological waste treatment
- quantities, quality, composition of wastes, in particular of natural organic wastes
- waste management concepts, collection systems, treatment methods, areas of application
- individual composting
- trends and system modifications
- synonyms used
- quality criteria for composts
- problem areas
- studies, surveys and preliminary overview

MATERIALS AND METHODS

The data required for this work were collected at two levels:

Level 1 - National reports
Level 2 - Description of plants and facilities

Exhaustive questionnaires were developed for both levels of the study and mailed to the corresponding contact persons, usually the national ISWA member.

While the first questionnaire was conceived as a guideline for the preparation of a national report - the already completed first draft for the Danish report was additionally enclosed - the second questionnaire comprised 374 questions relating to the installations, partly of a very detailed nature.

The extent of the ISWA book thus depended on the number and quality of the questionnaires that were returned. Also, only information submitted until copy deadline of the book could be included in this contribution.

The data submitted were processed and summarised in five chapters:

- Chapter 1: State of the Art and Future Perspectives
- Chapter 2: Summary of National Reports
- Chapter 3: National Reports
- Chapter 4: Plants and Facilities - Summary of Data
- Chapter 5: Plants and Facilities - Technical and Operational Data

The appendix contains colour illustrations of some of the plants and facilities presented.

RESULTS AND DISCUSSION

Quantities, extent of coverage and existing potentials
Currently, there are two collection concepts in use in the above mentioned countries: while separate biowaste collection was already introduced in Austria, Denmark, Germany, the Netherlands, Switzerland and partly also in Finland and Norway, domestic waste composting is still used in some parts of France, Italy and Spain.

Legal basis
Separate biowaste collection is the objective for almost all countries, including those where domestic waste composting is still in use.

Management of Urban Biodegradable Wastes

Table 1: Collection and treatment of natural organic wastes

Country	Population	Pop. density (inh./sq km)	Plants and facilities*) no.	Plant capacity**) (t/a)	Potential (t/a)
Austria	7.884,000	94	62	620,000	1.180,000
Denmark	5.150,000	120	112	280,000	850,000
(Finland)	(4.998,478)	(15)			
Helsinki	850,000	1,144	1	3,000	
France	56.000,000	102	76	1.900,000	10.000,000
Germany	80.594,000	226	500	5.000,000	9.000,000
Italy	57.739,000	192	33	1.298,000	8.420,000
Netherlands	15.000,000	367	22	1.415,000	1.700,000
Norway	4.300,000	13	2	19,000	570,000
Spain	38.900,000	77	24	2.544,000	6.800,000
Switzerland	6.943,000	168	161	350,000	920,000

*) excl. small-scale agricultural plants
**) incl. small-scale agricultural plants

Table 2: Specific production of natural organic wastes utilisation and potentials

Country	Potential per inhabitant and year (kg/inh.a)	Utilisation per inhabitant and year (kg/inh.a)	Coverage rate (%)
Austria	150	80	53
Denmark	170	40	24
(Finland) Helsinki	90	30	33
France	180	20	11
Germany	120	60	50
Italy	150	20	7
Netherlands	110	70	64
Norway	130	5	4
Spain	180	20	11
Switzerland	130	50	39

Table 3: Introduction of separate collection legal requirements

Country	Legally mandatory as of	Projected
Austria	1995	
Denmark		1996
(Finland) Helsinki	1993	
France	no information submitted	no information submitted
Germany	1993	
Italy		1988
Netherlands	1994	
Norway	no information submitted	no information submitted
Spain		late 1990s
Switzerland	1995	

The legal situation is rather complicated in some countries. On the one hand, there are several levels of legislation (federal republic - provinces, *Länder* and cantons in Austria, Germany and Switzerland), which can lead to complications above all in the field of environmental protection, an area difficult to adapt to political boundaries; on the other hand, decisions of one and the same legal body may have quite contradictory effects. For example, in Austria there may exist laws and ordinances which entail a dramatic increase of compost quantities while at the same time other legal requirements prohibit the use of compost as fertiliser.

Contaminants and disturbing substances, discussion of limit values

The high content of contaminants and disturbing substances to be found in mixed refuse is one of the most basic reasons why more and more countries are opting for the separate collection of natural organic wastes.

In the meantime, it has been established in several studies that separate collection is the only approach resulting in a low contaminant level and that there is no treatment method that actually permits separating these contaminants from the residual wastes. The heavy metal limits for compost vary from country to country:

Table 4: Comparison of heavy metal limits (the lowest values of the respective country)

Country	Parameter (ppm)						
	Cr	Ni	Cu	Zn	Cd	Hg	Pb
Austria	70	42	70	210	0.7	0.7	70
Denmark		30			0.8	0.8	120
France		200			8	8	800
Germany	100	50	200	440	1.5	1	150
Italy	50	50	100	300	3	2	100
Netherlands	50	10	25	75	0.7	0.2	65
Spain	750	400	1,750	4,000	40	25	1,200
Switzerland	100	30	100	400	1	1	120

Application of composts derived from separate biowaste collection

Compost is above all a soil conditioner and fertiliser. Due to the drastic reduction of the admissible limits for the content of heavy metals in some countries, it is possible that in the future only very small quantities of compost will be permitted e.g. in agriculture. There is already talk of reducing the admissible compost quantity to less than 5 t/ha.a (Austria, Denmark, Switzerland). However, in many areas the humus loss caused by erosion and digging is 4 - 6 t/ha.a.

Tests carried out in Vienna have demonstrated that e.g. in biological cultivation the optimum fertilising effect is in the range of 15 - 20 t/ha.a. Often, the natural (geogenic) heavy metal content of the soil or airborne emissions are not taken into account at all. In many cases, the utilisation of compost in reasonable quantities could lead to a reduction of the existing heavy metal concentration in the soil. Apart from these arguments, the application of 5 t/ha/a is practically not implementable.

Key problem areas

The odour problem

Of course, the danger of more manifest odour emission is greater for open facilities than for closed composting plants. Moreover, anaerobic fermentation has many advantages in this respect. However, even the most up-to-date and expensive technologies cannot effectively preclude odour emission during the fermentation process.

The fact that the odour problem can in many cases be defused by installing composting plants in suitable locations - thereby forgoing the deployment of complicated technologies - is often left out of consideration.

Marketing

Yet the interesting fact remains that compost is more and more often given away for free. This should, however, not astonish us too much. The competent local authorities are forced by the applicable laws to produce compost. It is evident that potential buyers take a wait-and-see position, assuming that the municipality or local authority will be forced to get rid of the compost sooner or later. In Germany, even negative pricing is not unheard-of.

Table 5: Prices obtainable for composts derived from organic wastes

	ATS/cubic metre
Austria	0 - 500
Denmark	0 - 200
France	0 - 400
Germany	0 - 300
Italy	0 - 200
Netherlands	0 - 100

Another phenomenon concerns the massive unwillingness of farmers to use compost. Apart from the actual additional costs involved in compost application, this may be to a certain degree triggered by anxiety. Recently, a general discussion erupted in Germany concerning the liability of compost producers for possible damage caused by compost application. There are plans to establish a compost guarantee fund in analogy with the model of the already existing sewage sludge guarantee fund. However, according to more sceptical voices, such a measure could destroy the still favourable reputation of compost as fertiliser for good.

Vienna has for several years been using a closed-cycle system. The compost produced is applied to the city-owned cultivated land. Vienna's agriculture is gradually shifting towards biological cultivation. In this way, it is possible to safeguard biowaste collection and composting while at the same time introducing valuable nutrients and accessory agents into the soil. In Vienna, you can already buy foodstuffs produced by Viennese agricultural enterprises using biowaste compost.

Health problems connected with the collection and treatment of biogenous wastes
Not too long ago, the press featured scare stories alleging that the use of organic waste containers (called 'Biotonnen' in Austria) can entail health hazards. In the meantime, many of these items have been largely disproved by scientific studies.

Further studies are required for many other areas including the safeguarding of jobs. The ISWA working group 'Biological Treatment of Waste' will deal more extensively with these issues in the future.

CONCLUSION

The state of the art presented and the overview of the existing and projected scientific studies and practical experiments show clearly that biological waste management is still experiencing a dynamic development phase. The following trends were identified:

- continuing high importance of natural organic wastes in waste management
- development of new perspectives for the biological treatment of residual wastes as an alternative to waste incineration (mechanical-biological treatment methods)
- dramatically increasing quantities of separately collected natural organic wastes (biowastes), effects on the development of collection and treatment systems
- introduction of separate organic waste collection in almost all countries studied
- anaerobic technologies and combined processes as effective methods for the future

REFERENCES

1. ISWA - Status and Trends for Biological Treatment of Organic Waste in Europe - Edited by W. Rogalski and J. Charlton, Published in Vienna 1995, Copyright ISWA
2. Biological Treatment of Organic Waste in Switzerland, Catherine Fischer, Johanna Lott - Fischer, Mauro Gandolla
3. Biologische Abfallbehandlung, Enzyklopädie der Kreislaufwirtschaft, Karl J. Thome - Kozmiensky

Section 4

Occupational Health

Emission of Spores of *Aspergillus Fumigatus* from Waste Containers and Piles

K. Messner and Ch. Mark, Institute of Biochemical Technology and Microbiology, Department of Mycology, University of Technology Vienna, Getreidemarkt 9, A-1060 Vienna, Austria.

ABSTRACT

The emission of spores of the ubiquitous mold fungus Aspergillus fumigatus from biowaste containers was determined, using a viable airborne spore sampler of the particle-impaction type. Measurements were performed at sites representative of both urban and suburban areas in Vienna. Spore emission from household waste bins (i.e. containing both organic and inorganic household waste) was also determined and compared to the spore concentration values measured at biowaste containers. The measurements showed that only in very rare cases elevated numbers of spores of Aspergillus fumigatus are emitted by biowaste containers. At most of them and at all of the household waste containers the spore counts were in the range of the naturally occurring background level. Around composting plants, the spore density is dependent on the treatment step during composting and the distance from the plant. After appr. 1400 m the spore count reaches the background level. An attempt to assess the health risk for users of biowaste containers is made.

KEYWORDS

Aspergillus fumigatus, aspergillosis, health, air sampler, spore count, waste container, biowaste, composting, risk assessment.

INTRODUCTION

The ubiquitous, opportunistic mold fungus *Aspergillus fumigatus* is increasingly recognised as a major pathogenic agent. The uptake of spores occurs primarily via the airways and lungs. A diameter, ranging from 2-4 µm, enables the spores to penetrate into the innermost regions of the lungs. Two kinds of diseases can be distinguished: allergies and aspergilloses. While allergies are caused by fungal spores, the pathogen in aspergilloses are growing fungal hyphae. Allergic responses are rhinitis, asthma and extrinsic or acute alveolitis, the latter is considered to be a typical occupational disease. The most important forms of aspergillosis are: allergic aspergillosis, mycetoma and invasive aspergillosis, respectively. The risk of acquiring an aspergillosis exists almost exclusively for patients with an impaired immune system as a result of an underlying disease (leukaemia, lymphoma, solid tumors, HIV) and their therapy (immune suppressant drugs, radiation). The background of allergies and aspergillosis is well documented [1-7]. However, the risk assessment is not always easy.

Aspergillus fumigatus grows within a broad temperature range reaching from appr. 15° to 55°C and is part of all fungal floras of organic composts. The human body temperature lies within its optimum growth range. Due to an increasing trend of separate organic waste collection and its centralised composting by the city administrations during the last years, people became increasingly concerned about *Aspergillus fumigatus* as a source of fungal infection. Contradictory statements reported in the press regarding the risks of infection involved added to the general confusion.

The objective of this paper is to present basic data on spore counts emitted by urban biowaste containers and the composting plants into the surrounding air. The investigations reported in this paper were launched by the municipal authorities of Vienna to enable a better assessment of the risks the citizens are exposed to by the collection and composting of biowaste. For comparison, spore counts of various human environments were investigated.

MATERIALS AND METHODS

Concentrations of viable airborne spores were measured using a particle impact collector, consisting of a ventilator to generate an air current that is directed through a sieve plate onto a contact plate containing oxgall-antibiotic agar, selective for *Aspergillus fumigatus* (agar 20g, peptone 10g, maltose 10g, oxgall 12g, 2mg chloramphenicol, aqua dest 1l). The air current is deflected by an angle of 90° leading to a deposition of particles onto the agar; where they are fixed by a thin film of water covering the medium. After exposure, the plates are removed from the air sampler and incubated at 45° C. These measures allowed *Aspergillus fumigatus* to grow rapidly, while restricting (at least for three days) practically all bacteria and fungi except some species of *Humicola* and *Mucor*. For identification of *Aspergillus fumigatus* colonies the criteria of Raper and Fennell were applied [8]. The detection limit of the spore samples except the ones taken at containers after reopening the lid was 0.033 spores per litre air. For the latter ones the detection limit was 0.011 spores per litre. The values below detection limit, termed 'bdl' in the tables, entered the calculation of the average values as zero.

Air-sampling at containers

Four types of containers were selected for investigations: 1) biowaste container in urban area (bwc-u), containing a higher percentage of organic waste from kitchens; 2) biowaste container in suburban area (bwc-s), containing mainly residues of gardening activities; 3) household waste container in closed rooms (hwc-r); 4) household waste container in the open (hwc-o).

The biowaste containers were emptied once a week and the household waste containers one to six times by the full service collection. Biowaste containers inside buildings had not been introduced for hygienic reason by the municipal authorities.

Air sampling was performed by placing the sampler at a distance of about 20 cm from the opening of the containers. Two samples were taken consecutively: Sample a: Emission of spores at deposition of waste; the lid of the container was lifted and immediately after deposition of waste, air was drawn through the sampler for 20 seconds, corresponding to a volume of 30 l. Sample b: Emission of spores only at lifting of the container lid; after closing the container, the lid was again lifted rapidly, thus generating an air current that would expel spores into the open. This time, 90 l of air were sampled.

In the first set of experiments 4 containers for type 1) (n=44), 5 containers for type 2) (n=50), 3 containers for type 3) (n=60) and 2 containers for type 4) (n=25) were chosen. As it was found that one container of type 1) emitted an elevated number of spores at two days, 14 randomly selected organic waste containers in urban areas were investigated additionally. The whole study is based on 193 air samples.

Air-sampling in the surrounding of composting plants
The air sampler was placed at various distances from the plant and the spore counts were determined according to the method described in the last chapter.

RESULTS AND DISCUSSION

Spore counts at containers
To be able to assess the risk of fungal infection by *Aspergillus fumigatus* for citizens when using biowaste containers, the results were not only expressed by mean values, but also by the maxima and minima and also by the living spore count$_{90}$ (lsc$_{90}$), showing the spore concentration (spores/m^3 air) which is not exceeded by 90% of the samples. This value does not take into account rarely occurring events of extremely high spore counts. The lsc$_{50}$ value expresses the spore concentration which is not exceeded by 50% of the values.

Table 1: Average values of spore concentrations of aspergillus fumigatus *in the surrounding air of waste containers (spores/m^3 air, bdl = below detection limit)*

	Disposal	Opening	Total
1) bwc-u	360	39	150
2) bwc-s	31	5.4	14
3) hwc-r	bdl	1.8	1.3
4) hwc-o	18	4.1	9.1
background level			7

Table 2: Average values of the four biowaste containers in urban areas (bwc-u) of table 1

Location	Spores/m^3 air
A	520
B	15
C	13
D	1.9

Tables 1 and 2 show that an elevated value of the spore counts was measured only with urban biowaste containers. To avoid misinterpretations it is of great importance to analyse this value in detail. The relatively high value, compared to the other types of containers, is caused by two measurements obtained at one single biowaste container within a time difference of two weeks. At the first measurement the spore count was 3500 spores/m^3 when waste was thrown into the container and 1700 at the second measurement. When opening the lid only, 790 and 130 spores, respectively, were counted per litre of air. All the other counts at the same container were in the range of uncontaminated air (7 spores/m^3).

Table 3: Maxima/minima of spore concentrations of aspergillus fumigatus *in the surrounding air of waste containers (spores/m^3 air; bdl)*

	MAXIMA	MINIMA
1) bwc-u	3500	bdl
2) bwc-s	230	bdl
3) hwc-r	11	bdl
4) hwc-o	67	bdl
Random samples	33	bdl

Table 4: lsc_{90} and lsc_{50} of the spore counts of aspergillus fumigatus *at waste containers (spores/m^3 air)*

	lsc_{90}	lsc_{50}
1) bwc-u	67	bdl
2) bwc-s	33	bdl
3) hwc-r	11	bdl
4) hwc-o	33	bdl

Table 5: Average values of spore concentrations of aspergillus fumigatus *in the surrounding air of 14 randomly chosen urban biowaste containers (spores/m^3 air)*

Disposal	Opening
2.4	4.4

It is clearly shown by table 5 that the spore counts obtained at most of the biowaste containers are within the range of the surrounding air. The average value for the surrounding air of all the test sites was 7 spores/m^3 air. That means that in most cases *Aspergillus fumigatus* did not develop in biowaste containers. Table 4 also demonstrates that at fifty percent, or more, of the containers the spore count was below limit of detection. 90% of the biowaste containers did not exceed a count of 66 spores per m^3 of air.

Nevertheless, it has to be mentioned that in exceptional cases very high counts are possible. To avoid these cases, an explanation should be found. The fact that all the other values of the same container were average clearly demonstrates that the content of the bin must be the main responsible factor. Household waste from gardens and probably also most of the plant residue coming from kitchens is not expected to heat up and enter the thermophilic phase until emptying of the bin after 7 days, giving rise to *Aspergillus fumigatus*. It can be assumed that either pre-composted material was filled in or very easily degradable waste like broths, favouring the growth of fungi, was filled into the container. The fact that nearby this container a restaurant is located may also speak for the second option.

Spore counts at composting plants and at distant areas

The biowaste containers are collected and transported to a place outside Vienna, where it is chopped to a smaller particle size, mixed with other organic waste materials in a drum and transported to a composting plant called 'Lobau'. There it is piled up to compost heaps.

During the rotting phase *Aspergillus fumigatus* dominates, reaching spore counts of appr. 2 million spores/g dry weight [9-10]. During this phase the compost heaps are turned to guarantee sufficient aeration [10]. When the compost temperature reaches appr. 53°C, *Aspergillus fumigatus* is killed. When it cools down, the compost becomes partly recolonized. After a resting phase, where the final decomposition takes place, the compost is sieved and packed for the end user.

The aim of this investigation was to measure the spore counts in the surrounding air at various distances from the composting plant, to be able to assess the risk of fungal infections for people in the vicinity of the composting plant. The air samples were taken at wind speeds of 0-40 km/h downwind from the composting plant.

Table 6: Average spore counts of aspergillus fumigatus *in the vicinity of the composting plant at various phases of composting (spores/m^3)*

distance from piles (m)	turning	resting	sieving
0-100	4500	74	87
100-400	1100	49	78
400-1000	670	7	36
1200-1400	7	7	7

Table 6 shows that above a distance of 1200 to 1400 m from the composting plant the spores were diluted and the average level of uncontaminated air is reached. Within this range the spore count is dependant on the activity and the distance.

Spore counts at various sites with expected elevated values

To obtain further data on spore counts of *Aspergillus fumigatus* 17 sites were investigated. Most of the sites, including leaf litter in woods and glass houses for tropical plants did not show higher counts than average. The highest counts were found when hay was present. Even in hay lofts the counts are very low as long as hay is not agitated. When hay was moved, the counts rose up to 4300 spores/m^3.

Two observations can be made from these results; i) When organic material is rotting at mesophilic temperature - as this is the case with leaf litter - other fungi than *Aspergillus fumigatus* are dominating. It usually only dominates at temperatures between 36 and 46°C. ii)

Surprisingly, as was the case in hay lofts, the spore counts of the surrounding air of compost heaps was also low as long as the compost was not turned. These two observations clearly explain why normally no spores are emitted by organic waste containers.

RISK ASSESSMENT

The alveolus in the lungs of an adult person have a diameter of 0.1-0.5 mm. Thus, the spores of *Aspergillus fumigatus* having an average diameter of 2.8 µm can easily penetrate the respiratory tract and will be deposited at their walls. At low spore counts, the spores are either phagocytised by the reticuloendothelial system or transported back into the trachea by the ciliated epithelium [11]. The breathing volume at resting phase is 0.5 litre, the maximum volume appr. 2.5 litres. If this volume is considered in relationship with the measured spore counts, it can be calculated that even at high spore counts only low numbers of spores are inhaled. At the highest spore density of 3500 spores/m^3, measured at biowaste containers, only 1.8 spores are inhaled at resting phase and 8.8 spores at maximum inhalation. Let us assume a manipulation time of one minute at the container and a number of 16 breaths per minute, we end up with a total of inhaled spores of 28, or 140 spores, respectively. Considering additionally that only 2 out of 193 samples at containers showed clearly elevated spore counts and that to acquire an allergic disease or an aspergillosis by *Aspergillus fumigatus* an exposure to high spore counts over a very long period of time is necessary, the health risk when using biowaste containers is almost zero for people with a sound immune system. Even for the risk group with an underlying disease the number of spores inhaled seems to be to low. Similar calculations and conclusions can be made for the risk assessment in the remote surrounding of composting plants.

For areas where people with suppressed immune systems are staying, as e.g. hospitals, reconvalescent centres or geriatric institutions, the use of biowaste containers should be avoided also because of a higher probability of a contamination of the air conditioning system [4]. Attention should be paid to the protection of workers employed at composting plants. One of the ways to assess the risk of aspergillosis when exposed to the spores of *Aspergillus fumigatus* is to examine the health conditions of people who have been working for a longer time or temporarily at sites where elevated spore counts prevail and to compare the intensity of exposure to the one to be investigated. Typical places with high spore counts of *Aspergillus fumigatus* are hay lofts and mushroom farms, where counts of up to 90,000 spores per cubic meter air are reported [12]. Typical occupational symptoms were reported for workers employed at these sites, such as farmer's lung and mushroom grower's lung. However, a final risk assessment has to be based on statistic medical data on the number of occasions of aspergillosis in areas of low and high spore densities. This paper is intended to deliver a tool for these studies by defining the concentrations of airborne spores of *Aspergillus fumigatus* at sites where people are concerned about being exposed to risks.

CONCLUSIONS

This study gives an overview of the number of spores of *Aspergillus fumigatus* emitted by organic waste material from its collection in biowaste containers to its final product - compost. Although a larger number of samples will be necessary to obtain statistically safe values, it seems to be a valuable tool for doctors to assess the risk of infections by *Aspergillus fumigatus* coming from biowaste containers and composting plants. It was clearly shown that the spore counts at most of the biowaste containers did not exceed the background spore count in the surrounding air. Around composting plants, the spore counts were elevated only during the turning phase, but reached the background level in a distance of up to 1,400 meters. The most striking conclusion which can be drawn from the results is that the health risk exerted by biowaste containers to the users is almost zero. Secondly, the risk of infections in the surroundings of composting plants seems to be rather restricted to the workers handling the compost than to people passing the zone of temporarily elevated spore counts.

Nevertheless, this article was written only by microbiologists. The final risk assessment has to be carried out by physicians.

REFERENCES

1. Tintelnot, K. & Seeliger, H. P. R. (1988) Aktuelle Gesichtspunkte zur Epidemiologie der Aspergillose. *Mycoses,* vol 31, no. 5, 245-254.
2. Weinberger, M., Elattar, l., Marshall, D. et al. (1992) Patterns of infection in patients with aplastic anemia and the emergence of *Aspergillus* as a major cause of death. *Medisine,* vol. 71, no. 1, 24-42.
3. Martino, P., Raccah, R., Gentile, G., Venditti, M., Girmenia, C., Mandelli, F. (1989) *Aspergillus* colonization of the nose and pulmonary aspergillosis in neutropenic patients: a retrospective study. *Haematologisa,* vol. 74, 263-265.
4. Klapholz, A., Salomon, N., Perlman, D. C. & Talavera W., F.C.C.P., (1991) Aspergillosis in the acquired immunodeficiency syndrome. *Chest,* vol. 100, 1614-1618.
5. Pursell, K. J., Telzak, E. E. & Armstrong D. (1992) Aspergillus species colonization and invasive disease in patients with AIDS. *Clin. Inf. Dis.* vol. 14, 141-148.
6. Moser M., Crameri R., Menz G. et al. (1992) Cloning and expression of recombinant *Aspergillus fumigatus* allergen l/a (rAsp f l/a) with IgE binding and type I skin test activity. *Joural of Immunology* vol.149, no2., 454-460.
7. Arruda L. K., Platts-Mills T. A., Fox J. W., Chapman M. D. (1990) *Aspergillus fumigatus* allergen 1, a major IgE-binding protein, is amember of the mitogillin family of cytotoxins. *Journal of Experimental Medicine* vol.172, no 5, 1529-1532.
8. Raper, K. B. & Fenell, D. l. (1965) The Genus *Aspergillus*. Baltimore. The Williams & Wilkins Co.
9. Klopotek, A. (1962) Über das Vorkommen und Verhalten von Schimmelpilzen bei der Kompostierung städtischer Abfallstoffe. Giessen: Inst. f. landwirtschaftl. Mikrobiologie derJustus Liebig Universität.
10. Millner, P. D., Bassett, D. A. & Marsh, P. B. (1980) Dispersal of *Aspergillus fumigatus* from sewage-sludge compost piles subjected to mechanisal agitation in open air. Appl. Env. Microbiol. vol. 39, 1000-1009.
11. Silbernagl, S. and Desopopoulos, A. (1988) Taschenatlas der Physiologie. Georg Thieme Verlag, Stuttgart.
12. Lacey, J. (1991) Frontiers in Mycology, Chapter 8. Honorary and general lectures from the Fourth International Mycological Congress. Regensburg 1990; Hawksworth, D.L. (ed); C-A-B- International.

Occupational Bioaerosol Exposure in Collecting Household Waste

N.O. Breum, B.H. Nielsen, E.M. Nielsen and O.M. Poulsen, National Institute of Occupational Health, Lersø Parkalle 105, DK-2100 Copenhagen, Denmark

ABSTRACT

Personal bioaerosol exposure in collecting household waste is correlated to governing parameters including quality of the waste, equipment at the households and type of equipment used for collecting the waste. It is difficult to generalise from exposure data on an individual waste collector to a large group of collectors. To solve this problem a work condition matrix was constructed using matrix elements characterised in terms of the governing parameters. Bioaerosol exposure data of a matrix element were obtained by personal sampling with the assistance of crews of waste collectors. This approach allows exposure of subgroups of waste collectors to be estimated on the basis of easily obtained data on the governing parameters. Except for garden waste the collection of mixed (non-separated) household waste kept in area containers caused the highest bioaerosol exposure level. For other types of waste and household equipment the median exposure levels were lower ranging 0.1-0.3 mg/m^3 (dust), 0.8×10^3-7×10^3 cfu/m^3 (bacteria) and 3×10^3-82×10^3 cfu/m^3 (fungi). It is concluded that the work condition matrix is useful in a generalised characterisation of waste collectors exposure to air contaminants.

KEYWORDS

Household waste collection, bioaerosol, exposure, dust, bacteria, fungi, work condition matrix, compostable waste.

INTRODUCTION

In recent years much attention has focused on recycling of household waste. The number of employees in waste collection and recycling is expected to increase and new technology is to be developed. To evaluate new technology in terms of exposure to air contaminants some base line information was recently reviewed for waste transfer stations, landfills, incineration plants and recycling plants [1], but data on exposure are sparse for waste collectors [2].

Personal bioaerosol exposure in collecting the waste is correlated to several governing parameters including quality of the waste, temperature, equipment at the households, type of truck used for the collection, and organisation of work. The multitude of governing parameters makes it difficult to generalise from exposure data on an individual waste collector to a large group of collectors. However, such generalisation is called for in epidemiological surveys on possible links between exposures and health effects. Therefore, the purpose of the present study was to solve this problem by characterising the exposure of subgroups of collectors according to their general work conditions, thereby establishing a crude work condition matrix. A matrix element is characterised in terms of the governing parameters, and measurements of bioaerosol exposure on waste collectors belonging to the matrix element provide an exposure description of the element.

MATERIALS AND METHODS

The work condition matrix

The governing parameters selected for the work condition matrix are listed in Table 1. Matrix elements already filled in with exposure data are indicated by a number referring to the list of references [3-7]. In the table some of the references are marked with a letter to indicate that the references include data on bioaerosol exposure for more than one set of governing parameters. Not all elements left open in the matrix are to be filled in - some elements are of no relevance and others may be considered of low priority in terms of number of workers subjected to these elements. Details of the governing parameters are given below.

Table 1: Structure of the work condition matrix

Type of truck	Household equipment	Type of household waste[*]		
		Mixed	Compostable	Garden
Basic compactor truck	paper sack	#3a	#5a	
	plastic sack			#7
	bin (appr. 0.1 m^3)		#5b	
	container (0.4-0.6 m^3)	#4a		
Modified compactor truck	container (0.4-0.6 m^3)	#4b		
Open truck	paper sacks		#5c	
Closed truck	paper sacks		#6a	
	bin	#3b	#6b	

[*] The number (without a letter) refers to the list of references

A basic compactor truck has a closed container for storage of the waste and at the rear a hydraulic lift for emptying bins and small containers into a magazine (the scoop) fitted to the container. When the scoop is full the waste is mechanically pushed into the container. A basic compactor truck can be modified by adding an air pollution control system, and the system included in the matrix had a sliced plastic curtain covering part of the air space above the scoop and a fan exhausted air from behind the curtain. The open truck had no technology dedicated to waste collection. The closed truck had a container fitted with a lift for loading of the waste from top of the container.

A multitude of household equipment is in use for collecting the waste. Paper sacks are often kept at fixed stands, and the waste collector has to carry the sack by hand to the truck [5]. However, a system is in use by which a cart is used for taking the sack to the truck [6]. Bins are in common use at single residences and often the waste collector has to carry the bin to the truck [3, 5], but a cart can be used for taking the bin to the truck [6]. At multi-storey houses, containers on wheels are often used for collecting the waste [4]. The garden waste is often collected in sacks of paper or plastic, but bundles of waste kept aside are also collected. Plastic sacks were included in the matrix, and in loading the waste into the scoop of the compactor truck the collector had to empty the sack to keep plastic away from the compostable waste [7].

Frequency in collecting household waste may be of importance for the potential of the waste to emit bacteria and fungi during handling. In Denmark, in general, mixed waste is collected once a week, compostable waste every second week, and garden waste once a month.

The field survey

The bioaerosol exposure level of a matrix element was obtained with the assistance of crews of waste collectors working under the specified set of governing parameters. For a 3-day period, or more, full shift personal sampling of all crew members (1-2 persons) was used for characterising the bioaerosol exposure. Dust (total particulate matter, TPM) was collected on 25 mm, 8.0 µm cellulose nitrate filters using closed-face Millipore field monitors with a 5.6 mm inlet operated at 1.9 L/min (1.25 m/s inlet velocity). The collected mass was determined by weighing the filter before and after the sampling. Closed-face Nuclepore field monitors operated at 1.0 L/min were used for collecting microbial samples on 25 mm 0.4 µm polycarbonate filters. Culturable bacteria and fungi were determined as described previously [8]. Briefly, samples on polycarbonate filters were resuspended, diluted and plated on Dichloran Glycerol agar (DG18 agar) and Nutrient agar with cycloheximide (50 mg/L) for viable counts of fungi and bacteria, respectively. The plates were incubated for 7 days at 25°C.

RESULTS

Exposure data of a matrix element were summarised in terms of the median and the range (Table 2). It is noted that the data for some elements should be considered preliminary and full details are reported elsewhere [3-7]. To assist in the view of the data medians versus reference # are plotted in Figure 1 (compostable waste) and Figure 2 (mixed waste). It is noted that microbial analysis was not performed for the samples selected from reference #3. Holding Figure 1 and Figure 2 together it is observed, except for garden waste, that the system of containers and a basic compactor truck for collecting mixed waste caused the highest exposure to dust, but a modified compactor truck reduced exposure to a level within the range (medians: 0.1-0.3 mg/m^3) of the other sets of governing parameters under consideration. The study of collecting garden waste yield the highest exposure to dust.

As for dust the picture from Figure 1 and Figure 2 on exposure to bacteria suggests, except for garden waste, the system of containers and a basic compactor truck for collecting mixed waste to cause the highest exposure, and again a modified compactor truck reduced exposure to a level at the middle of the range (medians: 0.8×10^3-7×10^3 cfu/m^3) of the other sets of governing parameters under consideration. Again the highest exposure was for garden waste.

Consistently exposure to viable fungi exceeded that of viable bacteria. Figure 1 and Figure 2 suggests, except for garden waste, the system of containers and a basic compactor truck for collecting mixed waste to cause the highest exposure to viable fungi, but again the modified compactor truck reduced exposure to a level at the middle of the range (medians:

3×10^3-82×10^3 cfu/m^3) of the other sets of governing parameters under consideration. The highest exposure was for garden waste.

Table 2: Exposure data of the work condition matrix

Ref. No.*	TPM	Bacteria	Fungi	Ref. No.*	TPM	Bacteria	Fungi
	mg/m^3	10^3 cfu/m^3			mg/m^3	10^3 cfu/m^3	
#3a	0.15; N=3** 0.1-0.2	-	-	#5b	0.3; N=7 0.2-0.5	6; N=7 1.6-7.6	82; N=10 20-313
#3b	0.3; N=21 0.2-0.5	-	-	#5c	0.2; N=6 0.1-0.5	7; N=6 3-40	28; N=6 12-58
#4a	0.5; N=6 0.3-0.6	20; N=6 4-120	250; N=6 30-340	#6a	0.11; N=6 0.06-0.16	2; N=6 <0.5-3	54; N=6 40-120
#4b	0.2; N=6 0.1-0.5	3; N=5 1-30	40; N=5 20-60	#6b	0.2; N=6 0.1-0.3	0.8; N=6 0.3-27	3; N=6 1-4
#5a	0.3; N=9 0.09-0.7	3; N=10 0.9-94	50; N=10 16-142	#7	0.8; N=6 0.6-1.2	60; N=6 20-280	370; N=6 80-660

* See Table 1. ** Median, number of observations, range

For all the field surveys the background bioaerosol concentration was at or below 0.08 mg/m^3 (dust), 0.7×10^3 cfu/m^3 (viable bacteria), and 0.8×10^3 cfu/m^3 (viable fungi). Air temperature throughout the surveys ranged 10-25 °C.

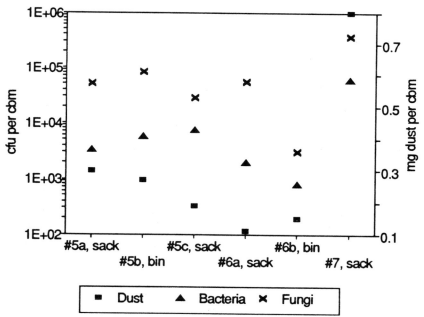

Fig. 1: Median exposure levels to bioaerosols in collecting compostable household waste versus reference # (see Table 1)

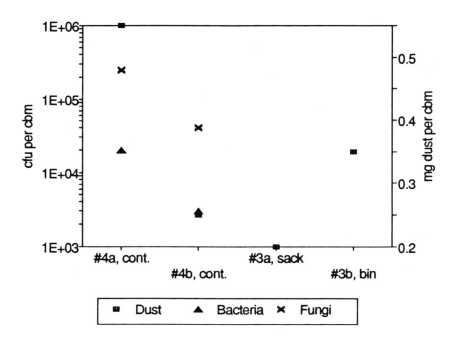

Fig. 2: Median exposure level to bioaerosols in collecting mixed household waste versus reference # (see Table 1)

DISCUSSION

Data on bioaerosol exposure in collecting household waste are sparse. Taking an approach of area sampling a study [9] on bioaerosol exposure in collecting mixed household waste kept in area containers reported concentrations ranging 1.4×10^4-10^5 cfu/m^3 (viable fungi) and 10^2-4×10^3 cfu/m^3 (viable gram-negative bacteria) close to the scoop of the compactor truck. To a degree data of the present study were consistent with those levels. However, for an exposure evaluation to be valid data on breathing zone concentrations are needed. Taking an approach of personal sampling a recent study [10] on collection of compostable household waste reported bioaerosol concentrations at or below 0.2 mg/m^3 (dust), 1.5×10^3 cfu/m^3 (bacteria), and 6.6×10^4 cfu/m^3 (fungi). The data obtained for the work-condition matrix [3-7] were consistent with those exposure levels.

After the collection of household waste and further down stream in the waste treatment process rather comprehensive data are available on occupational bioaerosol exposure. Fungal concentrations up to 10^6 cfu/m^3 were reported in tipping halls for waste transfer stations [11] and concentrations of bacteria and fungi numbered up to 7×10^4 cfu/m^3 in the tipping hall of an incineration plant [12]. At Finnish landfills [13] the concentrations in mechanically ventilated (no dust filter fitted to the systems) cabins of waste compactors were up to 4.9×10^4 cfu/m^3 (bacteria) and 7.3×10^4 cfu/m^3 (fungi). The obtained exposure levels for waste collectors (Table 2) were at or below the reported levels further down stream in the waste treatment process. However, it is noted that different sampling and analytical techniques were used for the studies mentioned so results are difficult to compare. For the sampling it is noted that airborne microorganisms are under stress and some are not very resistant to desiccation. Many bacteria are expected to die rapidly when aerosolised. In addition sampling onto membrane filters may also damage the organisms rendering them unable to grow on culture media [14]. In this paper only dust and viable microorganisms were used for characterising the bioaerosol exposure of a matrix element. Consequently the reported exposure levels to microorganisms should be considered biased towards a low exposure. The true exposure to live and dead microorganisms (total counts) can be measured by epifluorescence microscopy [3-8]. A full description of a matrix element in terms of exposure data, including total counts of microorganisms and endotoxin content of the dust, are reported elsewhere [3-7].

In literature no officially recognised occupational exposure limits seem available for airborne microorganisms. However, in compost handling a case [15] of acute ODTS (Organic Dust Toxic Symptom) was related to fungal concentrations ranging 1×10^6-5×10^8 cfu/m^3, but high concentrations were also observed for bacteria and endotoxin. For saw mill workers [16] an exposure up to 10^6 cfu/m^3 (fungi) was related to respiratory and ODTS-like symptoms. As observed from the exposure data (Table 2) of the work condition matrix full shift average exposure to microorganisms for all matrix elements were below the previously reported levels associated with respiratory symptoms and ODTS. However, it has to be emphasised that the concentration of air contaminants in the breathing zone of waste collectors fluctuate throughout the day and short term exposures may exceed the full shift average exposure level by an order of magnitude or more [7].

Local circumstances (wind velocity, temperature etc.) were comparable in the references used for the work-condition matrix. Therefore data on the filled in elements of the work-condition matrix support a hypothesis of a correlation between bioaerosol exposure and the selected governing parameters. However, the number of data is low and the spread in the data is large. It is noted that the list of governing parameters (Table 1) should not be considered exhaustive. Detailed statistical analysis of the obtained exposure data has pointed on organisation of work as an important governing parameter [4]. Such an observation was also made in another study (to be included in the matrix) on bioaerosol exposure in collecting mixed household waste [8]. Temperature is expected to be another governing parameter [5].

CONCLUSION

Bioaerosol exposure in collecting household waste is correlated to several governing parameters including quality of the waste, technical equipment for collecting the waste, and organisation of work. The work condition matrix is a useful concept in characterising waste collectors exposure to bioaerosols.

ACKNOWLEDGEMENT

The study is part of the Danish 1993-98 research programme Garbage Collection and Recycling, which is supported jointly by the Danish Ministry of the Environment and Ministry of Labour.

REFERENCES

1. Poulsen, O.M., N.O. Breum, N. Ebbehøj, Å.M. Hansen, U.I. Ivens, D. van Lelieveld, P. Malmros, L. Matthiasen, B.H. Nielsen, E.M. Nielsen, B. Schibye, T. Skov, E.I. Stenbaek and K.C. Wilkins (1995) Sorting and recycling of domestic waste. Review of occupational health problems and their possible causes. *Science of the Total Environment*, vol. 168, 33-56.
2. Poulsen, O.M., N.O. Breum, N. Ebbehøj, Å.M. Hansen, U.I. Ivens, D. van Lelieveld, P. Malmros, L. Matthiasen, B.H. Nielsen, E.M. Nielsen, B. Schibye, T. Skov, E.I. Stenbaek and K.C. Wilkins (1995) Collection of household waste. Review of occupational health problems and their possible causes. *Science of the Total Environment*, vol. 170, 1-19.
3. Breum, N.O., N. Ebbehøj, Å.M. Hansen, U.I Ivens, D. van Lelieveld, B.H. Nielsen, E.M. Nielsen, O.M. Poulsen, B. Schibby, E.I. Stenbæk and K.C. Wilkens (1995). Eksponeringer og helbredseffekter ved skift fra indsamling af usorteret husholdningsaffald til 4-strømsindsamling af kildesorteret husholdsaffald. (Collecting mixed versus source separated household waste in terms of personal bioaerosol exposure and health effects). Report from the National Institute of Occupational Health, Copenhagen, Denmark.
4. Breum, N.O., B.H. Nielsen, E.M. Nielsen and O.M. Poulsen (1995). Bioaerosol exposure during collection of mixed domestic waste - an intervention study on compactor truck design. Accepted for publication in *Waste Management & Research*.
5. Nielsen, E.M., N.O. Breum, N. Ebbehøj, U.I. Ivens, B.H. Nielsen, O.M. Poulsen, E.I. Stenbæk, C.K. Wilkins and U. Midtgård and N.O. Breum (1995). Eksponeringsvurdering samt helbredsstatus for skraldemænd beskæftiget med indsamling af komposterbart husholdningsaffald (An assessment of bioaerosol exposure and health effects for workers collecting compostable household waste). Report from the National Institute of Occupational Health, Copenhagen, Denmark.
6. Poulsen O.M., N.O. Breum, B.H. Nielsen and E.M. Nielsen (1994) Eksponering for luftforurening ved indsamling af grønt affald med Bates Combi henholdsvis Konfiskatsystemet (Bioaerosol exposure in collecting compostable household waste placed in paper sacks and containers, respectively). Report from the National Institute of Occupational Health, Copenhagen, Denmark.
7. Breum N.O., E.M. Nielsen and B.H. Nielsen (1995) Bioaerosol exposure in collecting garden waste, recyclable materials and waste for incineration. Accepted for publication in *Annals Agricultural Environmental Medicine*.
8. Nielsen E.M., B.H. Nielsen and N.O. Breum (1995). Occupational bioaerosol exposure during collection of household waste. *Annals Agricultural Environmental Medicine*, vol. 1, 53-59.
9. Ducel, G., J.J. Pitteloud, C. Rufener-Press, M. Bahy and P. Rey (1976). Importance de l'exposition bacte'rienne chez les employe's de la voirie charge's de la leve'e des ordures. *Soz.- und Präventivmedizin*, vol. 21, 136-138.
10. Sørensen, C. and H. Bjørnstrup (1993). Indsamling af madaffald fra husstande i København (The collection of compostable household waste in Copenhagen). Report No. 29/1993 from the Environmental Protection Agency, Copenhagen, Denmark.
11. Crook, B., S. Higgins and J. Lacey (1987) Airborne Micro-Organisms Associated with Domestic Waste Disposal. Report from AFRC Institute of Arable Crops Research, Rothamsted Experimental Station, Harpenden, Herts, England.
12. Rahkonen, P (1992). Airborne contaminants at waste treatment plants. *Waste Management & Research*, vol. 10, 411-421.

13. Rahkonen, P., M. Ettala and I. Loikkanen (1987). Working conditions and hygiene at sanitary landfills in Finland. *Annals Occupational Hygiene*, vol. 31(4A), 505-513.
14. Thompson, M.W., J. Donnelly, S.A. Grinshpun, A. Juozaitis and K. Willeke (1994). Method and test system for evaluation of bioaerosol samplers. *Journal Aerosol Science*, vol. 25(8), 1579-1593.
15. Weber, S., G.J. Kullman, E. Petsonk, W.G. Jones, S.A. Olenchock, W. Sorenson, J. Parker, R. Marcelo-Baciu, D.G. Frazer and V. Castranova (1993). Organic Dust Exposures From Compost Handling: Case Presentation and Respiratory Exposure Assessment. *American Journal Industrial Medicine*, vol. 24, 365-374.
16. Eduard, W., P. Sandven and F. Levy (1993). Serum IgG Antibodies to Mold Spores in Two Norwegian Sawmill Populations: Relationship to Respiratory and Other Work-Related Symptoms. *American Journal Industrial Medicine*, vol. 24, 207-222.

Work Related Symptoms and Metal Concentration in Danish Resource Recovery Workers

Torben Sigsgaard and Jens C Hansen, Steno Centre of Public Health, Department of Environmental and Occupational Medicine, University of Aarhus, Denmark
Per Malmros and Jakob U Christiansen, Directorate of National Labor Inspection, Copenhagen, Denmark

ABSTRACT

This study is a cross sectional study of recycling workers and controls. The purpose of the study was to identify adverse health effects of modern waste handling including recycling of materials from household waste. The study-group comprises 40 garbage handling, 8 composting, 20 paper sorting workers from all over Denmark and as controls 119 drinking water supply workers from Copenhagen. We have earlier described that recycling workers are prone to respiratory tract, mucosal, eye and skin symptoms. This paper describes the metal content of the blood from recycling workers, and also the gastrointestinal and musculoskeletal symptoms among these workers. We found a higher amount of Cadmium among the garbage handling workers of 3.6 (1.7) µg/l and 2.1 (1.6) µg/l among smokers and non-smokers respectively vs. 1.7 (4.3) µg/l and 0.6 (0.5) µg/l among the smoking and non-smoking controls, $p < 0.05$. Probably as a consequence of batteries contaminating the waste. Nausea more than once a month was more frequent among the garbage handling workers. Vomiting or diarrhoea in relation to work was experienced by 27% of the garbage handling workers compared to 5% among the controls, $p < 0.05$. Finally we found a higher rate of musculoskeletal symptoms from the shoulders 27% and the hands 16% among the garbage handling workers compared to 12% and 5% among the controls ($p < 0.05$). This study emphasises the need to address the working environment when recycling facilities are being planned.

KEYWORDS

Mucosal symptoms, musculoskeletal symptoms, organic dust, metals, endotoxin, bacteria, molds, dust, recycling, garbage.

INTRODUCTION

The treatment of waste from urban areas and especially recycling has been shown to have adverse health effects on people, who are occupationally exposed to these environments.

During the last 15 years, several studies have shown increased prevalence rates of gastrointestinal symptoms among workers in the sewage treatment-plants and among sewer workers. Symptoms have been characterised as acute episodes of self-limited diarrhoea [1-4]. Resource recovery and composting plants have also been shown to induce gastrointestinal symptoms among employees [5,6]. One report have shown that a newly built resource recovery plant exposed the workers to heavy metals [5]. So far no reports of adverse ergonomical effects has been published in this area.

This study of garbage handling and recycling workers in Denmark was performed to investigate the working environment and to see if the presently employed workers suffered any work-related symptoms. In an earlier paper [7] we have described the respiratory and allergic symptoms, and this paper will deal only with gastrointestinal and musculoskeletal disorders. Furthermore the blood content of heavy metals among the workers were investigated and this will also be reported here.

MATERIALS AND METHODS

Plant types

The survey includes all Danish resource recovery systems in 1991. As controls, blue collar workers from the drinking water supply industry of Copenhagen were chosen. The plants were subdivided into three categories: paper sorting where only paper was processed, compost plants, (chopping and composting garden waste only), and garbage handling plants where pre-separated fractions of mixed house-hold waste were treated by various methods, mechanical or hand sorting for subsequent reuse. The residue was either composted or combusted.

One garbage handling worker did not wish to participate, 99% of the recycling workers and 81% of the controls participated. Thus the study comprises 40 garbage handling, 8 composting, 20 paper sorting workers from all over the country and as controls 119 drinking water supply workers from Copenhagen.

Table 1: Personal and smoking characteristics of Danish recycling workers according to the different plant types mean (standard deviation in parantheses).

Plant type	Control	Paper sorting	Compost	Garbage handling
Men	119	20	8	40
Women	-	-	-	4
Mean age	44.5(11.3)	[1]37.2 (14.3)	37.5 (15.6)	[1]37.7 (11.9)
Mean time in current job (months)	146.2(120.7)	[1]62.0 (95.5)	[1]9.5 (10.9)	[1]33.5 (59.7)
Employment time in current branch (years)	12.4(10.0)	10.7 (13.3)	[1]0.8 (0.9)	[1]3.1 (5.1)
Smoking: Pack years	15.6(17.5)	9.2 (20.6)	22.8 (29.2)	12.8 (12.7)

[1] $P< 0.05$ Mann Whitney U-test group vs. Control.

Hygienic survey

Work zone samples of total dust were measured by methods specified by the National Labour Inspectorate, Denmark [8] over a work-shift with a sampling time of approximately 7 hours. Dust samples were collected on 37 mm 8µ pore size membrane filters enclosed in two-piece filter holders and weighed to an accuracy of 10^{-5}g. Samples were taken from representative workers at every job type.

Endotoxin content was determined using the limulus lysate method [9] in the dust from the filters. A lysate with a sensitivity of 0.06 EU (cape cod) was used.

Airborne bacteria and fungi were collected with impingers over 3 hours at a flow rate of 0.9 l/min. Tryptone soya agar was used for total count, Drigalsky agar for Gram-negative bacteria and modified Rose Bengal agar for fungi.

Metal analysis

Sera and blood from the participants were analysed for heavy metals. Mercury, lead and cadmium were analysed on whole blood samples by atomic absorption-spectrophotometry (Perkin Elmer 1100B). For Pb and Cd a graphite furnace (HGA300) was used and for mercury a mercury/hydride system (MH520). Analytical quality control was performed on an internal daily blood standard and a certified reference blood sample, Seronorm Trace Elements, whole blood I and II, Nycomed AS Oslo.

The other trace metals were measured by PIXE (particle induced X-ray emission) analysis [10]. Yttrium was used as internal standard.

Interview and diagnosis

The questionnaire has been reported earlier [7] and will only be described briefly in this paper.

The questionnaire was designed to identify respiratory, gastrointestinal and skin symptoms. Questions of cough and earlier diseases was taken from British Medical Research Council questionnaire of respiratory symptoms [11]. This was supplemented by questions about asthma and allergy.

Questions about gastrointestinal symptoms concerned lack of appetite, nausea, changes in stool quality, and vomiting or diarrhoea in relation to work.

Questions on ergonomical problems during the last 12 months and during the last 7 days from neck, upper extremities, back and lower extremities. Questions on medical attention for the symptoms were included. The questions were adopted from the standardised Nordic questionnaire for the analysis of musculoskeletal symptoms [12,13].

A complete occupational history was obtained. This was especially detailed for periods of employment in the industry allowing us to link these periods to the hygienic measurements.

Statistics

Tabulation, graphical analysis and analysis of the data in group means were carried out using SPSS statistical package for continuous measurements [14] To compensate for multiple comparison Duncan's test with a confidence interval of 0.05 was used in the oneway analyses. For dichotomised data a x^2 MH test was used with a confidence interval of 0.05.

RESULTS

The personal characteristics of the workers are given in Table I. Mean age and mean employment time in current job were significantly higher of among water supply workers compared to the resource recovery workers.

Table 2 summarises the hygienic survey. Dust was found in all plants with the lowest concentration in the water-supply plant. The mean concentration of 0.74 mg/m^3 in the garbage handling plants was significantly higher than the 0.42 mg/m^3 found in the water supply-plants. Sampling of microorganisms revealed great variations in the total count. Far greater counts of total microorganisms and molds were found in the resource-recovery plants compared to the very few found in the water supply plants. Significantly higher amounts of total bacteria were found in compost plants as well as garbage handling plants, compared to paper sorting plants. Significantly higher counts of Gram negative bacteria were found in the composting and garbage handling industries compared to the water supply industry. The concentrations of endotoxins were low in all plants but the concentrations in the garbage handling plants were significantly higher than in the paper sorting plants. In the garbage sorting plants we found the highest concentrations of microorganisms in the reception hall where the garbage is dumped from trucks (0.2 - 27.6·10^4 cfu/m^3) and during hand sorting (0.2 - 14.8·10^4 cfu/m^3) In the composting plants the concentrations of microorganisms were highest in association to aeration of piles (1.0 - 83.6·10^4 cfu/m^3). The highest concentration measured in this survey was in a plant during indoor aeration of a compost pile.

Table 2: Hygienic survey of the Danish.

	Plant type			
	Control	Paper sorting	Compost	Garbage handling
Total dust mg/m^3	0.42 (0.25)	0.83 (0.57)	0.62 (0.57)	0.74 (0.77)
Colony forming units/m^3 x 10^3	0.08 (0.04)	4.7 (5.9)*	54.4 (77.1)*,**	46.1 (125.5)*
Endotoxins ng/m^3	2.5 (3.9)	1.3 (1.5)	0.8 (1.1)	2.5 (4.4)**
Molds cfu/m^3 x 10^3	0.01 (-)	5.2 (5.4)*	26.9 (67.5)	14.3 (31.0)*
Numbers of samples	6	18	13	50

* t-test groups vs. Control, $p < 0.05$.
** t-test groups vs. paper sorting, $p < 0.05$.

Table 3 shows the results of the metal analysis. It is seen that all the concentrations of metals are within the normal range. The smoking paper sorting and recycling workers had significantly lower Pb-levels compared to the control workers 2.7, 3.9 and 5.8 µg/100ml respectively. Mercury was highest among the non smoking paper sorting workers 2.9 (2.5) µg/l vs. 2.3 (2.2) and 2.3 (3.1) among the controls and the garbage handling workers respectively. For Cadmium a higher level was associated to smoking in all groups. However, there was a significant increase in the Cd level in smoking as well as non-smoking garbage handling workers of 3.6 (1.7) and 2.1 (1.6) µg/l compared to 1.7 (4.3) and 0.6 (0.5) among controls and 2.3 (1.6) and 1.8 (0.6) among paper sorting workers, as seen in figure 1.

Selenium showed lower concentrations among the recycling workers compared to controls. However, this difference was only significant among the smokers. For the rest of the trace metals only minor differences were found. Interestingly we have been able to measure strontium in the serum of smokers in all groups.

Table 3: Heavy mental content in the blood among Danish recycling workers according to plant type

Mean (SD) Heavy metal	Plant type							
	Control		Paper Sorting		Compost		Garbage handling	
	Smokers N = 65	Non-smokers N = 47	Smokers N = 8	Non-smokers N = 12	Smokers N = 4	Non-smokers N = 4	Smokers N = 25	Non-smokers N = 15
Pb µg/100 ml	5.8 (2.9)	5.4 (2.9)	¹2.7 (2.5)	4.9 (4.0)	-	-	¹3.9 (2.7)	3.4 (1.6)
Hg µg/l	2.3 (2.2)	2.0 (1.6)	2.3 (1.8)	¹2.9 (2.5)	-	-	2.3 (3.1)	1.4 (1.2)
Cd µg/l	1.7 (4.3)	0.6 (0.5)	2.3 (1.6)	1.8 (0.6)	-	-	3.6 (1.7)^A	¹2.1 (1.6)
Fe	1.61 (0.55)	1.51 (0.98)	1.59 (0.39)	1.50 (0.36)	1.12 (0.23)	(0.22) 1.29	1.52 (0.49)	1.54 (0.51)
Cu	0.87 (0.17)	0.83 (0.16)	¹0.69 (0.10)	0.78 (0.09)	0.83 (0.13)	0.93 (0.07)	³0.84 (0.15)	0.80 (0.01)
Zn	0.77 (0.13)	0.82 (0.18)	0.76 (0.02)	0.83 (0.19)	0.67 (0.1)	0.81 (0.1)	0.76 (0.14)	0.77 (0.13)
Se	0.11 (0.02)	0.12 (0.02)	¹0.1 (0.01)	^{A}0.12 (0.02)	¹0.09 (0.01)	0.11 (0.01)	¹0.10 (0.02)	0.11 (0.02)
Br	4.49 (1.15)	4.19 (1.28)	¹3.56 (0.95)	4.04 (0.84)	¹2.48 (0.71)	3.34 (1.02)	¹3.88 (1.18)	3.92 (0.91)
Sr	0.01 (0.00)	-	0.01 (0.03)	0.02 (0.01)	-	-	0.01 (0.02)	-

[1] P < 0.05 Duncan oneway group vs. control. [2] P > 0.05 Duncan group vs. garbage handling.
[3] P > 0.05 Duncan group vs. paper sorting.
[A] P > 0.05 t-test smoker was as non-smoker.

Gastrointestinal symptoms, nausea more than once a year and ever having experienced vomiting or diarrhoea in relation to work, were significantly more common in the garbage handling industry than in the controls (see table 4). The latter symptoms were also common among garden compost workers although not to a significantly higher level. Almost all workers had a normal appetite. However, we found a non-significant increase in the frequency of loose stomach among the composting and garbage handling workers.

Table 4: Gastrointestinal symptoms among Danish recycling workers according to plant type n (%)

		Plant type		
Symptoms	Control	Paper sorting	Compost	Garbage handling
Always normal appetite	112 (94.1)	14 (70.0)	8 (100)	38 (86.4)
Nausea more than once a month	2 (1.7)	1 (5.0)	1 (12.5)	8 (18.2)
Working stomach more than once a month	8 (6.7)	1 (5.0)	0 (-)	4 (9.1)
Loose stomach more than once a month	14 (11.8)	1 (5.0)	2 (25.0)	8 (18.2)
Ever experienced vomiting or diarrhoea in relation to work	6	0 (-)	2 (25.0)	[1]12 (27.3)
N	119	20	8	44

[1] P < 0.05 Fishers exact test group vs. Control.

Table 5 shows the frequency of musculoskeletal disorders according to type of plant. It is seen that the most prevalent symptom is low back pain which occurs among 36%, 40%, 63% and 39% of the controls, the paper sorting workers, the composting workers and the garbage handling workers respectively. No significant differences were found between the groups. Among the garbage handling workers we found a significantly increased prevalence of shoulder and wrist/hand problems which occurred among 27% and 16% of the garbage handling workers compared to 12% and 5% among the controls. When the analysis was restricted to males we still found an excess frequency of 20% and 15% among the garbage handling workers. However, the differences were only borderline significant in this analysis (0.05 < p < 0.1, Fishers exact test).

DISCUSSION

The mean age and length of employment was found to be higher in the paper sorting industry than in the garbage industry. This might be explained by the newness of the branch. Another possibility is a high turnover among the newly employed garbage workers. The latter explanation is currently being investigated.

The total dust measurements were lower than the results from a resource recovery plant in USA [5], and also lower compared to compost plants in USA and Sweden [15,16]. The total number of microorganisms however, were comparable to concentrations in the above mentioned plants. The endotoxin levels were low compared to other industries with exposure to organic dust [17-20], but comparable to other compost plants [15]. However, they were substantially lower than what was found at a resource recovery plant with a high prevalence of respiratory symptoms [21]. We have not measured the heavy metal content in the air of the plants. from another study [5] it is known that there might be exposure to heavy metals in the sorting facilities. In the former study no Cd was found in the samples, but the authors state that this must be due to shortcomings in their measuring technique.

Table 5: Ergonomical symptoms within the last 12 months the last 7 days and medical attention according to plant type

	Control		Paper sorting		composting		Garbage handling	
Has had problems within the last 12 months from:	N	%	N	%	N	%	N	%
Neck	23	19.3	4	20	1	12.5	10	22.7
Med	12	10.1	1	5		0	4	9.1
7D	10	8.4	3	15		0	4	9.1
Shoulders	14	11.8	4	20		0	^{A}12	27.3
Med	3	2.5	1	5		0	3	27.3
7D	9	7.6	2	10		0	4	9.1
Elbow	7	5.9	2	10		0	4	9.1
Med	1	0.8	1	0		0	1	2.3
7D	2	1.7	2	5		0	3	6.8
Wrist/hand	6	5.0	2	10		0	^{A}7	15.9
Med	2	1.7		0		0	3	6.8
7D	1	0.8	2	10		0	3	6.8
Upper back	14	11.8	4	20	1	12.5	7	15.9
Med	7	5.9		0	1	12.5	2	4.5
7D	8	6.7	2	10		0	3	6.8
Lower back	43	36.1	8	40	5	62.5	17	38.6
Med	19	6.0	2	10	1	12.5	3	6.8
7D	17	14.3	2	10	1	12.5	5	11.4
Hips	7	5.9	2	10		0		4.5
Med	2	1.7		0		0		0
7D	5	4.2	1	5		0		0
Knee	22	18.5	4	20	2	25	9	20.5
Med	1	0.8	1	5	1	12.5	3	6.8
7D	7	5.8	2	10	1	12.5	2	4.5
Feet	1	0.8	^{B}3	15	1	12.5	2	4.5
Med		0		0	1	12.5		2.3
7D		0	1	5	1	12.5	1	0
	119		20		8		44	

A = p < 0.05 Control vs. garbage handling workers, Chi^2, B = p < 0.05 Control vs. paper sorting workers, Chi^2, Med = Has sought medical attention for the problems, 7D = Has had these problems in the past 7 days.

The metal content of the blood of all the workers investigated in this survey shows values within the normal range [22-25]. However, there was an effect of smoking as well as job on the concentration of Cd with the recycling workers having the highest blood concentrations of Cd. This might reflect an occupational exposure, although much has been done in Denmark to get rid of batteries containing heavy metals from the waste. There has been an ongoing campaign with a maximal impact possibly after this survey. The Cd-concentrations were, however, small and will not cause any health effects on the workers. Still emphasis must be on the reduction of the exposure to Cd in the garbage handling industry.

The lead concentrations were highest among the control workers. For this we have no obvious reason, although a few of the control workers were skilled auto-mechanics with some ongoing hobby garage work. Another possibility would be lead water-piping which has not been used in the Copenhagen area for many years. An interesting finding in our material is the detectable level of Sr although at low concentrations, we found consistently mean serum concentrations from 0.006 to 0.012 µg/l among smokers. This means, that cigarette smoke may contain some amounts of Strontium.

The frequency of musculoskeletal problems were at the level of what is normally encountered among male workers in Denmark [26]. We found an excess of symptoms from the upper extremities among the garbage handling workers and paper sorting workers compared to the controls as well as among other male workers. This is probably an effect of the high amount of manual handling of materials in the group of garbage handling workers. The frequencies of symptoms were lower when we restricted the analysis to only male workers and were only borderline significantly increased. However all the women in the study were employed in manual handling of materials, often with the extremities in extreme positions. Hence the increase in symptoms which was found is probably not entirely an effect of sex.

CONCLUSION

We have earlier shown that garbage recycling workers are prone to skin, respiratory tract, upper airway, mucosal and eye symptoms [7].

This study shows that garbage handling workers are prone to higher concentrations of Cd in their serum compared to controls.

We also found higher amounts of gastrointestinal symptoms among the garbage handling workers and compost workers.

Finally we have found an overall low prevalence of musculoskeletal symptoms. However, we found an increase in the prevalence of musculoskeletal symptoms from the upper extremities among the garbage handling workers compared to controls probably as a result of a high amount of manual handling in the garbage handling industry.

Together with the former study, this survey emphasises the need to address the working environment, when recycling facilities are being planned.

ACKNOWLEDGEMENTS

This study was supported by the Danish Labour Protection Fund 1988, the Danish Department of Environment & The University of Aarhus Research Fund.

REFERENCES

1. Clark C.S., Bjornson H.S., Schwartz-Fulton J., Holland J.W. & Gartside P.S. (1984) *J Water Pollut Control Fed* 56, 1269-1276.
2. Clark C.S. (1987) *J Water Pollut Control Fed* 59, 999-1008.
3. Clark C.S. & Linnemann Jr C.C. (1986) *CRC* 16, 305-326.

4. Dean R.B. (1978) Environ Health. Eff. Res. Series. 1-11.
5. Diaz L.F., Riley L., Savage G. & Trezek G.J. (1976) *Compost. Science* 18-24.
6. Lundholm M. & Rylander R. (1980) *J Occup Med* 22, 256-257.
7. Sigsgaard T., Malmros P., Nersting L. & Pedersen C. (1994) *Am. Rev. Respir. Dis.* 149, 1407-1412.
8. Wilhardt P. (1991) in: *Vejledning i arbejdshygiejniske undersøgelser - Luftforurening*, National Institute of Occupational Safety & Health, Copenhagen.
9. Goto H. & Rylander R. (1987) *J. Lab. Clin. Med.* 110, 287-291.
10. Hertel N. (1986) in: *Nuclear instruments and methods in physical research* (University of Aarhus, Aarhus, pp. 58-60.
11. British Medical Research Council (1965) *Lancet* i, 775-779.
12. Kuorinka I., Jonsson B., Kilbom Å., Vinterberg H., Biering-Sørensen F., Andersson G. & Jørgensen K. (1987) *Appl Erg* 18, 233-237.
13. Andersson G., Biering-Sørensen F., Kilbom Å., Kuorinka I., Vinterberg H., Hermansen L. & Jonsson B. (1984) *Nordisk Medicin* 99, 54-55.
14. Norusis M.J./SPSS INC. (1988) in: SPSS/PC+ V2.0 Base manual for the IBM PC/XT/AT and PS/2, SPSS INC.
15. Clark S.C., Rylander R. & Larsson L. (1983) *Appl. Environ. Microbiol.* 5, 1501-1505.
16. Clark C.S. (1986) *Am J Ind Med* 10, 286-287.
17. Clark S., Rylander R. & Larsson L. (1983) *Am Ind Hyg Assoc J* 44, 537-541.
18. Olenchock S.A., Christiani D.C., Mull J.C., Ting Y.T. & Lian L.P. (1983) *Appl. Environ. Microbiol.* 46, 817-820.
19. Olenchock S.A., May J.J., Pratt D.S., Piacitelli L.A. & Parker J.E. (1990) *Am J Ind Med* 18, 279-284.
20. Rylander R. & Morey P. (1982) *Am Ind Hyg Assoc J* 43, 811-812.
21. Sigsgaard T., Bach B. & Malmros P. (1990) *Am J Ind Med* 1, 92-93.
22. Tarp U., Thorling E.B. & Hansen J.C. (1990) *Nutrition Research* 10, 1171-1176.
23. Helgeland K., Haider T. & Jonsen J. (1982) *Scand. J. Clin. Lab. Ivest.* 42, 35-39.
24. Nygaard S.-P., Otteosen J. & Hansen J.C. (1977) *Danish Medical Bulletin* 24, 49-51.
25. Möller-Madsen B., Hansen J.C. & Kragstrup J. (1988) *Scand. J. Dent. Res.* 96, 56-59.
26. Nord-Larsen M., ¥rhede E., Nielsen J. & Burr H. (1992) in: *Lønmodtagernes arbejdsmiljø 1990 - bind 1. Sammenhænge mellem arbejdsmiljø og helbred*, Arbejdsmiljøfondet, Copenhagen.

Occupational Health Problems in Waste Recycling

Per Malmros, Danish Working Environment Service, Landskronagade 33, DK-2100 Copenhagen Ø, Denmark

ABSTRACT

In recent years, action plans concerning the environment have been issued by governments in Denmark and many other industrialized countries. The overall objective is to reduce quantities of and impact on the environment from all types of waste. As a result of these activities, many experiments with recycling of waste, have been carried out through the recent years. in Denmark. These waste recycling activities have unexpectedly resulted in a number of serious working environment problems among the employees collecting, handling and sorting the waste. Several employees have become seriously ill, and have had to leave the job, and be given social pension, and it has been necessary to change the technical installations and the work organization. In particular, the isolation of the biodegradable fraction of the rubbish has given rise to new problems in the working environment, regarding collection, sorting, and processing (composting and biogasformation) of this fraction.

KEYWORDS

Health, safety, working environment, microorganisms, biodegradable waste, recycling, endotoxins, fungi, guidelines, regulation.

INTRODUCTION

In recent years, it has become a common understanding, that we must limit the burdens we place on nature and curb the growing consumption of natural resources if the natural environment is to be preserved for future generations.

The ever greater volumes of waste are creating increasing problems - it is becoming steadily more difficult to find places for disposal sites without jeopardising the environment.

The overall objective of the action plan 1993 - 1997 is to reduce quantities of all types of waste and their impact on the environment. Since the generation of waste is inherent in the activities of our society, the aim is to make optimal use of the resources of waste - first of all of materials, then of the energy resources, so that the impact on the environment, including the working environment, is reduced to the extent possible.

OCCUPATIONAL HEALTH PROBLEMS IN RECYCLING ACTIVITIES

As a result of the above-mentioned politics, many experiments with recycling and composting of organic waste, have been carried out through the recent years in Denmark, and several large waste sorting facilities, have been established throughout the country of Denmark.

These waste recycling activities have unexpectedly resulted in a number of serious working environment problems among the employees handling the waste. Several employees have become seriously ill, and have had to leave the job, and have received disability pensions.

Because of these experiences, changes in design of equipment used for collection, and in the technical arrangement of the treatment facilities, have been made, and it has, at the same time, proven necessary to pay a lot of attention to the actual organization and performance of the work.

The Danish experiences presented in the following, deals primarily with recyclable materials derived from household waste, or waste from industries very much resembling household waste. It must be born in mind, however, that waste can also contain hazardous materials. Demolition waste for example in some cases contains asbestos, and household waste may contain chemical toxic substances, e.g. solvents. It is the Danish experience that chemical waste have to be dealt with separately.

The results, briefly reviewed in this paper, focuses on the recycling of materials derived from household waste, or waste of a similar composition. The recyclable materials from households are paper and cardboard, metal (maybe in the future plastics) and the biodegradable fraction. The latter fraction is often called 'the green fraction', and it is collected for composting or biogas formation. It is sometimes refered to as 'the organic fraction' although both paper and plastics are also organic.

Problems relating to collection and treatment of the biodegradable fraction, have received a lot of attention. It is very important to the Danish waste planning that the this fraction is removed from the waste-stream to be incinerated, because of a lack of incineration capacity. It is part of the action plan for the external environment, to increase the utilization of the biodegradable waste as much as possible.

The second point is that the trend to separate the waste at the source, is a result of working environment problems at the sorting plants. This emphasizes the necessity of evaluating waste management as an integrated system.

THE COLLECTION SYSTEM

In Denmark the waste is traditionally collected at the producers doorstep, and the dustman usually delivers the dustbin or a waste sack at the refuse collection truck, usually equipped with a comprimation device, and the waste is then transported to incineration or to the dumpground.

The above mentioned isolation af the organic fraction of the rubbish has given rise to new problems in the working environment, regarding collection of this fraction. However, only limited information exists on possible occupational health problems related to such new systems.

Occupational accidents are very frequent among waste collectors. Recently, the DWES-Registry of Occupational Accidents and Diseases, have analyzed reports on accidents and diseases for garbage collectors for the period of 1989-1993. The results clealy shows, that garbage collectors have high incidence rates of work related accidents, that is a 6 times increased risk as compared to the entire work force. Moreover the reports on accidents are increasing over the last 5-year period. The diseases shows the same patern. The frequency is twice that of the entire workforce. Based on current knowledge, it apears that the risk factors should be considered as an integrated entity, i.e. technical factors as poor accessibility to the waste, and poor ergonomic design af the equipment. This may act in concert with high working rate etc.

Musculoskeletal problems are also common among waste collectors. The DWES-registry of Occupational Accidents and Diseases have received reports on diseases twice that of the entire workforce, and the annual number of reports have been doubled in the period 1989-1993. 60% of the reports of diseases relates to the musculoskeletal system. Exess risk of chronic bronchitis has been reported, and Danish data indicates an excess risk for pulmonary problems among waste collectors compared with the total work force(1). High incidence rates of gastrointestinal problems, irritation of the eye and skin, and perhaps symptoms of ODTS (Organic Dust Toxic Syndrome = influenza-like symptoms, cough, muscle pains, fever, fatigue, headache) have been reported among workers collecting the biodegradable fraction of the domestic waste.

In the following, a Danish example concerning the collection of the biodegradable fraction shall be presented. In the late 80s a composting plant was established in the northern part of Sealand. It was decided to collect the organic fraction in bins instead of waste sacs, due to the very wet fraction, and it was decided to collect the waste once per fortnight, instead as once a week as before. Very quickly, and in 3 summer-seasons in a row, many of the workers collecting the organic fraction reported symptoms; Diarrhoea, vomiting, nausea, fatigue, headache and general indisposition. The Danish Working Environment Service (DWES) organized an enquiry, and out of 26 workers reporting to DWES, 11 reported symptoms, primarily during the summertime. The results are summarized in table 1.

Table 1: Working environment problems due to collection of biodegrable waste from house-holds

AFAV (composting Plant in the Northern part of Sealand). Collection of Organic Waste from Households	
Summer 90:	4 cases of Headache, Fatigue, Nausea
90/91	11 out of 26 reports include symptoms when collecting biogradable fraction.
April 91:	4 reported cases
June 91:	2 reported cases
Summer 92:	4 reported cases
Summer 93:	reported cases, not verified.

These symptoms are attributed to microorganisms, primarily bacteria and fungi, growing in the rubbish(1). The growth is enhanced during the hot season, and the amount of fluid, in Denmark often called percolate, in the dustbin, increases for several reasons. Firstly because of the composition of the garbage in the summertime is more humid, secondly because of the removal of the paper-fraction which otherwise retains much of the water, and thirdly because of the separation into two waste streams, means that the dustbin is emptied only once per fortnight, instead of originally once a week.

So, in more general terms, the challenge is to design collection material i.c. trucks, dustbins, waste sacs, carts and so on, which completely limits the exposure to microbial aerosols, that is, allows the work to be performed inclosed via contained way. Concerning the ergonomical problems, the rules in Denmark imply that the garbage should be transported on wheels (i.e. carts to transport the waste sack or the dustbin) instead of being carried on the back. Access for the collector shall be free, and it shall be possible to roll the cart on a smooth surface.

PROBLEMS IN THE WORKING ENVIRONMENT AT SORTING PLANTS

In order to reduce the strain on the environment from deposits of waste in landfills and combustion at incineration plants, several governments throughout the industrialized world have planned greatly increased recycling af domestic waste. New recycling facilities have been built and the number of workers involved in waste sorting and recycling will increase in the years to come.

Cross sectional studies on occupational exposures and/or health problems in the waste sorting and recycling industry have been carried out in several countries (2)(3)(4)(5)(6). These studies have shown, that workers in this industry have an exess risk of work related problems such as musculoskeletal problems, pulmonary diseases, ODTS-symptoms (cough, chest-tightness, dyspnoea, influenza-like symptoms such as chills, fever, muscle ache, joint pain, fatigue and headache), gastrointestinal problems, and irritation of the eye, skin and mucous membranes (8). The odds ratio (OR) for different pulmonary diseases and symptoms among workers at waste sorting plants, paper sorting plants, and garden composting plants as compared with water supply workers serving as a reference population has been established (6)(7). Workers at waste sorting plants had significantly higher risk of chest-tightness and ODTS. In the interpretation of these results it must be born in mind that these kind of studies often underestimates the risks because of the healthy worker selection. Many observations in Denmark seems to indicate that this factor is very dominant in these facilities.

The causality of the occupational health problems among workers at waste recycling plants is still not known, but there is no doubt that these problems are very much connected to high exposures to airborne dust containing bacteria, fungi and microbial toxins, i.e. endotoxins and glucans.

To illustrate how serious the problems can be at a waste sorting facility, the experiencies from at danish recycling plant, shall be shortly reviewed in the following. In recent years the problems relating to sorting of garbage in Denmark were first described at a plant performing mechanical and manual sorting of industrial and household waste.

This plant was build in the spring of 1986, and was considered to be the most modern installation in Europe, designed for maximum recycling. Often the wish to obtain very high percentages in recycled material, leads to building of plants designed for treatment of mixed waste.

After only three months of functioning, the first case of occupational illness appeared, and until the summer of 1991, 10 cases of occupational illness have been registered, and it must be underlined that the group of workers exposed to organic dust, amounts to only 15 people. The diagnosis were predominantly Asthma bronchiale, but also Chronic bronchitis and toxic and allergic alveolitis have been recorded. The findings from this specific plant are summarized in table 2.

Table 2: The occupational health-status of the 4S plant

4S plant status 1991. Plant started medio 1986	
Initial Symptoms:	
August 1986:	Upper Airway Symtoms
	Allergic Alveolitis
	Monday fever
Diagnose:	8 Asthna Bronchiale
	1 Chronic Bronchitis
	1 Allergic Alveolitis
Results:	8 Notifications of occupational diseases pr. March 1987
	1 Report as pr. September 1988
	1 Report as pr. June 1991

Exposed group of workers approximately 15. 7 out of 9 had to change job. Only 2 out of 7 were free of symptoms after approx. 2 years away from the plant.

Consequently, a study on 8 sorting plants and 4 composting plants was carried through (6). This study showed that occupational health problems was a general risk with these activities, and that the garbage recycling workers were prone to problems with skin, respiratory tract, upper airway mucosal, eye and gastrointestinal diseases. The conclusions are summed up in table 3.

Table 3: Conclusions from the DWES experiences

(8 garbage-, 4 compost plants) 15-20 cases of asthma 20-30 cases of ODTS (Organic Dust Toxic Syndrome)	
Symtoms significantly increased, compared to a control group of water trestment workers:	-Chest tigtness
	-Influenza feeling
	-Itching eyes at work
	-Sore or itching throat
	-Gastrointestinal symptoms
Observations of ergonomical problems	
Observations of noise and vibrations	
Observation og draught/cold	

Symptoms in the recycling industry

There are many different symptoms attributed to microbiological exposure. The problems are not attributed to infections, but to toxic and possibly allergic reactions, and it is apparent, that even though the source may be the same, i.e. microbes, the symptoms differs, depending on the work performed. These observations, which are essential in the preparation of guidelines and regulation of the technical demands and the actual performing of the work, are in a non-scientific and tentative way, shown schematic in table 4.

Table 4: A summarized presentation of symptoms related to waste handling

Condition of the garbage	Dry/Dusty Paper/Industrial	Wet/ Aerosol Household
Kind of Work	Treatment (sorting, composting)	Collection of waste
Symptoms	Headache, Fatigue, Fever, Chills, Muscular pains, Painful Joints, Loss of appetite	
	Shortness of breath	Nausea
	Chest tightness	Vomiting
	Cough	Diarhoea
	Asthma Attack	
Illness	Asthma, Allergic, Toxic alveolitis(ODTS), Cough and Chronic Bronchitis	Gastro In-testinal Illness
Skin:	Itching and rashes	

The Danish system for materials recovery

The Danish experiences point to the conclusion, that increased recycling activities, pose the possibility of problems to the workers at the facilities. It is our understanding that the risk of meeting these problems is prominent in all recycling activities, and even more so, when the recycling percentage goals are very high.

The solutions enforced in Denmark rests on certain specific assumptions. The future systems shall be flexible, with possibility of local, i.e. municipal solutions, without compromising workers safety. It is also clear, that these systems must be planned and put into action, with the cooperation of all the parties involved that is, authorities, employees, trade unions and so on.

The Danish system in this way rests upon the experiences concerning the working environment, and the action plan concerning 1993-1997 for garbage and recycling.

The Danish guidelines concerning the working environment are summarized in table 5.

Table 5: The danish experiences concerning the working environment

Work Environment Guidelines	
Waste Received:	High Quality (i.e. well defined and free of organic contamination if it is to be handsorted)
	Control Systems (Technical possibility of rejecting waste of bad quality)
Technical Instalations:	Generation of aerosols to be minimized(encapsulation)
	Ventilation systems
	Cleaning Systems should be present

Ergonomics:	Filters on trucks and for personal should be present protection
	Machines and technical aids designed properly and adjustable to the user. Possibility of changing places at sorting lines.
Ergonomics:	Well designed bins for collection
	Possibility of transporting the refuse on wheels.
Working Routines:	Safety Instructions known and present
	Personal Protection Systems
	Maintenance instructions and routines.
Training:	A 'Must' !
	Programmes for every step in the systems are necessary
	Hygiene
	Personal, cleaning fascilities present.
	Occupational, daily cleaning routines.

CONCLUSIONS AND FUTHER WORK

The data from Denmark and other sources currently available on aerosol exposure indicate that workers at transfer stations, landfills and incineration plants may have elevated prevalence of pulmonary disorders and gastrointestinal problems.

Manual sorting of unseparated household garbage may have high risk of pulmonary diseases, toxic alveolitis, gastrointestinal problems, and irritation of the eye and mucous membranes of the upper airway. At plants sorting and recycling source separated household waste, the workers may have an increased risk of gastrointestinal symptoms and irritation of the eye and skin. Workers sorting clean paper and board do not appear to have specific problems.

Workers at compost plants experience ODTS-like symptoms, cough, chest tightness, flue-like symptoms such as chills, fever, pains in muscles and joints, fatigue and headache, gastointestinal problems, skin and eye irritation. Moreover reports on asthma, alveolitis and bronchitis exist.

It has not been possible to find scientific data on working environment problems related to biogas production from household waste. However, WES receives reports from the Danish plants, and it seerms more than likely, that the problems exist, and that the pattern ressembles the ones at composting plants.

No scientific data exists regarding sorting of return-bottles, but reports from the local inspectors in DWES, indicates that problems indeed exists. These problems are microbial problems due to contamination with organic remains, ergonomical problems due to monotonous, repetitive work and increased risk of cutting damages. More over there are problems due to draught and cold.

Further work

More basic work is needed in the coming years. To cope with the problems raised by the aim of recycling 50% of the garbage by the turn of the century, and still being able to design systems taking local (human, economical, historical etc) aspects in consideration, we find it necessary to give more elaborated standards about the performance of the systems.

We have therefore established a 5-year programme ending in 1998, with the aim to produce the necessary knowledge to give information concerning the dose-response relationship concerning the most important parameters.

The programme has an administrative part, and a research part, closely interlinked. The programme is outlined in table 6.

Table 6: Danish working environment service programme

Administration:	Guidelines, rules and regulations for collection systems, 2-stream and 4stream syatems. Guidelines and regulations for waste treatment plants. Instructions of the Local Authorities. Establishment of values of acceptance.
Research:	New and standardidzed methods for measuring and evaluating the most important exposures, primarily microbiological and ergonomical. Establishment of exposure - response relationsships.

REFERENCES

1. Poulsen, O.M., Breum, N.O., Ebbehøj, N., Hansen, Å.M., Ivens, U.I., van Lelieveld, D., Malmros, P., Matthiasen, L., Nielsen, B.H. Nielsen, E.M., Schibye, B., Skov, T., Stenbæk,E.I., Wilkins, K.C. (1995) 'Review on Sorting and Recycling of Household Garbage, Occupationel Problems and their possible causes'. The Science of the Total Environment 170: 1-19.
2. Mansdorf, S.Z., Golembievsky, M.A., and Fletcher, M.W. (1982):'Industrial Hygiene Characterization and Aerobiology of ressource Recovery Systems'. Morgantown: NIOSH
3. Crook, B..,Higgins, S., and Lacey, J.(1987):'Airborne Micro-organisms Associated with Domestic Waste Disposal, Harpenden, Hertz.,AL5 2JQ:Crop Protection Division, Afrc. Institute of Arable Crops Research, Rothamstead Experimental Station.1-119
4. Rahkonen, P. and Ettala, M.,(1987):'Working Condition and Hygiene at Sanitary Landfills in Finland. Ann. Occup. Hyg; **31**(4A):505-513.
5. Rahkonen,P.(1992):'Airborne Contaminants at Waste Treatment Plants. Waste Management and Research; **10**:411-421.
6. Malmros, P., Nersting, L., Petersen, C., and Sigsgaard, T,:'Arbejdsmiljøforhold ved Genanvendelse af Affald. Miljøprojekt Nr.161. Miljøstyrelsen.
7. Sigsgaard, T., Abel, L., Donbæk, L. and Malmros, P.,(1994):'Lung Function Changes Among Recycling Workers Exposed to Organic Dust'.Am. Joru. Ind. Med.;**25**:69-72
8. Poulsen, O.M., Breum, N.O., Ebbehøj, N., Hansen, Å.M., Ivens, U.I., van Lelieveld, D., Malmros, P., Matthiasen, L., Nielsen, E.M., Nielsen, E.M., Schibye, B., Skov, T., Stenbæk, E.I., Wilkins, K.C., (1995) 'Review on collection of Household Garbage. Occupational Health Problems and thei possible Causes. The Science of the Total Environment 168: 33-56.

Section 5

Anaerobic Processing

Anaerobic Digestion Processes for the Energetic Valorisation of the Green Fraction of Municipal Solid Waste and the Recovery of Volatile Fatty Acids

P.Pavan*, A.Musacco*, P.Battistoni**, J.Mata-Alvarez***, F.Cecchi****, *University of Venice - Dept. of Environmental Sciences - Calle Larga S.Marta 2137 - Venice - Italy, **University of Ancona - Dept. of Materials and Earth Sciences - Via Breccie Bianche - Ancona - Italy, ***University of Barcelona - Dept. of Chemical Engineering - Marti i Franquez 6 pl. - Barcelona - Spain, ****University of L'Aquila - Dept. of Chemistry, Chemical Engineering and Materials - Monteluco di Rojo - L'Aquila - Italy

ABSTRACT

The aim of this work is to present the results of several pilot projects which study various anaerobic digestion processes and applications. The first study is related to energy recovery from the anaerobic digestion, AD, process applied to the source sorted organic fraction of municipal solid waste, SS-OFMSW, in single and two phase approaches in the thermophilic range of temperature. The separation of the phases (Hydrolytic-acidogenic from methanogenic) allows a better control of the process (concentrations of VFAs in the reactor were less than 1 gHAc/l in each two phase run studied, compared with 17.3 gHAc/l in the single phase one) and three-times larger biogas productions were obtained (0.72 vs. 0.23 m^3/Kg in feed). The second study is on the acidogenic fermentation of a mixture of MS-OFMSW (mechanically sorted) and SS-OFMSW in the mesophilic range of temperature using a plug-flow reactor in order to obtain an effluent rich in short-chain compounds. These light organic fractions can be extracted from the effluent using a kerosene/trioctil phospine oxide mixture and be used as liquid fuel additives. Several operative conditions were studied in the mesophilic range, obtaining concentrations of VFAs in the effluent up to 23 g/l.

KEYWORDS

Anaerobic fermentation, anaerobic digestion, volatile fatty acids, lactic acid, biogas, mesophilic, thermophilic, municipal solid waste, liquid fuel additives.

INTRODUCTION

Growing interest in environmental protection, especially regarding wastewater and solid waste treatment, has lead to research into strategies that associate the efficacy with a low environmental impact. Municipal wastes, in particular those obtained using separate collection, require treatments in which the high biodegradability can be exploited to achieve a beneficial use of the resources. In this field, useful results can be obtained by adopting the anaerobic digestion processes. The aim of this work is to present the results of various pilot studies that look into the possibility of using anaerobic process applications. The first experiment is linked to the energy recovery from the biomass, comparing the results obtained in different experiments using a single phase and a two phase approach. The second study is concerning the production of liquid fuel additives through the acidogenic phase of anaerobic digestion.

MATERIALS AND METHODS

Fig.1 shows the pilot plant used for the studies on the two phase experiments (part A= shredding and fermentative treatment; part B= methanogenic phase). The reactors were completely mixed and had a working volume of 0.8 (fermenter, n.6 in Fig.1) and 1 m^3 (methanogenic reactor, n.7 in Fig.1).

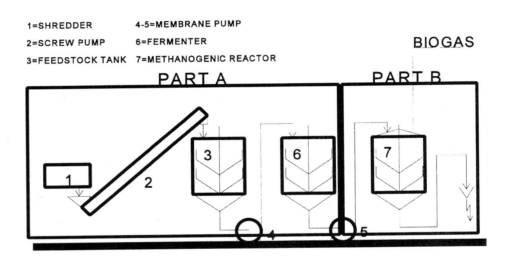

Fig.1. Fermenter and methanogenic completely mixed reactor used in the gas recovery experiments

Fig. 2 shows the plug flow reactor used for the experiments on the chemicals production. The working volume was 0.3 m^3. The digesters were monitored following Standard Methods procedures [1]. Temperatures, flow rates and CH_4/CO_2 ratios in the biogas were monitored by using infrared on-line probes and a PC for data aquisition. Volatile fatty acids, VFA, were monitored through gas chromatographic method and lactic acid through ionic chromatography. Other details are reported in [2].

The (source separated) organic fraction of MSW used the experiments was acquired from the supermarkets and canteens of the city of Treviso (SS-OFMSW). The mechanically separated MSW was from the mechanical sorting plant of S.Giorgio di Nogaro (UD) (MS-OFMSW). Main characteristics are reported elsewhere [2].

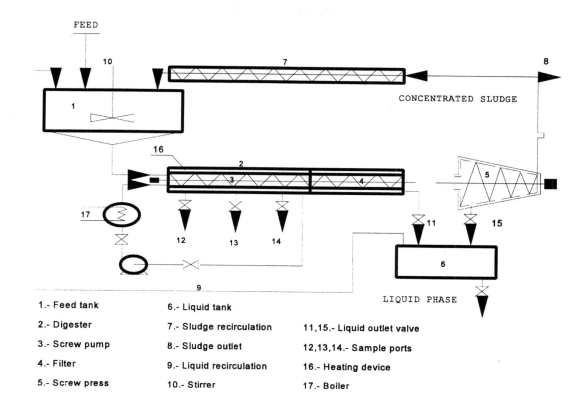

Fig.2. Plug flow reactor used in the chemical recovery acidogenic fermentation experiments

RESULTS AND DISCUSSION

The recovery of energy - single and two-phase approach

The high biodegradability of SS-OFMSW constitutes serious disposal problems. Due to the high water content, the heating value is low and disposal in landfills invariably leads to the generation of leachate. Anaerobic digestion exploits these wastes in a conservative way, resulting in considerable energy retrieval, waste stabilization and a reduction of volume before disposal [3]. In particular, the separation of the acidogenic phase from the methanogenic allows one to optimize the yields of the different bacterial consortia, stabilizing the process and improving the biogas production.

The experiment is presented in two parts: In the first the behaviour of the process using the single phase approach is presented. In the second the same substrate was treated using the separate phase AD process. The operative conditions studied, as well as the feed characteristics and the process yields obtained during the experiments are reported in table 1.

The single phase process has shown few problems, mainly linked to the high biodegradability of the substrate. The VFA content in the digester reached an average value of 17 gHAc/l, which is extremely high when compared with levels found elsewhere in digesters. The specific gas production (SGP) was also very low (0.23 m^3/KgTVS).

Table 1: Comparison of the results obtained using the single phase and the two phase process for gas recovery.

	Single Phase	Two Phases			
Run	S	3	5	6	4
Operative conditions					
T, °C	55.1	53.7	54.9	55.1	54.9
HRT, days	20.5	13.7	13.1	12.5	9.5
OLR, kgTVS/m^{3}d	5.3	4.1	4.8	4.7	6.9
Digester characteristics					
TS, g/kg	124.8	30.9	26.3	20.4	31.2
TVS/TS, %	50.1	61.3	62.8	55.6	63.5
VFA, gHAc/l	17.3	0.944	0.264	0.212	0.554
pH	7.2	8.1	7.9	7.9	8.1
TA(4), gCaCO3/l	16.8	10.2	9.3	8.9	9.2
Yields					
GPR, m^3/m^{3}d	1.2	2.3	3.3	2.5	5.1
SGP, m^3/kgTVS	0.23	0.57	0.68	0.53	0.72
CH4, %	59.9	57.9	57.3	56.9	56.2

The results obtained with the two-phase process (using a second reactor of 1 m^3) show a better situation. HRT of 13 d and 9.5 days were used. The VFA content in the digester was always under 1 gHAc/l and the pH near 8, without any chemical control. The specific gas production was at least twice the one obtained in the single phase process. In particular in the condition where the lowest HRT was used, the SGP reached values about three times larger (0.72 m^3/KgTVS). Considering this gas production, an amount of SS-OFMSW per inhabitant per day from 200 to 400 g w.w., and an average TS-TVS composition of this type of waste (12%TS, 0.85 TVS/TS), it was estimated that a production of 15-30 l of biogas per day can be achieved per inhabitant.

Chemicals from fermentation of MSW

The second study deals with the exploitation of the acidogenic phase of the AD process to recover chemicals. In fact, the short chain compounds, such as VFA and lactic acid, can be extracted from the effluent of the fermenter using a mixture of kerosene and trioctil phospine oxide and then esterificated, obtaining liquid fuel additives [4]. The study was investigated using a pilot scale plug-flow reactor (see Figure 2) in the mesophilic range (37 C), applying different operative conditions [5]. The main results are reported in Table 2.

The study of the reactor performance has been done considering the system as a black box; that is, comparing the values of the parameters determined for the inlet and outlet sludge.

The TVFA (Total Volatile Fatty Acids) concentration in the outlet sludge obtained when working at HRT of 2 days with sludge recirculation was 11.8 g/l, while in the case of run 2 and 3 it was 18.3 and 23.1 g/l respectively. It is observed that a HRT=2 days is not sufficient for an appropriate level of substrate degradation. This is supported by the fact that TVFA production and yields increased with retention time from run 1 to run 3. In contrast, applying HRT higher than 6 days seems inappropriate considering the literature results. The TVFA concentration obtained applying a HRT= 4 days is comparable with that reported by Antonopoulos and Wene [6] who, working at 8 days HRT obtained 10-15 g/l of TVFA. Albin et al. [7], digesting agro-industrial wastes reached a maximum TVFA concentration in the effluent of 18.9 g/l, with a HRT of 6 days and organic load rates, OLR, of 10 kgVS/m^{3}d. Working at the same retention time, TVFA concentration was slightly higher in this study (18.3 g/l at HRT = 4 days) although the OLR employed was more than 3 times greater (42.3 kg VS/m^{3}d and 38.5 kg VS/m^{3}d respectively).

TVFA concentration obtained in the sludge samples from the first sampling valve of the tubular reactor (see Figure 2, and Table 2) were quite high for the three periods, and increased with retention time from 9 g/l in period 1 (time taken for sludge to travel from inlet to first valve: 8 h), 12.8 g/l in period 2 (time to travel to first valve: 16 h), to 13.4 g/l in period 3 (time to travel to first valve: 24 h). This was because acidogenic bacteria are fast growing, with minimum doubling time around 30 minutes [8] and they are capable of fermenting part of the soluble fraction of the organic refuse to produce mixed VFA in a short interval of time.

Due to the high OLR applied under all conditions studied, the availability of soluble substrate was never a limiting factor and only a small amount of the soluble substrate was fermented between inlet and first sampling valve. This suggests that an inocculation stage should be considered in order to reduce the retention time in the reactor [9].

Alkalinity and pH were always within the optimal ranges: pH was between 5.0 and 5.7 in the runs considered, but never at inhibition levels for the acidogenic process, that is below 4.5 [10]. TA (difference of Total Alkalinities measured at pH 4 and 6) was always high enough (in the range 14.7 - 19.5 gCaCO$_3$/l) to consider the system well buffered.

The following Table 3 shows single volatile fatty acid concentrations in the outlet sludges during runs 1, 2 and 3. It can be seen that at 2 and 4 days HRT acetic acid is the favoured product of the acidogenic process, while at a higher retention time (6 days - run 3), it is more towards the higher acids, and particularly butyric acid.

The distribution of the individual acids obtained with this pilot plant is in agreement with the findings of a number of investigations, such as those carried out by Wulfert and Weiland [11], Hanaki et al. [12] and Zhang and Noike [13]. As it could be expected, they found that at short retention times, acetic acid tended to be the major product of the acidogenic process, with low concentrations of propionic and butyric acids. With increasing retention time, the percentage of butyric acid in the total VFA production increased considerably, as well as propionic acid concentration, which indicated a major production of longer VFA with higher retention times.

Table 2: Operative conditions and main results obtained during acidogenic a.d. Experiments

RUN	1	2	3
Operative conditions			
HRT, days	2.0	4.0	6.0
OLR, kgVS/m^{3}d	85.2	42.3	38.5
Feed characteristics			
TS, g/kg	248.3	244.9	285.5
TVS, g/kg	159.0	159.2	160.4
pH	6.11	6.19	6.06
TVFA, g/l	2.645	7.651	4.361
Sample valve 1			
pH	4.77	5.68	5.69
ΔTA, gCaCO$_3$/l	12.3	9.0	13.0
N-NH$_4$, mg/l	40.6	177.6	100.7
TVFA, g/l	9.061	12.782	13.372
Sample exit reactor			
TS, g/kg	255.2	304.0	303.6
TVS, g/kg	153.9	188.2	152.7
pH	5.08	4.98	5.72
ΔTA, gCaCO$_3$/l	14.7	17.2	19.5
N-NH$_4$, mg/l	101.3	81.2	486.7
TVFA, g/l	11.780	18.265	23.110

Table 3: Single VFA concentrations obtained in the outlet sludge of the acidogenic plug-flow reactor in the three runs considered

RUN	1	2	3
Acetic acid, g/l	8.156	15.346	8.176
Propionic acid, g/l	0.927	1.217	3.668
iso-Butyric acid, g/l	0.205	0.299	0.451
Butyric acid, g/l	1.651	1.099	5.741
iso-Valeric, g/l	0.284	0.161	0.632
Valeric acid, g/l	0.556	0.143	4.406
Total VFA, g/l	11.780	18.265	23.110

The production of methyl or ethyl esters of C_2 - C_6 by the isolation and esterification of the acids obtained by the anaerobic digestion of a mixture of OFMSW in a plug-flow reactor, appears particularly attractive considering the final single VFA concentrations obtained. Conditions of period 1 and 3 seem to be appropriate ones for obtaining high acetic, propionic and butyric acid concentration. These acids could consitute the raw material for obtaining VFA esters with high octane number. In fact, the motor octane numbers of methyl acetate, methyl propionate and methyl butyrate are sensibly higher than the usual MTBE octane number (109, 107 and 104 compared with 97 - [4]). Such compounds could be advantageously used as additives for gasoline.

CONCLUSIONS

The comparison from the single and two phase approach of the AD process applied to the SS-OFMSW shows that the phase separation allows to by-pass the problems connected with the high biodegradability of the substrate. The high levels of VFA (17.1 g/l observed in the single phase process) are reduced to less than 1 g/l in all the operative conditions of the two phase process. The stability of the process also leads to the improvement of yields, obtaining a SGP about three times larger than the one observed in the single phase.

Acidogenic fermentation of the OFMSW can be carried out at high solids in plug-flow reactors, reaching considerable yields in terms of VFA: up to 23.1 g/l with an HRT of 6 days. Considering the results obtained, reactor operation at higher retention time does not seem appropriate from an economical point of view, because of the required reactor volume.

The acids could be the raw material for valuable products like methyl- and ethyl esters with octane numbers superior to the MTBE octane number.

No particular attention needs to be paid to the pH control. pH never drops, in the operative conditions range adopted, below the values critical for this process (4-4.5).

REFERENCES

1. APHA, (1985). Standard Methods for the examination of water and wastewater. 14th edition, 1193pp.
2. Cecchi, F., Pavan, P., Mata-Alvarez, J., Bassetti, A., Cozzolino, C., (1991) Anaerobic digestion of municipal solid waste: thermophilic vs. mesophilic performance at high solids. *Waste Manag. & Research*, 9, 305-315.
3. Cecchi, F., Vallini, G., Mata-ALvarez, J. (1989). Anaerobic digestion and composting in an integrated strategy for managing vegetable residues from agro industries or sorted organic fractions of municipal solid waste. Int. Symp. on waste manag., Istambul, 25-27 Sett., 53-60.
4. D'Addario, E., Pappa, R., Pietrangeli, B., Valdiserri, M. (1993). The acidogenic digestion of the organic fraction of municipal solid waste for the production of liquid fuels. *Wat. Sci Technol.*, 27, .
5. Sans, C., Mata-Alvarez, J., Cecchi, F., Pavan, P. (1993). Acidogenic fermentation of urban waste for the production of valuable products. 6th Mediterranean congress of chemical eng., Barcelona, P-5.2.51, p.590.
6. Antonopoulos, A., Wene, E.G. (1988). Bioconversion of Municipal Solid Waste and recovery of short chain organic acids for liquid fuel production. Energy and Environmental System Division, Energy from Municipal Waste Program.
7. Albin, A., Ahlgrim, H.J., Weiland, P. (1989). Biomethanation of solid and semi-solid residues from harvesting and processing of renewable feedstocks. Proceedings 5th EC Conference on Biomass for Energy and Industry. Lisbon, Portugal, 9-13 October.
8. Mosey, F.E. (1983). Mathematical modeling of the anaerobic digestion process: regulatory mechanisms for the formation of short chain volatile acids from glucose. Water Sci. Technol., 15, 209-32.
9. Sans, C., Mata-Alvarez, J., Cecchi, F., Pavan, P., Bassetti, A. (1995). Volatile fatty acids production by mesophilic fermentation of mechanically sorted urban organic wastes in a plug-flow reactor. Bioresource Technology, 51, 89-96.
10. Zoetemeyer, R.J., Cohen, A., Van den Heuvel, R.C. (1982). pH influence on anaerobic dissimilation of glucose in an anaerobic digester. Water Res., 16, 303-11.
11. Wulfert, K., Weiland, P. (1985). Two-phase digestion of distillery slops using fixed bed reactors for biomethanation. 3rd E.C. Conference on Energy from Biomass. pp. 271-80.
12. Hanaki, K., Matsuo, T., Nagase, M., Tabata, Y. (1987). Evaluation of effectiveness of two-phase anaerobic digestion process degrading complex substrate. Water Sci. Technol., 19, 311-22.
13. Zhang, T.C., Noike, T. (1991). Comparison of one-phase and two-phase anaerobic digestion processes in characteristics of substrate degradation and bacterial population levels. Water Sci. Technol. 23, 1157-66.

Codigestion of Manure With Organic Toxic Waste in Biogas Reactors

B.K. Ahring, H. Garcia, I. Mathrani and I. Angelidaki, *The Anaerobic Microbiology/ Biotechnology Research Group, Institute of Environmental Science and Engineering, Technical University of Denmark, 2800 Lyngby, Denmark*

ABSTRACT

Codigestion of manure together with organic toxic waste containing either tetrachloroethene (PCE) or aniline was studied in continuously stirred tank reactor experiments. The process could be adapted to a feed concentration up to 200 ppm PCE, and the concentration of PCE and it's degradation products in the effluent were normally below the detection limit (1.0 ppm). PCE loss *via* the biogas was not a major removal route; PCE was not detectable in the biogas (detection limit 1.0 ppm). Aniline was not inhibitory to the process at up to 500 ppm. However, aniline was not entirely degraded and was present in the reactor effluent.

INTRODUCTION

Many large-scale biogas plants have been built in Denmark. In these plants, it is common to codigest of manure with organic industrial waste from food-processing industries. This practice improves plant economy by increasing the gas production compared to plants operated solely on manure, and by the waste fees obtained for treatment of the industrial wastes. Lipid-containing wastes can give alot of biogas if they are fed to biogas reactors properly, allowing microbial adaptation to the waste. Laboratory experiments have shown nearly 100% conversion of the lipid wastes added to reactors treating manure [1,2].

Waste and wastewaters from chemical industries often contain solvents and low concentrations of other toxic or polluting xenobiotic chemicals. For example, wastes containing n-butanol, traces of tetrachloroethene (PCE), isobutylacetate, isobutanol, acetate, aniline, methanol and other organic chemicals are produced. Today, problematic chemical wastes are normally treated using expensive, physico-chemical methods or by incineration. Therefore, there has been great interest from many industries to develop new, cheaper and environmentally friendly methods for disposal of these wastes.

The biodegradation of many solvents is part of the natural carbon cycle. Considerable research on aerobic but less on anaerobic biodegradation has been done [3,4]. Anaerobically, degradation of chemical compounds can take place as result of anaerobic respiration with nitrate and sulphate, and by fermentation with methane and carbon dioxide as the main products [5].

Co-metabolic activities, both aerobic and anaerobic, can also degrade xenobiotic chemicals [6-10]. During detoxification through cometabolism, additional substrates are needed to support microbial growth (i.e. as a carbon and energy sources), and are stimulatory to detoxification activities [11-14]. Many studies have documented that addition of a supplemental carbon source stimulates anaerobic detoxification reactions [12,14]. Nies and Vogel [13] studied dechlorination of polychlorinated biphenyls in sediments and found that the relative extent and rates of dechlorination were greatest when methanol, acetone, glucose or acetate were present. When no supplemental carbon was added dechlorination activity was insignificant. Similarly, we showed [11] that addition of carbohydrates significantly increased dechlorination of pentachlorophenol in granular sludge.

Polychlorinated compounds such as PCE and tetrachloromethane are only dehalogenated under strictly anaerobic conditions [6,10]. Under methanogenic conditions, the chlorine atoms are reductively dechlorinated and replaced with hydrogen atoms. Hydrogen is supplied by many different, easily degradable organic substrates, such as sugars, methanol [7], acetate [10] and benzoate [9]. Mononuclear aromatic compounds can be degraded in the absence of oxygen if they possess a carboxyl, hydroxyl or an amino substituent, such as in alanine [9].

The positive effects of codigestion of different types of wastes, such as codigestion of manure together with industrial wastes from the food processing industry, has been previously demonstrated [1]. In the present study, we investigated the degradation of toxic wastes containing PCE or aniline in anaerobic reactors treating manure.

MATERIALS AND METHODS

Reactor Experiments
The experiments were performed in 4.5 l, automated, water jacket-heated lab-scale reactors with a working volume of 2 l [15]. The reactors were fed automatically every 6 hours, with a hydraulic retention time (RT) of 15 days. The reactor temperature was $55^\circ C$. All reactors were operated with the same conditions for approx. one month before the experiments were started.

Substrates
Cattle manure, obtained from a Danish biogas plant, was the main substrate for the reactors. The aniline-waste contained, among other things, isobutylacetate, isobutanol, acetate, barbituric acid and aniline. The overall chemical composition was C_6H_7N and its water solubility was approx. only 1 ppm. The other toxic waste contained PCE and had an overall chemical composition of C_2Cl_2. This waste was also only slightly soluble in water (100 ppm).

Experimental Design
Five reactors were used. One reactor served as control and was fed only cattle manure. Two reactors, PCE,1 and PCE,2 were fed manure and different concentrations of PCE-waste. Two reactors, ANI,1 and ANI,2 were fed manure and different concentrations of aniline-waste. For part of the experimental period a second control reactor was run with manure and PCE waste from which the PCE had been removed. Table 1 shows the experimental plan.

Table 1: Experimental plan

	Control	PCE,1 reactor	PCE,2 reactor	ANI,1 reactor	ANI,2 reactor
Feed	Manure	Manure + PCE-waste	Manure + PCE-waste	Manure + anil.waste	Manure + anil.waste
Conc. of toxic waste (g/l)		0.71→7.10	1.43→14.3	0.71→7.1	1.79→17.9
		1.79→3.57[#]	3.57→7.14[#]		
Conc. of toxic* component (ppm)		20→200	50→500	50→500	100→1000
		50→100[#]	100→200[#]		

*The concentrations were calculated from the specifications given by the waste producer
[#]Phase 2 - after reactors PCE,1 and PCE,2 were reinoculated from the control reactor

The toxic waste was added to the manure and was mixed thoroughly. The feeding flasks were closed to minimize evaporation and nitrogen gas was added to the head-space of the feeding flasks.

The PCE reactors were run in two phases. After phase 1 the two PCE reactors were restarted with 50% reinoculation from the control reactor. The concentration of the toxic waste was stepwise increased as indicated in table 1. When a new concentration of PCE waste was added, the waste was added to the feed flask and the reactor to immediately reach the new concentration. However, in the second step for aniline, from 50 to 500 ppm and 100 to 1000 ppm for reactors ANI,1 and ANI,2, respectively, aniline-waste was only added to the feed.

Analysis

Gas production in the reactor experiments was measured using an automated water-displacement gas meter. Gas composition was measured using gas chromatography with thermal conductivity detection. Volatile fatty acids (VFA) and aniline were analyzed using gas-liquid chromatography with flame ionization detection. PCE and its dechlorination products were detected and quantified using mass spectrometer with gas chromatography system for separation.

RESULTS AND DISCUSSION

PCE-degradation

Addition of 20 or 50 ppm PCE to the manure reactors did not inhibit the process (Fig.1). The methane yield of all the reactors was 0.2 to 0.3 l/gVS, with slightly higher yields for the PCE reactors (Fig.1a). When the addition of the PCE-waste was increased 10 times to 200 and 500 ppm for reactors PCE,1 and PCE,2 respectively, the methane yield decreased: the methane yield in reactor PCE,2 dropped to below 0.1 l CH_4/gVS within one week, while it took approx. two weeks before the methane yield in the reactor PCE,1 dropped to the same level. Similarly, the total VFA concentration increased from below 3 g/l to over 6 g/l, indicating that the process was disturbed (Fig.1b). VFA accumulation resulted in a pH decrease despite the high buffer capacity of manure [16].

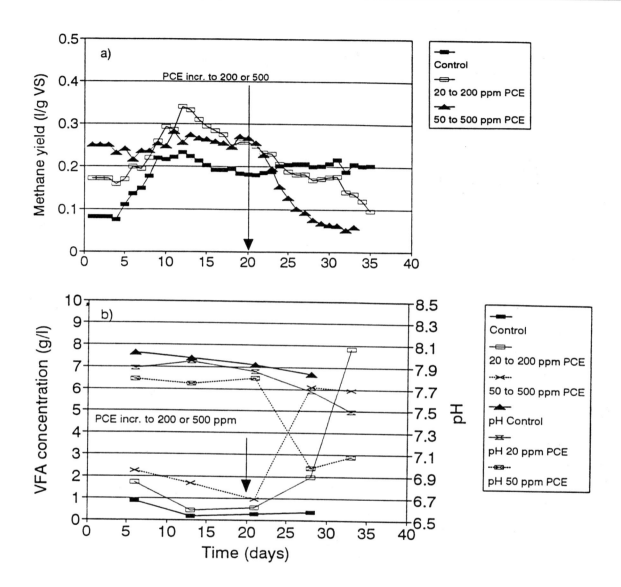

Fig. 1: Codigestion of manure together with PCE-waste

Addition of PCE-waste from which the PCE had been removed did not cause inhibition, indicating that the toxicity was due to PCE (data not shown).

In the second phase, after 50% reinoculation of the PCE reactors with material from the control reactor, the process accepted PCE concentrations up to 200 ppm without any signs of inhibition (Fig.2). The methane yields of reactors PCE,1 and PCE,2 were stable and at the level of the control reactor, approx. 0.2 l CH4/gVS. The VFA concentration stabilized at below 2 g/l, indicating that the process was in balance.

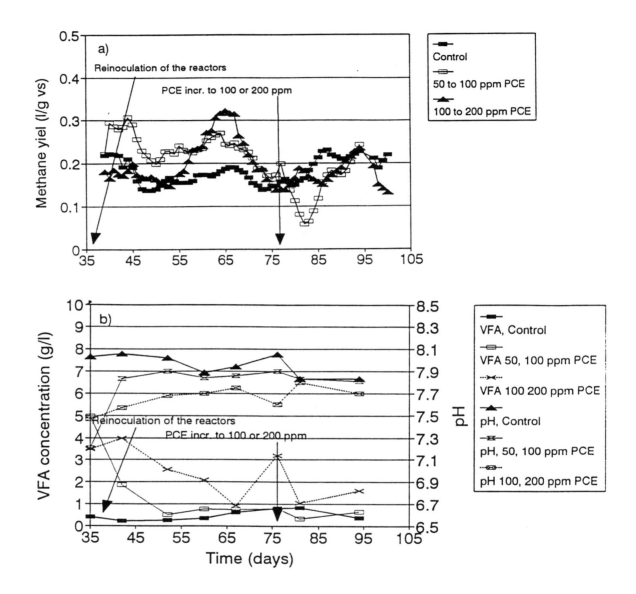

Fig. 2: Codigestion of manure together with PCE-waste. Phase 2, the reactors were 50% reinoculated with digested manure from the control reactors

Fifty percent reinoculation resulted in dilution of the toxic components and intermediates and the process was then able to handle an increase to 200 ppm, contrary to phase 1, where the same concentration resulted in process failure. This indicated some degree of adaptation of the reactor biomass. It has previously been shown, that adaptation can result in a process more resistant to inhibition [15].

In Fig.3 the measured concentrations of PCE in the feed and reactor are shown. The PCE concentrations in the feed varied and were somewhat lower than expected from the quantity PCE of added. The reason could be sampling error due to the nonsoluble character of the PCE-waste. Another explanation could be evaporation of the PCE from the feed bottle.

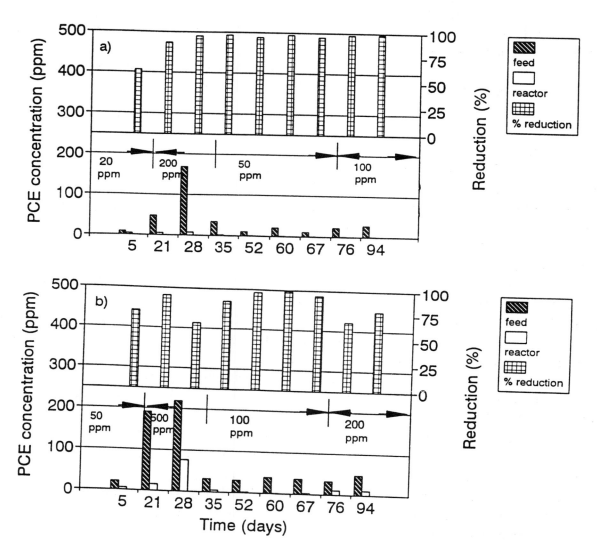

Fig. 3: PCE concentrations and PCE reduction: a) reactor PCE,1 and b) reactor PCE,2

The concentration of PCE in the reactors was almost always below the detection limit (1.0 ppm), indicating effective removal of the toxic compounds during anaerobic digestion. PCE reduction was close to 100% and measurement of intermediates showed only low concentrations of DCE and no vinyl chloride, indicating full dechlorination (detection limits for both were 1.0 ppm). PCE in the biogas was never above the detection limit (1.0 ppm).

Anaerobic reactors are closed systems, thus PCE that evaporated from the reactor, would end up in the biogas and would finally will be incinerated. From an environmental point of view, it is important that the liquid effluent concentration does not contain the toxic compound. This requirement seems to be fulfilled in the present study. The high effectiveness of the process, and the low effluent PCE concentrations justify treatment of PCE-containing waste in anaerobic reactors. Adaptation of the system, with slow stepwise addition of the toxic waste seems to be necessary to avoid inhibition and disturbance of the anaerobic process.

Aniline degradation

Addition of aniline-waste at 50 and 100 ppm aniline did not disturb the anaerobic process (Fig.4). The methane yield of reactors ANI,1 and ANI,2 was slightly higher than the 0.2 l CH_4/gVS of the control reactor. At day 20 the aniline concentration was increased 10 fold to 500 and 1000 ppm in the feed of reactors ANI,1 and ANI,2, respectively. After approx. 3 retention times (40 days) reactor ANI,2 was inhibited and the methane yield dropped to approx. 0.03 l CH_4/gVS, while ANI,1 stayed uninhibited (Fig.4a).

In reactor ANI,2 the VFA level increased to almost 8 g/l, indicating that the process was out of balance. The pH dropped from approx. 8 to almost 7 as a result of VFA accumulation (Fig.4b). Inhibition in reactor ANI,2 (Fig.4b) occurred approx. 16 days after the increase in the feed concentration. The delay was probably due to the slow increase of aniline concentration in the reactor compared to the feed.

The above experiments showed that the anaerobic process could tolerate an aniline concentrations of up to 500 ppm. However, addition of aniline-waste with an aniline concentration of 1000 ppm severely inhibited the process.

Fig. 5 shows the concentrations of aniline in the feed and reactor. As for PCE, the aniline concentrations measured were lower than the aniline added. Aniline transformation in the reactors was, however, much lower than with PCE. High concentrations of aniline were found in the effluent of the reactor (especially ANI,2), showing that aniline was not fully degraded. Further adaptation of the process or introduction of microorganisms that could degrade aniline may have increased aniline degradation.

CONCLUSIONS

The anaerobic process could be adapted to degrade PCE-waste with PCE concentrations up to 200 ppm. The concentration of PCE and it's dechlorination intermediates in the reactor effluent was below the detection limit (1.0 ppm). The use of anaerobic treatment as a means of destruction of such toxic wastes may be a feasible process in the future.

Aniline, at a feed concentration up to 500 ppm, was tolerated by the anaerobic process. Aniline was, however, not satisfactorily destroyed in the reactors, especially at higher concentrations. Further adaptation of the process was probably required.

Codigestion of toxic organic waste with manure or other types of non-toxic waste is a potentially efficient method of disposal of toxic organic wastes. Further experimentation is, however, required before application can develop for full-scale anaerobic treatment. Establishment of the degradability of toxic wastes, as well as toxicity limits and concentration requirements loading and adaptation are all parameters which should be determined prior to process design.

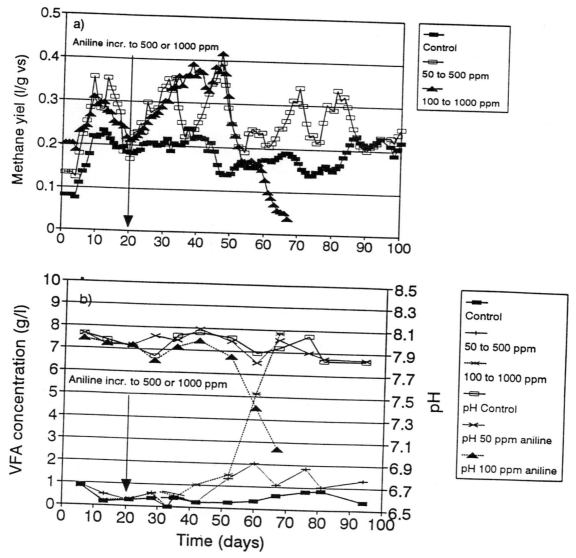

Fig. 4: Codigestion of manure together with aniline-waste

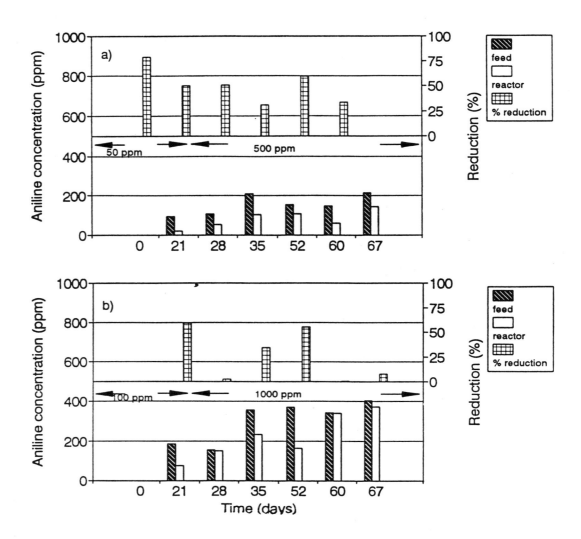

Fig. 5: Aniline concentrations and aniline reduction: a) reactor ANI,1 and b) reactor ANI,2

REFERENCES

1. Ahring B.K., Angelidaki I., and Johansen K. (1992). Anaerobic treatment of manure together with industrial waste. *Wat. Sci. Tech.* No 25, 311-318.
2. Angelidaki, I., S.P. Petersen, and B.K. Ahring. (1990). Effects of lipids on thermophilic anaerobic digestion and reduction of lipid inhibition upon addition of bentonite. *Appl. Microbiol. Biotechnol.* No 33, 469-472.
3. Fannin K.F., Conrad J.R., Srivastva V.J., Chynoweth D.P., and Jerger D.E. (1986). Anaerobic processes. *J. Water Pollution Control Fed.* No 58, 504-510.
4. Vega J.L., Clausen E.C., and Gaddy J.L. (1988). Performance of immobilized cell reactors. In: Handbook on anaerobic fermentation. Erickson L.E., and Yee-Chak Fung (1988). New York.

5. Sleat R., and Robinson J.P. (1983). Methanogenic degradation of sodium benzoate in profundal sediments from a small eutrophic lake. *J. Microbiol.* No 140, 141-152.
6. Egli C., Tschan T., Scholtz R., Cook A.M., and Leisinger T. (1988). Transformation of tetrachloromethane to dichloromethane and carbon dioxide by *Acetobacterium woodii*. *Appl. Environ. Microbiol.* No 54, 2819-2824.
7. Freedman R., and Gosset J.M. (1989). Biological reductive dechlorination of tetrachloroethylene and trichoroethylene to ethylene under methanogenic conditions. *Appl. Environ. Microbiol.* No 55, 2144-2151.
8. Schink, B. (1988). Principles and limits of anaerobic degradation: Environmental and technological aspects. In: A.J.B. Zehnder (ed), Biology of anaerobic microorganisms. Chapter 14. A Wiley-Interscience Publication. John Wiley & Sons.
9. Schink, B. (1990). Anaerobic news on phenols and aniline. Wat. Poll. Res. Rap. no. 25. Proceedings of the Workshop "Organic Micropollutants in the aquatic environment". Copenhagen.
10. Vogel T.M., and McCarty P.L. (1985). Biotransformation of tetrachloroethylene to trichloroethylene, vinyl chloride, and carbon dioxide under methanogenic conditions. *Appl. Environ. Microbiol.* No 49, 1080-1083.
11. Hendriksen H.V., Larsen S., and Ahring B.K. (1992). Influence of a supplemental carbon source on anaerobic dechlorination of pentachlorophenol in granular sludge. *Appl. Environ. Microbiol.* No 58 365-370.
12. Kuhn, E.P. Townsend G.T. and Suflita J.M. (1990). Effect of sulfate and organic carbon supplements on reductive dehalogenation of chloroanilines in anaerobic aquifer slurries. *Appl. Environ. Microbiol.* No 56, 2630-2637.
13. Nies, L., and Vogel T.M. (1991). Identification of the proton source for the microbial reductive dechlorination of 2,3,4,5,6-pentachlorobiphenyl. *Appl. Environ. Microbiol.* No 57, 2771-2774.
14. Suflita, J.M., J. Stout, and Tiedje J.M. (1984). Dechlorination of 2,4,5-trichlorophenoxyacetic acid by anaerobic microorganisms. *J. Agric. Food Chem.* No 32, 218-221.
15. Angelidaki I. and Ahring B.K. (1993). Thermophilic digestion of livestock waste: the effect of ammonia. *Appl. Microbiol. Biotechnol.* No 38, 560-564.
16. Angelidaki I. and Ahring B.K. (1994). Anaerobic thermophilic digestion of manure at different ammonia loads: effect of temperature. *Wat. Res.* No 28, 727-731.

A Mathematical Model for Dynamic Simulation of Anaerobic Digestion of Complex Substrates: Focusing on Codigestion of Manure with Lipid Containing Substrates

I. Angelidaki, L. Ellegaard, and B.K. Ahring, The Anaerobic Microbiology/Biotechnology Research Group, Institute of Environmental Science and Engineering, Building 115, The Technical University of Denmark, 2800 Lyngby, Denmark.

ABSTRACT

A mathematical model describing the combined anaerobic degradation of a complex organic material, such as manure, and a lipid containing additive, has been developed. The model includes an enzymatic hydrolytic step and 6 bacterial steps and involves 14 chemical compounds. Free ammonia, acetate and long-chain fatty acids constitute the primary modulating factors in the model. Simulations showed that stepwise increase of lipid addition, resulted in substantial increase of the methane production.

KEY WORDS

Mathematical model, simulations, anaerobic, lipids, long-chain fatty acids

INTRODUCTION

Anaerobic digestion is widely used for waste treatment. In Denmark, large biogas plants have been built to treat manure in addition to wastes from slaughter houses food industries and source-sorted household wastes.

Livestock manure is a complex substrate containing undissolved and dissolved organic matter such as polysaccharides, lipids, proteins and volatile fatty acids (VFA) as well as a number of inorganic compounds of importance for the chemical environment. Lipids are one of the most energetic organic compounds with a high specific biogas potential. Manure has only a low content of lipids (2-4% of the total organic content) [11]. Consequently, addition of wastes with a high lipid content into a biogas reactor can result in significant increase of the biogas production. Codigestion, especially with waste with high lipid concentration, has therefore, been employed in most manure based biogas plants in Denmark in the recent years.

Although lipids are a rich substrate for biogas production, they can also inhibit the biogas production [2,5,6]. It has been shown that long-chain fatty acids, which is an intermediate of lipid degradation, is responsible for the inhibition [2]. It has been shown in batch experiments with synthetic media, that long-chain fatty acids can be inhibitory at relatively low concentrations [2,7,8].

Many mathematical models have been described previously. Only a few models containing a separate description of carbohydrates, lipids and proteins have been reported [10, 12]. However, mathematical models containing a detailed description of the interactions occurring during the lipid degradation has not previously reported.

A mathematical model focusing on codigestion of waste with carbohydrates as main organic component, such as manure together with lipid containing additives has been developed. Dynamic changes after addition of lipid to digestions running at steady state with manure are presented.

MODEL DESCRIPTION

The model used is based on a previously described model [4] (Fig.1a), which was extended to include lipid and long-chain fatty acid (LCFA) degradation (Fig.1b). The extended model involves one enzymatic process (hydrolysis of undissolved organic matter), and six bacterial groups: glucose fermenting acidogens, lipolytic bacteria, LCFA degrading acetogens, propionate degrading acetogens, butyrate degrading acetogens and aceticlastic methanogens. Stoichiometry of all the conversions, except for lipid and LCFA have previously been described in [4]. Equilibrium relationships of ammonia, carbon dioxide, and pH, as well as gas phase dynamics and temperature effects have been included in detail. Free ammonia inhibition of the aceticlastic step and acetate inhibition of the acetogenic steps is included in the model. VFA inhibition of the initial enzymatic hydrolytic step is also included. Furthermore, inhibition caused by LCFA on all steps of the process except hydrolysis, has been introduced.

As model lipid glycerol triolate (GTO) is used. Oleate is the most abundant type of LCFA in many vegetable oils. GTO is hydrolysed to glycerol and oleate by glycerol fermenting acidogenic bacteria (Fig. 1b). Glycerol is further degraded to propionate. This step is not kinetically included in the model, i.e., this step is assumed to occur instantly from GTO to propionate.

The oleate formed as intermediate is degraded by LCFA acetogenic bacteria to acetate and H_2. Oleate exerts inhibition on all the bacterial stages of the anaerobic process. As long as the oleate concentration is kept low, lipid degradation occurs uninhibited. However, if oleate accumulates over a certain level, the entire process is inhibited [2].

Substrate: <u>Manure</u> consists of a complex mixture of dissolved and undissolved organic matter and important inorganic components. In the model the primary substrate is represented as soluble (s) and insoluble (is) carbohydrate units, with the basic formula $(C_6H_{10}O_5)_s$ and $(C_6H_{10}O_5 \cdot nNH_3)_{is}$ respectively [4]. The insoluble fraction includes organic bound nitrogen which is only released during hydrolysis. The elemental composition of the insoluble fraction represents the carbohydrate and protein content of manure. The total organic content of the manure used for simulations was 43 g/l, consisting of 34 g/l soluble and insoluble carbohydrate units, 7 g/l VFA and 2 g/l lipids. In addition 2.88 g/l inorganic CO_2 is included. The CO_2 concentration was experimentally determined, by acidifying a sample of undigested manure to pH 4 and measuring the gas evolution. The cation (Z^+) concentration was set to 6.75 g/l. The composition of the undigested manure used for the simulations is as shown in Table 1.

Table 1: Cattle manure composition used in model[a].

Component	Concentration
Insoluble organic matter	28.9 g/l
Soluble organic matter as:	
- $(C_6H_{10}O_5)_s$	5.1 g/l
- Glycerol trioleate	2.0 g/l
- Acetate	4.5 g/l
- Propionate	2.3 g/l
- Butyrate	0.2 g/l
Total organic	43.0 g/l
Dissolved NH_4^+-N	2.5 g-N/l
Organic bound NH_3-N	1.2 g-N/l
Phosphorus	0.55 g-P/l[b]
CO_2	2.88 g/l[d]
Z^+	6.75 g-K/l[c]

[a] From ref. 2; [b] Based on cattle manure analyses from a Danish biogas plant; [c] Determined iteratively to obtain a certain pH level. [d] Experimentally determined from cattle manure

STOICHIOMETRY

Stoichiometry for the hydrolysis of the insoluble carbohydrate, the soluble carbohydrate fermentation, VFA degradation and methanogenesis are as previously described [4]. GTO ($C_{57}H_{104}O_6$) is hydrolysed to glycerol ($C_3H_8O_3$) and LCFA ($C_{18}H_{34}O_2$) by glycerol degraders (eq.1), which then convert glycerol to propionate ($C_3H_6O_2$) (eq.2) [9]. As the rate of glycerol conversion is relatively high, and the concentration of glycerol therefore small and of little importance, the dynamic description of glycerol was omitted. The degradation of GTO is derived by combining, a) the GTO lipolysis to oleate and glycerol (eq.1) and b) the glycerol degradation to biomass ($C_5H_7NO_2$) and propionate (eq.2):

$$C_{57}H_{104}O_6 + 3\ H_2O \longrightarrow C_3H_8O_3 + 3\ C_{18}H_{34}O_2 \tag{1}$$

$$C_3H_8O_3 + 0.04071\ NH_3 + 0.0291\ CO_2 \longrightarrow 0.04071\ C_5H_7NO_2 \\ + 0.9418\ C_3H_6O_2 + 1.09305\ H_2O \tag{2}$$

which results in the overall GTO degrading reaction:

$$C_{57}H_{104}O_6 + 1.90695\ H_2O + 0.04071\ NH_3 + 0.0291\ CO_2 \longrightarrow \\ 0.04071\ C_5H_7NO_2 + 0.941843\ C_3H_6O_2 + 3\ C_{18}H_{34}O_2 \tag{3}$$

Degradation of the resulting oleate by LCFA degrading acetogens is derived by combining the LCFA step (eq. 4), together with the hydrogen utilizing step (eq.5):

$$C_{18}H_{34}O_2 + 15.2398 \ H_2O + 0.2500 \ CO_2 + 0.1701 \ NH_3 \longrightarrow \qquad (4)$$
$$0.1701 \ C_5H_7NO_2 + 8.6998 \ C_2H_4O_2 + 14.500 \ H_2$$

$$14.500 \ H_2 + 3.8334 \ CO_2 + 0.0836 \ NH_3 \longrightarrow 0.0836 \ C_5H_7NO_2 \qquad (5)$$
$$+ 3.4139 \ CH_4 + 7.4997 \ H_2O$$

which results in the overall oleate degrading reaction:

$$C_{18}H_{34}O_2 + 7.7401 \ H_2O + 4.0834 \ CO_2 + 0.2537 \ NH_3 \longrightarrow \qquad (6)$$
$$0.2537 \ C_5H_7NO_2 + 8.6998 \ C_2H_4O_2 + 3.4139 \ CH_4$$

The resulting yield coefficients of the steps included in the model expressed as g per g bacteria synthesized, are summarized in table 2.

Table 2: Yield coefficients (g/g-biomass) used in the model.

	$(C_6H_{10}O_5)_s$	GTO	Oleate	NH_3	Acetate	Propion.	Butyr.	CH_4	CO_2
Acidogenesis	-12.86	-	-	-0.15	3.54	2.93	3.08	-	2.41
GTO degr.	-	-192.16	183.90	-0.15	-	15.15	-	-	-0.28
Oleate degr.	-	-	-9.84	-0.15	18.21	-	-	1.91	-6.27
Propion. degr.	-	-	-	-0.15	8.01	-10.57	-	1.51	1.01
Butyrate degr.	-	-	-	-0.15	15.37	-	-11.92	0.96	-3.30
Methane prod.	-	-	-	-0.15	-24.14	-	-	6.08	16.72

GROWTH KINETICS

For the hydrolytic step a first order reaction rate was applied, as described in [4].

Acidogenic and lipolytic step:
For these steps, Monod growth kinetics were applied with non-competitive inhibition by LCFA, acidogenic (eq.7) and lipolytic (eq.8):

$$\mu = \mu_{max}(T) \left(\frac{1}{1 + \frac{K_s}{[(C_6H_{10}O_5)_s]}} \right) \left(\frac{1}{1 + \frac{[LCFA]}{K_{i,LCFA}}} \right) \qquad (7)$$

$$\mu = \mu_{max}(T) \left(\frac{1}{1 + \frac{K_s}{[GTO]}} \right) \left(\frac{1}{1 + \frac{[LCFA]}{K_{i,LCFA}}} \right) F(pH) \qquad (8)$$

where $\mu_{max}(T)$ is the temperature dependent saturation growth rate, $[(C_6H_{10}O_5)_s]$ or [GTO] the substrate concentration and K_s the Monod saturation constant, [LCFA] is the LCFA concentration and $K_{i,LCFA}$ the LCFA inhibition constant. F(pH) is a pH modulating function as described in [4].

LCFA acetogenic step:
For the LCFA acetogenic step Monod kinetics were assumed with a Haldane type substrate inhibition by LCFA:

$$\mu = \mu_{max}(T) \frac{1}{1 + \frac{K_{s,LCFA}}{[LCFA]} + \frac{[LCFA]}{K_{i,LCFA}}} F(pH) \qquad (9)$$

Propionate, butyrate acetogenic step:
For the propionate and butyrate degrading step Monod growth kinetics were assumed with non-competitive inhibition by acetate, [1,10] and LCFA:

$$\mu = \mu_{max}(T) \frac{1}{1 + \frac{K_s}{[A]}} \frac{1}{1 + \frac{[HAc]}{K_{i,HAc}}} \frac{1}{1 + \frac{[LCFA]}{K_{i,LCFA}}} F(pH) \qquad (10)$$

where $K_{i,HAc}$ and $K_{i,LCFA}$ is the inhibition constant for acetate and LCFA respectively; [A] is either the propionate or the butyrate substrate concentration [HAc] is the acetate concentration and F(pH) a pH modulation function.

Aceticlastic step:
For this step Monod growth kinetics with non-competitive free ammonia inhibition and inhibition caused by the LCFA were assumed. Furthermore, pH modulation was included.

$$\mu = \mu_{max}(T) \frac{1}{1 + \frac{K_s}{[HAc]}} \frac{1}{1 + \frac{[NH_3]}{K_{i,NH3}}} \frac{1}{1 + \frac{[LCFA]}{K_{i,LCFA}}} F(pH) \qquad (11)$$

where $K_{i,NH3}$ is the inhibition constant for ammonia.

Gas and liquid were assumed to be in quasi-stationary equilibrium and the distribution of the volatile components between gas and liquid phase was determined as previously described [4].

KINETIC CONSTANTS

The kinetic constants for the enzymatic and bacterial steps used in the model are presented in table 3. Values for kinetic constants are as previously described [4]. Kinetic and inhibition constants for LCFA degradation was from [2,3]. Kinetic constants for GTO degradation was indirectly estimated in order to satisfy experimental observations concerning inhibition dynamics of GTO and oleate [2,5]. For all bacterial steps a constant death rate with a value of 5 % of the saturation growth rate (μ_{max}) was used.

Table 3: Kinetic constants used in model

Group	μmax (d^{-1})	K_s (g/l)	K_i (g/l)	K_i[b] (g/l)
Hydrolytic	1.0	-	0.33 (VFA)[a]	-
Lipolytic	2.86	0.020 (GTO)	-	0.2[c]
LCFA-degraders	0.57	0.040 (ol.)	-	0.2[d]
Acidogenic	5.0	0.500 (glc)	-	0.2[c]
Propionate degr.	0.54	0.259 (HPr)	0.96 (HAc)	0.2[c]
Butyrate degr.	0.68	0.176 (HBt)	0.72 (HAc)	0.2[d]
Methanogenic	0.60	0.120 (HAc)	0.26 (NH_3)	0.2[d]

[a] VFA as acetate; [b] LCFA inhibition, experimentally determined, by batch experiments on synthetic media inoculated with 5% digested manure at 55 °C; [c] inhibition constant for non-competitive inhibition; [e] inhibition constant for Haldane-type inhibition

RESULTS

The effect of introducing a lipid (GTO) containing additive to a reactor, previously operated at steady state conditions with cattle manure and a retention time of 15 days, were simulated. In simulations the GTO content of the manure fed to the reactor was suddenly increased at day 7.

When 3 g/l of GTO was added, the biogas production rate (methane and carbon dioxide) gradually increased, after an initial disturbance, to reach a new steady state level after approx. 30 days (Fig.2a). A peak of oleate (LCFA) is seen immediately after the introduction of GTO, due to accumulation of oleate, until a sufficient increase in the population of LCFA degraders occurred. Although the peak only lasted a few days, and reached only moderate concentrations, the LCFA inhibition sparked a long lasting accumulation of VFA, until steady state conditions were reestablished, with VFA levels slightly below the level before the addition of GTO (Fig.2b). pH slightly decreased from 8.05 to approx. 8.00 due to the VFA accumulation, but stabilized again at the same level as before the GTO addition.

Addition of 8 g/l GTO to the process resulted in a dramatic process failure after a short time. The biogas production rate dropped to almost zero within the first week (Fig.3a). All the acids increased (Fig.3b). pH fluctuated with a final decrease due to the accumulation of VFA. LCFA (oleate) increased shortly after the GTO addition to a very high level compared to the inhibition level ($K_{i\,LCFA}$ 0.2 g/l), which is the reason for the total process failure. A sufficient amount of LCFA degraders was not present to counteract the accumulation of LCFA.

However, when the GTO addition was increased stepwise, the process managed to adapt, and higher final concentrations could be tolerated without process inhibition. In a simulation (Fig.4) the GTO addition was increased to 4 g/l at day 7 and then further to 8 g/l at day 17 and finally to 12 g/l at day 27. The biogas production increased, as the organic load increased, and stabilized at a level approx. double as high as before the GTO addition (Fig.4a). The methane content of the gas produced was higher than before the addition of GTO, due to the more reduced form of GTO compared to the basic organic composition of manure. The VFA concentration rose at each increase of GTO, but finally stabilised on a level somewhat lower than before the GTO addition (Fig.4b).

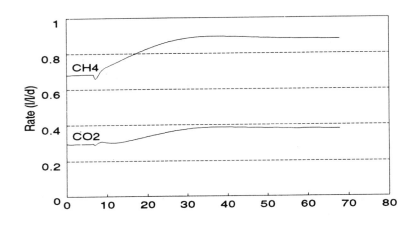

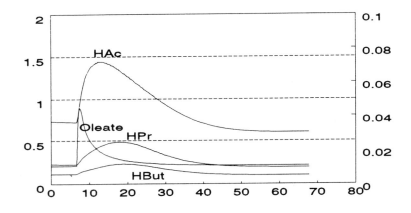

Figure 2: Simulation with digestion of cattle manure and addition of 3 g/l GTO at day 7

The LCFA concentration increases after the first GTO addition from approx. 0.01 to 0.08 g/l. However, at the next elevations of the GTO concentration LCFA accumulated to a much lower level. The lower LCFA accumulation observed after the second and third increase of the

GTO feed concentration, is due to the higher number of LCFA degrading acetogenic bacteria established during the first step increase.

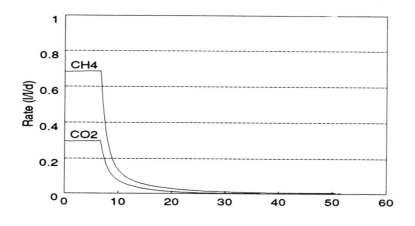

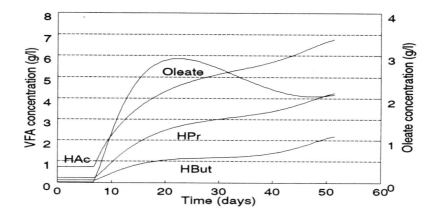

Figure 3: Simulation with digestion of cattle manure and addition of 8 g/l GTO at day 7

DISCUSSION

The key factor determining the fate of the process after introduction of lipid, is the kinetic balance between LCFA production by hydrolysis (the lipolytic step) and LCFA consumption by the LCFA acetogenic step. The higher rate of the former step results in a LCFA accumulation, until a sufficient population of LCFA acetogens has been established. The simulations described clearly demonstrate the need of cautiousness when introducing or changing the addition of lipid containing waste, as has been noticed on several occasions on full scale biogas plants. Furthermore, the simulations demonstrate the importance of an adaptation of the process to lipids, when the content of lipid is low in the main substrate as is the case for manure. During adaptation, development of an active LCFA degrading culture takes place, which is responsible for keeping the concentration of LCFA below toxic levels.

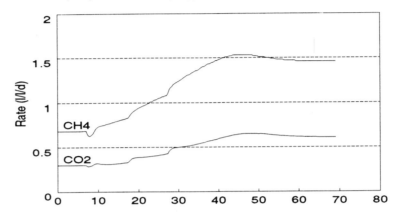

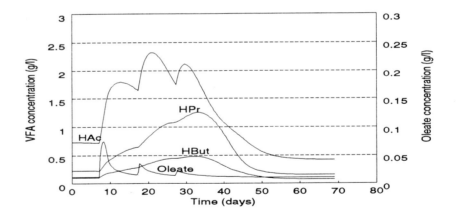

Figure 4: Simulation with digestion of cattle manure and stepwise increase of GTO addition 4 g/l at day 7, 8 g/l at day 17 and 12 g/l at day 27

It is interesting to note the lower steady state concentration of VFA, which is obtained when the substrate contains lipid (GTO). This means a slightly higher biogas yield is obtained from the manure, when codigested with lipids. The reason for this phenomenon is a decrease of the ammonia level, due to the higher bacterial activity when lipids are included. The increased synthesis of bacterial biomass consumes nitrogen, and lower the concentration of ammonia, which is an important modulating factor of the biogas process. Whether this synergistic effect of codigestion is achieved under practical conditions, depends on many factors, as small disturbances of the pH balance might augment or counteract the above phenomena. Lipids are usually added in the form of a waste product containing other elements than lipid, which might well also influence the chemical environment of the process.

It should however, be underlined that the presented results are only simulations, Therefore, the results should be interpreted as qualitative indication of process performance and not as absolute indications of concentrations that could cause inhibition. Inhibition concentrations of LCFA would depend on many factors according to the conditions of digestion.

CONCLUSIONS

Lipids are a good and rich substrate for biogas production, but at the same time intermediates from lipid degradation are potential inhibitors for the anaerobic processes. It is therefore important to include the interactions occurring during lipid degradation, e.g. when modeling and simulating the processes.

ACKNOWLEDGEMENTS

This work was supported by grants from the Center for Process Biotechnology.

List of symbols:

μ:	growth rate
$\mu_{max}(T)$:	temperature dependent saturation growth rate
$[(C_6H_{10}O_5)_s]$:	concentration of soluble carbohydrates
[GTO]:	glycerol trioleate concentration
[LCFA]:	long-chain fatty acids concentration
[HAc]:	acetate concentration
[NH_3]:	unionized ammonia concentration
[A]:	propionate or the butyrate substrate concentration
K_s:	Monod saturation constant
$K_{i,HAc}$:	inhibition constant, for inhibition caused by acetate
$K_{i,LCFA}$:	inhibition constant, for inhibition caused by LCFA
$K_{i,NH3}$:	inhibition constant, for inhibition caused by unionized ammonia
F(pH):	pH modulating function

REFERENCES

1. Ahring B.K., Westermann P., (1988). Product inhibition of butyrate metabolism by acetate and hydrogen in a thermophilic coculture. *Appl Environ Microbiol* 54,2393-2397
2. Angelidaki I., Ahring B.K., (1992). Effects of free long chain fatty acids on thermophilic anaerobic digestion. *Appl Microbiol Biotechnol* 37,808-812
3. Angelidaki I., Ahring B.K. (1995). Establishment and characterization of an anaerobic thermophilic (55°C) enrichment culture degrading long-chain fatty acids. *Appl Environ Microbiol*. In press.
4. Angelidaki I., Ellegaard L., Ahring B.K. (1993). A mathematical model for dynamic simulation of anaerobic digestion of complex substrates, focusing on ammonia inhibition. *Biotechnol Bioeng* 42,159-166
5. Angelidaki I., Petersen S.P., Ahring B.K. (1990). Effects of lipids on thermophilic anaerobic digestion and reduction of lipid inhibition upon addition of bentonite. *Appl Microbiol Biotechnol* 33,469-472

6. Hanaki K., Matsuo T., Nagase M. (1981). Mechanisms of inhibition caused by long chain fatty acids in anaerobic digestion process. *Biotechnol Bioeng* 23,1591-1610
7. Koster I.W., Cramer A., (1986). Inhibition of methanogenesis from acetate in granular sludge by long-chain fatty acids. *Appl Microbiol Biotechnol* 53,403-409
8. Rinzema A., Alphrnaar A., Lettinga G. (1989). The effect of lauric acid shock loads on the biological and physical performance of granular sludge in UASB reactors digesting acetate. *J Chem Tech Biotechnol* 46,257-266
9. Schauder R., Schink B. (1989). Anaerovibrio glycerini sp. nov., an anaerobic bacterium fermenting glycerol to propionate, cell matter, and hydrogen. *Arch Microbiol* 152,473-478
10. Siegrist H., Renggli D., Gujer W. (1992). Mathematic modelling of anaerobic mesophilic sewage sludge treatment. *Inter Sympos on Anaerob Digest of Solid Waste* 43-53(Abstract)
11. Zehnder A.J.B. 1988. Biology of anaerobic microorganisms. Zehnder A.J.B. (ed.) Publ. John Wiley Sons
12. Vavilin V.A., Vasiliev V.B., Ponomarev A.V., Rytow S.V. (1994). Simulation model "methane" as a tool for effective biogas production during anaerobic conversion of complex organic matter. *Bioresource Technol* 48,1-8

Innovative Pretreatment to Optimize Anaerobic Conversion of Biowaste

Heijo Scharff and Thijs Oorthuys, Grontmij Consulting Engineers, P.O.Box 203, 3730 AE De Bilt, The Netherlands

ABSTRACT

A mechanical waste separation plant operated by VAGRON in Groningen (Netherlands) produces a wet and a dry fraction from household waste. At present the wet fraction is composted and the dry fraction is landfilled. Due to government regulations these treatment methods cannot be continued in the near future. This paper describes the development, testing and evaluation of a pretreatment system for the wet fraction. Pilot tests show that the pretreatment can separate inert materials from the wet fraction. Reuse of waste can be promoted by application of the inert materials in civil engineering works. Anaerobic digestion of remaining organic material is facilitated due to the removal of inert materials that have caused mechanical problems in several full-scale fermentation plants in Europe.

KEYWORDS

Waste management, mechanical separation, recycling, anaerobic digestion.

THE VAGRON MECHANICAL SEPARATION PLANT

The mechanical separation plant operated by VAGRON in Groningen was built in 1987 and treats 80,000 tonnes of household waste per annum. The input of the VAGRON plant is separated into a dry fraction, or refuse derived fuel (RDF), a wet organic fraction (WOF) and several ferrous metal fractions. A schematic representation of the VAGRON plant is given in Figure 1. The compositions of Groningen household waste and the products of the VAGRON plant are presented in Table 1.

The heart of the VAGRON mechanical separation plant is its sieves. Separation is carried out in two consecutive stages to obtain a rapid separation between the different components. This prevents contamination of the different fractions and reduces the size of equipment by early removal of material from the process. The waste is transported through the separation plant by means of conveyor belts. During this process, ferrous metal is removed from the fractions by overhead magnets. All fractions are collected in containers. Because of its proven reliability and flexibility, dry mechanical separation is a very suitable system for waste disposal schemes requiring maximum recycling of solid waste.

Refuse derived fuel

The mechanical separation plant produces 55 to 60% dry fraction or refuse derived fuel. Application of refuse derived fuel in specific high efficiency incineration plants both reduces the total mass to be incinerated and increases the effiency of the incineration plant [1].

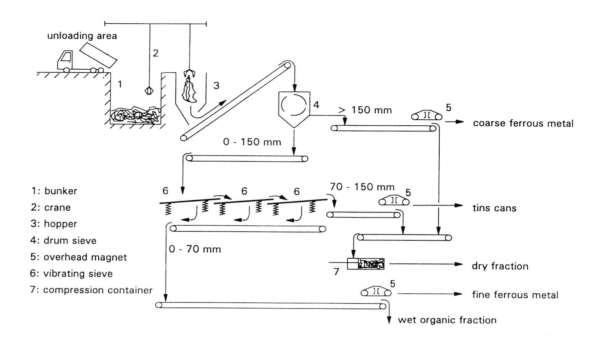

Fig. 1: Schematic representation of the VAGRON mechanical separation plant.

Table 1: Groningen household waste and vagron products

Component	Household waste [%]	RDF [%]	WOF [%]
glass	7-9	<1	14-18
stone	5-7	<1	10-14
metals	3-5	<1	<1
paper/cardboard	26-31	39-45	2-6
plastics	9-13	34-32	<1
textile	3-5	4-8	<1
sand	3-5	<1	8-11
v.f.g. & u.r.*	35-39	6-14	56-60

*vegetable, fruit and garden waste and unidentifiable remainder

Metals

Several recyclable ferrous metal fractions (see Figure 1) are separated (3 to 5% of the input), the largest of which contains mainly tin cans. This fraction is the most attractive material for reuse and is sold for Dfl 25 (DKK 90) per tonne. A coarse ferrous metal fraction contains various items like chairs and bicycle frames. A fine ferrous metal fraction contains mainly crown-corks and jar lids. All fractions are reused in the metal industry.

Non-ferrous metals are mostly small sized items and can mainly be found in the wet organic fraction. Even without prior treatment of the wet organic fraction separation of non-ferrous metals (approximately 1% of the wet organic fraction) by means of an Eddy current separator has proven succesfull in a pilot test (1 tonne wet organic fraction per hour). The metal industry has offered to pay Dfl 500 (appr. DKK 1800) per tonne non-ferrous metals. Further separation into components like copper, lead, brass, aluminium, etc. prior to recycling is carried out by the metal industry.

Wet organic fraction

The remaining material (37 to 42% of the input) is called the wet organic fraction. Until recently the wet organic fraction was composted in the open air, but present Dutch regulations on odour emission require indoor composting. Moreover since 1994 more stringent regulations have been implemented with respect to the heavy metal content of compost. In the meantime less contaminated compost from source separated biowaste became available. Therefore the end product of the wet organic fraction can no longer be sold as compost. At the moment the material is used as a temporary cover material on landfill sites. In the near future alternative methods are required for the treatment of the wet organic fraction.

The wet organic fraction has two major components: (bio)degradable organic matter and inert material like sand and glass. The latter does not facilitate the treatment of the first component. Several full-scale waste fermentation plants in Europe have experienced problems with blocking of pipes or reduction of reactor volume by build-up of sand in the system. Also excessive wear and tear of pumps, valves and pre- and post-treatment equipment has occurred. It is preferable that the inert material is separated before further treatment of the wet organic fraction. The inert material is a promising product to be used in civil engineering structures and could thus promote recycling.

INNOVATIVE PRETREATMENT SYSTEM

In cooperation with suppliers of equipment an innovative wet separation system has been developed. The system is based on proven technology as applied in sand dredging, soil remediation and waste treatment [7]. Figure 2 gives a schematic representation of the intended full-scale system. An expected mass balance is given based on the VAGRON waste composition, the results of pilot plant tests and intended optimization of the waterbalance.

First the wet organic fraction is mixed with process water and sedimentation of coarse inert material occurs in an upflow separator (1). The coarse material is removed from the separator by means of a submerged screw conveyor (2). In a counter-current wash drum (3) remaining organic particles are removed from the coarse inert material. The overflow of the first separator is transported to a drum sieve (4). The process water in the meantime washes

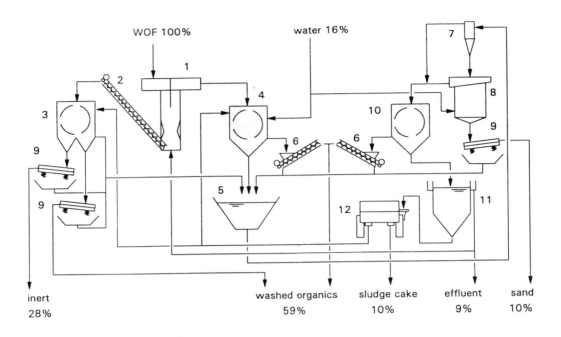

Fig. 2. Schematic representation of the wet pretreatment system (explanation see text)

smaller particles from the coarse organic fraction. The process water and smaller particles fall through the sieve into a reservoir (5). The coarse organic fraction is dewatered in a screwpress (6). The sequence of the first two steps enables optimal use of process water. From the reservoir the water containing small particles is pumped to a cyclone (7). In the cyclone both inert and organic particles of 0.06 to 2 mm are separated. The mixture of particles falls into a fluidized bed separator (8). Based on measurement and control of the specific weight of the medium in the separator a partition is made between organic and inert material. The inert material is mainly sand and is dewatered on a vibrating screen (9). The organic material is separated from the process water in a drumsieve (10) and is dewatered in a screwpress. All organic material is combined for further treatment. The water from the drum sieve (10) containing particles smaller than 0.06 mm is pumped to a sedimentation tank (11). From the bottom of the sedimentation tank a slurry is pumped to a centrifuge (12). Particles are continuously removed from the slurry and concentrated in a sludge cake to avoid build up in the process water. The sludge cake can either be landfilled or combined with the organic material for further treatment. Some equipment needs to be fed with 'clean' water and for this purpose either surface water or effluent from a wastewater treatment plant can be used.

Pilot test results

A pilot test treating approximately 100 kg wet organic fraction per hour during two weeks was performed at the VAGRON mechanical separation plant. The particle size of the wet organic fraction was smaller than 40 mm. The test proved that it was possible to separate inert materials from the wet organic fraction by applying the new washing technique. During the first test several problems were encountered. In the first pilot plant the wash drum was not present. The inert material contained approximately 10% organic matter. Reuse of such material in civil engineering works is not possible. Also the sedimentation tank and the centrifuge were not present. Suspended solids could not be removed from the process water and therefore no water was recirculated. Feeding to the drumsieve (4) was obstructed due to a very narrow entrance. A lot of additional water had to be used to flush the waste into the drumsieve. To operate the drumsieve and to obtain clean sand a waterflow of approximately 3 m^3/hour was required. During the test tap water was used. A 'clean' water requirement of 30 m^3 per tonne waste is unacceptable, therefore a major part of the water must be recirculated.

In the second pilot plant the drumsieve was adapted and a centrifuge was included. Semi-technical pilot tests treating approximately 1 tonne wet organic fraction per hour were carried out during two months. The tests focused on treatment of <35 mm wet organic fraction. Additionally <70 mm wet organic fraction, <60 mm source separated biowaste and <180 mm source separated biowaste were treated. It was observed that the source separated biowaste containing items up to 180 mm could not be processed. This test was discontinued after several hours of trial and error. The results of the tests are summarized in Table 2. Due to spilling of waste, water or products the balance of input and output did not match completely. This is indicated in the last column of Table 2.

The capacity of the pilot plant was lower than expected. The main problem was the funnel shaped feeding unit of the upflow separator (1). The separator was frequently blocked by uneven feeding. Thus lumps of material formed and hampered contact between the waste and the process water. The coarse inert separator required regular feeding. A few times obstruction of the drum sieve (4) occured. This was caused by material spilt on the chain of the driving mechanism. The fluidized bed separator (8) was a critical part of the equipment. The organic load of the material entering the separator was high compared to the material that is usually treated in this type of equipment in sand dredging. The fluidized bed separator required frequent attention in order to adjust the waterflow to the actual input. The fine organic material separated in the fluidized bed separator contained a lot of fats. The fat caused clogging of the screen of drum sieve (10). When this occurred process water entered the screwpress (6) and caused spilling of water. The screen had to be cleaned regularly. Improvements have been designed based on this experience. They will be implemented in the full scale plant that will start operation in 1997. The centrifuge performed very well. It had to be cleaned less frequently than expected. Without adding polymer a sludgecake with an average total solids content (TS) of 40% was produced.

Table 2: Results of the pretreatment system

week	oper-ation hours	product-ion time (%)	capacity (kg/h)	Input waste (kg)	water (kg)	Output inert (kg)	sand (kg)	sludge (kg)	water (kg)	organics (kg)	balance (kg)
				wet organic fraction <35 mm							
39	7.0	49	860	2,960	42,000	870	370	280	41,670	1,640	120
40	29.7	55	960	15,610	165,600	3,690	920	1,130	164,410	10,820	220
41	34.2	83	980	27,680	264,600	7,920	2,900	2,360	263,550	15,550	0
42	33.3	72	980	23,270	236,600	5,690	2,220	1,900	234,980	14,910	170
43	30.9	74	920	21,110	219,600	6,680	1,800	1,910	218,920	11,400	10
44	29.1	69	1,000	20,030	168,200	7,780	1,510	2,190	166,320	10,160	270
45	31.8	50	1,060	17,070	221,200	3,130	2,490	2,310	219,680	10,740	-80
46	9.9	62	980	5,940	97,700	990	730	650	97,020	4,180	70
total	205.9	67	970	133,680	1,415,500	36,750	12,940	12,740	1,406,550	79,400	790
				wet organic fraction <70 mm							
47	10.5	58	830	5,040	198,000	1,630	670	500	197,430	2,690	120
				source separated biowaste <60 mm							
48	9.8	74	940	6,830	111,000	590	1,270	600	110,120	5,110	140

Although the waterbalance was improved by using a centrifuge still 11 m³ of tap water per tonne wet organic fraction was used during treatment. The 'clean' water is required to prevent clogging of nozzles in the wash drum as well as the drum sieve and to obtain clean sand from the fluidized bed separator. The water mainly needs to be free of suspended solids. Therefore effluent of a wastewater treatment plant can be used for this purpose. In the Netherlands the wastewater of a fermentation plant always needs to be treated in a wastewater treatment plant. When the pretreatment system, fermentation plant and wastewater treatment are combined more water can be recirculated. The net input of water to the pretreatment system will be confined to the amount of water that is lost in the products of the pretreatment system. The results from the pilot tests indicate that this amount is 0.07 to 0.13 m³ per tonne wet organic fraction or source separated biowaste.

The source separated biowaste contained less stones and glass than the wet organic fraction. Due to the large amount of garden waste in the source separated biowaste a significant amount of sand was produced. Again a washed organic material was produced containing hardly any inert material.

Product quality

After washing the coarse inert fraction contained 1 to 2% organic matter. A sand fraction with an organic matter content less than 3% was produced. The low organic matter content of both sand and glass and stone fractions allows application in road construction. Glass and stone can also be used in concrete as a replacement for gravel. Contractors have shown interest in the materials. They expect to be able to pay Dfl 5 (DKK 18) per tonne sand and Dfl 2 (DKK 7) per tonne of glass/stone fraction. The compositions of the input and the products of the wet pretreatment system are summarized in Table 3.

Table 3: Composition of pretreatment input and output

Component	WOF (%)	Inert (%)	Sand (%)	Sludge (%)	Organics (%)	Process water	(mg/l)
glass	17	56	<1	<1	<1	COD	18,600
stones	13	42	<1	<1	<1	BOD	6,700
metals	<1	<1	<1	<1	<1	Kjeldahl-N	520
paper/cardboard	3	<1	<1	<1	5	N-tot	550
plastics	<1	<1	<1	<1	<1	P-tot	30
v.f.g. & u.r.	55	2	3	40	86	Suspended solids	2,100
sand	11	<1	97	<1	<1	Total solids	18,900
other	<1	<1	<1	60	7	Sum Cu, Ni, Pb, Zn	0.0009
total	100	100	100	100	100		

Due to binding of heavy metals to clay and humus particles in the sludge cake, it was expected that the heavy metal content of the washed organic material could be improved. The end product of digestion of this material may be applicable as compost depending on local regulations. Heavy metal analyses were carried out on one sample of sludge cake and washed organics that were aerobically composted. The results of these heavy metals analyses and the values of the Dutch regulations are presented in Table 4.

Table 4: Heavy metals

Component	Sludge cake	Composted washed organics	Compost limits up to 31-12-1994	Compost limits from 01-01-1995
total solids (TS) (%)	46	52	-	-
volatile solids (VS)(% of TS)	30	65	>20	>20
Chromium (mg/kg TS)	40	35	200	50
Nickel (mg/kg TS)	15	20	50	20
Copper (mg/kg TS)	45	150	300	60
Zinc (mg/kg TS)	350	320	900	200
Arseni (mg/kg TS)	7	2	25	15
Cadmium (mg/kg TS)	1.2	1.2	2	1
Mercury (mg/kg TS)	0.4	0.6	2	0.3
Lead (mg/kg TS)	190	130	200	100

The analyses do not support the theory that heavy metals can be removed by means of the sludge cake. The results do however indicate that application of both materials as compost might have been possible in the Netherlands up to 31-12-1994. It should be noted that the sludge cake was not composted. The heavy metal content will increase due to the loss of part of the volatile solids (VS) and thus total solids (TS). Because of the limited number of samples analysed further investigations are required.

FERMENTATION OF WASHED ORGANIC FRACTION

After the wet separation, the total solids content of washed organic material is approximately 30% of which 80% were volatile solids. The washed organic material was fermented in laboratory scale batch tests in both wet (10% TS) and dry (30% TS) fermentation tests. Both thermophilic (55°C) and mesophilic (35°C) temperature conditions were imposed. The conditions and results of the fermentation tests are presented in Table 5.

Table 5: Fermentation tests

Batch number	1	2	3	4	5
TS in reactor (%)	30	30	10	10	10
Temperature (°C)	55	35	35	35	35
Inoculum/waste (-)	0.9	0.5	0.8	0.5	0.1
Retention time (days)	14	17	21	21	125
Methane production m^3 STP*/kg VS	0.29	0.25	0.28	0.27	0.33
Surplus water composition					
g COD/l	22.5	40.0	6.3	10.0	12.0
g BOD/l	5.3	10.0	2.5	3.2	3.3
g Nkj/l	2.1	4.0	1.1	1.0	0.9
pH	7.3	7.5	7.3	7.5	7.5

*STP: standard temperature and pressure = 0°C and 1013 mbar

The tests indicate that thermophilic conditions and wet fermentation tend to give a slightly higher biogas production as compared to dry mesophilic fermentation. Peres et al. [8] have given a summary of the methane production from municipal solid waste of several European and American fermentation systems. They report a specific methane production varying from 0.23 to 0.39 m^3 per kg VS. The observed gas production during the fermentation tests of washed wet organic fraction are in the same range. By removing inert material the VS content of the input is increased. Applying a similar TS content in the reactor means that also the organic load of the system is increased. The fermentation tests indicate that this has neither a positive nor a negative effect on the methane production. Optimization of anaerobic treatment of biowaste by means of the wet pretreatment system is not to be found in a higher methane production. Optimization is achieved by the removal of inert materials and thus in less mechanical failures and smaller reactor volumes.

Application of a ratio of inoculum to fresh waste of 0.1 is not sufficient to start the fermentation process. In batch 5 rapid acidification occurred. Only after 1 month of daily pH correction a stable pH value of more than 7 was reached. Then the methanogenic activity slowly started to rise. It took 125 days to reach a methane production comparable to the other batches.

Efficient removal of sand, glass and stones by the pretreatment system was confirmed by the fact that no sedimentation layer could be observed on the bottom of the batch reactors.

The filtrate of the reactor contents after fermentation indicated that COD and nitrogen concentrations in surplus water from dry fermentation are considerably higher than from wet fermentation. A difference in the amount of surplus water could not be investigated during the tests. In the Dutch situation where wastewater treatment is a prerequisite a higher Kjeldahl-N load is a disadvantage. The BOD/N ratio of the surplus water from batches 1 to 4 varies from 2.5 to 3.2. This indicates that for efficient nitrogen removal additional BOD may be required.

CONCLUSIONS

Large scale tests with the innovative pretreatment system have shown the technical feasibility of the separation of inert materials from the wet organic fraction of household waste. Application of the pretreatment system removes the material that does not generate biogas as a product with recycling potential. It can reduce the required volume of the fermentation tank in both wet and dry fermentation systems without reducing the total biogas production. Finally it prevents blocking and excessive wear and tear of equipment in the fermentation plant. It must be concluded that all fermentation systems can benefit from the innovative pretreatment system.

REFERENCES

1. Grontmij Consulting Engineers; Upgrading of products from waste separation - Optimisation of energy generation from refuse derived fuel (in Dutch); Dutch National Research Program Reuse of Waste; NOH 8924; 1989.
2. Environmental Impact Statement; Fermentation plant (Valorga) for VFG-waste (in Dutch); Waste Authority Central Brabant; 1992.
3. Dactech Milieu bv; Treatment of Dutch VFG-waste in a Dranco Plant (in Dutch); Dutch National Research Program Reuse of Waste; NOH 9249; 1992.
4. BTA GmbH; Vergärung von Biomüll und Nassmüll; Ergebnisse aus 5 Jahren Entwicklung und Versuchsbetrieb der BTA-Anlage Garching; 1991.
5. MAT GmbH; Realization of the BTA Technology in a Big Scale Plant for Processing and Disposal of Organic Waste; document presented in Irvine (UK); 1994.
6. Avecon; The Waasa Process; documentation of the ASJ fermentation plant (Waasa, Finland); 1993.
7. DWA Amsterdam; Full scale sand separation for sewage sludge (in Dutch); H_2O (231, 650-652); 1991.
8. Peres, C.S., Sanches, C.R., Matumoto, C., Schmidell, W.; Anaerobic biodegradability of the organic components of municipal solid waste (OFMSW); Proceedings of the Sixth International Symposium on Anaerobic Digestion; May 12 - 16, 1992, Sao Paulo, Brazil.

Wet Mechanical Pretreatment of Organic Wastes - the Full Scale Experience

P. Hochrein, W.L.Gore & Assoc. GmbH, Postfach 1149, D-85634 Putzbrunn, Germany

ABSTRACT

Wet Mechanical Pretreatment (WMP) of organic wastestreams uses pulping technology adapted from the paper industry to transform contaminated organic wastes into a purified organic feedstock for biological processes. With more than three years of full-scale and eight years of pilot operation, it is now a proven and reliable technology.

WMP may be used either for preparing feedstocks for quality compost production or high-performance anaeriobic digestion. As pretreatment for residual residential wastes, WMP segregates inert, combustible and biodegradable materials and thus allows optimized stabilization of each fraction for residual repositories.

WMP acts as highly flexible, completely mechanized and automated interface between difficult wastestreams and sophisticated biological treatment processes. Contaminant removal from the organic portion exceeds 99 percent with a yield of biodegradable matter of 95 percent for subsequent biological treatment. The new generation of waste pulpers offers the option of treating wastes in completely contained equipment. Hence, odor emissions and occupational biohazards are minimized.

KEYWORDS

Wet mechanical treatment, mechanical processing, waste pulping, pulper, composting, anaerobic digestion, contaminant removal, odor, emission, process optimization

PROCESS DESCRIPTION

WMP centers around waste pulping, a process derived from paper technology. Waste pulping is done in a batch process in order to maximize organics yield and minimizing reject volume. In a waste pulper, the feedstock is blended with water which is usually recirculated in the overall process. The waste pulper is fitted with a vertical axis impeller inside a cylindrical or octagonal vessel. The impeller causes a strong toroidal flow inside the pulper with high velocity gradients and strong hydraulic shear forces. These shear forces selectively comminute biodegradable matter and leave plastics and other biologically inert substances intact. When the wastes have been sufficiently pulped, a pulp of 5 to 15% TS is withdrawn from the pulper through a screen at the bottom of the vessel retaining the coarse contaminants. These contaminants are then resuspended in process water. They are removed by means of a rake catching materials of low density such as plastic and by a lock of gate valves at the bottom edge discarding sedimented heavies

The rake fraction is dewatered to at least 45 percent TS with a hydraulic dewatering press. The pulp from the waste pulper is run through a specifically designed hydrocyclone setup using a Circulating Bed Cylone. This apparatus extracts a mineral fraction with minimal organic content using only minute amounts of water. The purified pulp contains hardly anything but organic matter: the VS/TS ratio is between 80 and 90 percent, whereas organic wastestreams typically have a VS/TS-ration betwee 60 and 70 percent.

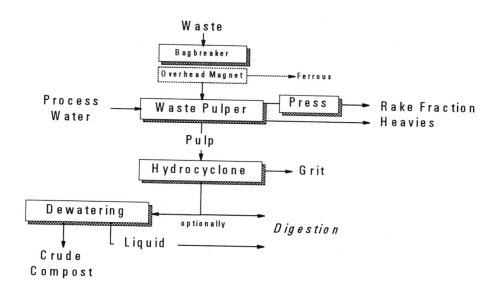

Fig.1. Flowchart depicting essential components of Wet Mechanical Pretreatment

Purified pulp may then go various directions: it may be dewatered into crude compost for aerobic composting, or it may be suscepted to anaerobic digestion for biogas and compost production. In the first case a solid-liquid separation such as a decanter or a dewatering press is the last step of WMP.

If a solid-liquid separation is used to produce crude compost, the liquid phase should be digested for biogas production. The digested liquid may be used as process water for pulping, depending on the kind of environment that WMP is used in.

PILOT SCALE APPLICATION

In 1986, the pilot plant of BTA (Biotechnische Abfallverwertung GmbH & Co. KG) plant in Garching near Munich started operation. At this time, the Wet Mechanical Preteatment was a novel approach to tackle the problem of selecting biodegradable matter from contaminated wastestreams.

During the first three years of the Garching pilot, Wet Waste from a wet/dry separated collection system were processed. These Wet Wastes contained some 20 to 30 percent of contaminants such as plastic, glass, metals and compound materials. The pulp produced proved to be a feedstock well suited for the subsequent high-performance biogasification. Later on, the wet/dry-collection approach was dropped throughout Germany, because the macroscopic as well as heavy metal contamination of the organics was inacceptable. [1]

From 1989 to 1991, source separated biowaste was processed and the WMP was optimized for yield and/or throughput.

From October 1991 to March 1992, combined treatment of biowaste, McDonald's restaurant wastes, market wastes and food wastes from cantinas was successfully tested. The Garching WMP proved that these extremely different wastestreams could easily be transformed into a homogeneous, clean pulp for digestion. [2]

Several other trials at Garching were conducted with residual residential wastes, commercial wastes, excavated landfill material and rake material from public sewage treatment. Many of these experiments were minutely documented with handsorting analyses of both input and products of the WMP-stage. [3]

Both yield of organic matter into the pulp as well as contaminant removal consistently exceeded 90 percent for this wide variety of wastestreams without any mechanical changes to the equipment.[3]

From these pilot experiences, a wide range of applications for WMP has been concieved. It also implies that this technology is capable of handling ill-defined or very variable wastestreams, acting as interface between difficult wastes and sophisticated biological processes.

PRETREATMENT FOR ANAEROBIC DIGESTION: HELSINGØR

In September 1991, North-Sealand Biogasplant was started up. Equipped with two 20 m3 BTA waste pulpers, NBA was designed to treat 80 Mg per day of source separated residential organic wastes estimated to contain less than 2,5 percent contaminants. The process guarantee implied that these 80 Mg were to be processed in 21 batches at 65 minutes each, translating into a Waste Loading Rate (WLR) of 175 kg/m3 pulper/h.

The waste volume was considerably lower than projected. TS of the wastes was around 40 percent and contamination was in the range between 5 and 15 percent and the bagbraker did not reach the designed throughput resulting in prolonged feeding times. Manually operated, the pulpers were not run at their full capacity. The high level of contamination caused prolonged rake operation and delays, because a single rake serves both pulpers in turns. Also, the effect of wear in the pumps was considerably higher than expected.

The **elapsed WLRs** were calculated for each batch from operator recordings. They reflect the performance reached under conditions considerably worse than assumed for the process guarantee: The feeding rate of the bag breaker used was much lower than projected, the unpredictedly high contamination of the wastestream required increased raking time, the waste pulpers were not filled and fed to their capacity and poor pump maintenance prolonged the time needed for withdrawing the pulp from the vessel.

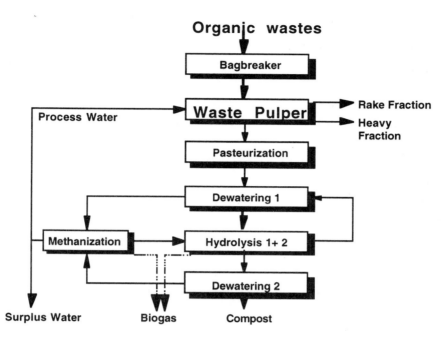

Fig. 2. Waste Pulper as pretreatment for source separated biowaste at North Sealand Biogasplant (NBA), Helsingør, DK.

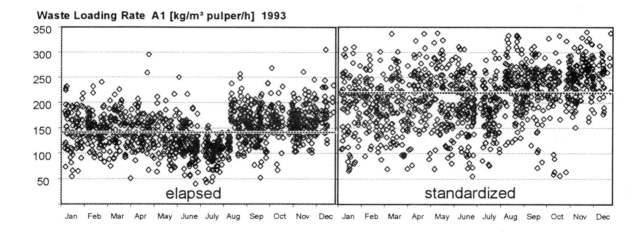

Fig. 3: Waste Loading Rates for pulper A1 at NBA in 1993. [5]

Standardized WLRs were computed to make theses measurements comparable to the process guarantee: A feeding time of 10 minutes, a complete filling to 20 m3 and 10 rake passes per batch according to 5% contaminants were assumed. This calculation does **not** take into account the idle times of the pulpers that are due to the manual operation mode used. These idle times may be avoided using an automatic operation mode.

When comparing elapsed and standardized WLRs in Figure 3, the standardized WLRs appear to be more scattered. This may be due to a variety of reasons: 1. Due to the inconsistent manual operation, some batches were run for a very long time or equipment was idling, 2. The last batch of each day was often only half filled with waste, 3. There were batches with very low waste loading due to an improper feeding mode, 4. Operator recordings appeared to be inconsistent at times, but the errors were not reproducable at the time of evaluation. Usually, these causes resulted in very low elapsed WLRs. Correcting feeding times and rake passes does not proportionally affect these errors, which may be an explanation for increased scatter. On the other hand, it was safeguarded that operators did not record any times lower those actually elapsed.[6] Therefore, standardization is on the safe side when comparing actual performance to the guarantee.

Table 1: Operational results of wet pretreatment at nba in 1993 [5]

	Waste processed Mg	Batches No.	WLR [kg/m3/h]	standardized WLR [kg/m3/h]	average TS of Pulp [%]
Pulper 1	4916	1219	145	220	6,8%
Pulper 2	4310	1100	135	220	6,9%

These results indicate that the waste pulpers at NBA are capable of exceeding the guaranteed performance of 175 kg/m3/h by 30 percent.

In spring 1994, the operators of NBA ceased the manual recording of the process data for the wet pretreatment because performance was considered satisfactory, even before the process guarantee of BTA expired in September 1994. [6]

Optimization experiments by the author conducted at NBA showed that an Improved Pulping Cycle is capable of increasing **WLR beyond 250 kg/m3/h for the given wastestream**. The objective of these experiments was to maximize WLR without compromising selectivity of the separation process.

This was done by increasing the amount of wastes fed per batch to the very capacity of the equipment. In order to sustain maximum capacity of the equipment, maintenance was improved as well as there were changes throughout each step of the batch process.

PRETREATMENT FOR COMPOST PRODUCTION: BADEN-BADEN

Near the City of Baden Baden, a 20 m3 BTA-waste pulper, a hydrocyclone and a sludge decanter at the local sewage treatment plant started operation in April 1993. The 5500 Mg annually of source separted biowastes from the City of Baden-Baden are processed into a homogenous, fluffy feedstock that is composted together with separately collected yard and brush wastes in open windrows. The liquid from the decanter is fed into the sludge digestors of the sewage plant for increased gas production.

After a year of introducing source separated collection in the region, full waste volume is being delivered since April 1994. Since then, all the biowaste collected has been processed at the plant.

The biowaste feedstock contains some 3 percent contaminants together with 70 percent moisture and produces considerable odors. WMP almost halves the volume of material going into windrow composting. This crude compost is mechanically well digested and easily composted due to its consistent solids content of 65 percent. Thorough handsorting analyses in Febuary indicate that contaminant content in the crude compost is around 0.05 percent, meaning that the efficiency of the fully automated contaminant removal is better than 99 percent! [7]

The results of optimization are reflected in Table 2. There, the monthly tonnages and the respective percentages as compared to the feedstock are given. Between August 1994 and Febuary 1995, several improvements were implemented:

1. A redesigned impeller was installed in the waste pulper improving selectivity of the pulping process. This reduced the amount of organic matter in the rejects and in turn increased the yield of crude compost. The content of contaminats in the compost was also reduced.
2. Total Solids content of the pulp was increased by improving the automatic algorithm of the feeding mode. This reduced liquid consumption and output per Mg of waste.
3. A hydraulic dewatering press was installed for the rake fraction. By dewatering the rake fraction from 20 to 45 percent TS, reject mass was reduced even further.

Table 2: Monthly mass balances before and after process optimization [7]

	August 1994		Febuary 1995 optimized	
Waste	477 Mg	100%	385 Mg	100%
Rejects	89 Mg	18%	27 Mg	7%
Liquid	1506 m3	309%	1120 m3	291%
Crude Compost	197 Mg	40%	218 Mg	57%

Furthermore, the batch sequence has been optimized and automated in a way that a WLR of 200 kg/m3/h is consistently reached, even though the small bag breaker chosen for this installation needs between 30 and 40 minutes to feed one batch. Assuming a 15 minute feeding time would translate into WLRs around 300 kg/m3/h.

The finished product from co-composting Crude Compost and yard waste meets all requirements for unrestricted application in Certified Organic Farming according to German „Bioland"-standards. [10]

The odor emissions of the crude compost are identical to yard waste, so that odor emissions from the open windrow composting site are not increased. The figures given are

German Odour Units (OU) according to Guideline of the German Society of Engineers (Verband Deutscher Ingenieure - VDI) and comparable to Dilutions to Threshold (D/T)-values used in the U.S.A.

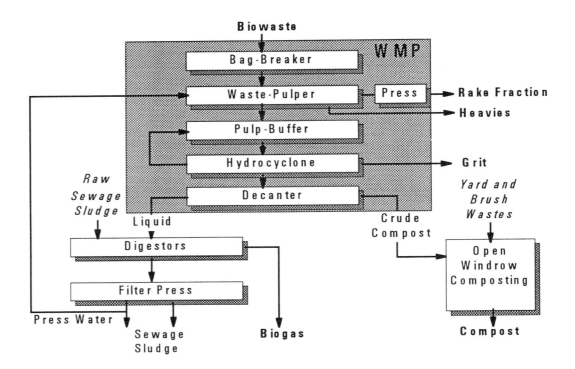

Fig. 4: WMP installation in Baden Baden

Table 3: Measurement of odor emissions from wet mechanical processing without biofilter [8]

	Source Emission [OU/s]	Time-factorized emission [OU/s]
Waste on tipping floor	1860	470
Crude Compost	10	160
Machine Room		560
Total WMP-Plant		1190

Odor emissions of the WMP-facility amount to less than 15 percent of the odors emitted from the activated sludge sewage treatment and may such be considered negligible within this environment.

In order to compare WMP with the odor emissions of other biowaste treatment schemes, the specific odor flow per Mg of annual capacity of the plant is computed.

Table 4: Specific odor emissions of biowaste composting with wmp and with conventional dry pretreatment and composting

	Specific emissions before or w/o biofilter [OU/s/Mg]	Specific emissions after biofilter ($\eta = 85\%$) [OU/s/Mg]
WMP Baden-Baden, open windrows, no air managment nor biofilter	0.23	
Dry pretreatment and box composting with biofilter, open curing [9]	1.34	0.40
Dry pretreatment, aerated static pile, fully enclosed, with biofilter [9]	1.58	0.25

It is evident that the biowaste processing scheme with WMP and windrow composting produces the least primary odor emissions of these concepts. Thus, expensive end-of-pipe air management and biofilters are obsolete, keeping the technical efforts at a reasonable minimum.

The liquid going into the digestors enhances biogas production as well as the degradation of the sewage sludge. No increase in the amount of sewage sludge has been observed throughout 1994. Gas production increased by some 20 percent. Utilized with the existing cogeneration units, the whole biowaste processing scheme from tipping floor to the end product balances its electric energy consumption with what it produces. There are some 150 kWh/Mg waste of excess heat produced from cogeneration.

PROCESS OPTIMIZATION

Optimizing Wet Mechanical Processing plants is a complex task and allows anything but general solutions. As with almost any waste management facility, WMP-optimization starts out by determining as many „enviromental" factor as possible: Times, volumes, qualities and variablity of waste delivery; permitted operating hours and limiting factors in the process environment, such as availability of process water or constrained downstream processing and storage capacities. Knowledge about the economics of the operation is essential for motivating the investments needed to implement improvements. Customer demands for product quality need to be met on a sustained basis.

But the single most influential factor for plant performance is at the same time the „softest": motivation of both staff and the political background. Only if the operating personell is excited about what they are doing and about the equipment they are given at hand, the sytem will work to its best. This also means regarding staff as valuable source of information and ideas concerning process and equipment. The most efficient way to optimize such a process is to help staff develop and at the same time accept the ideas.

For the optimizing engineer, there is a multitude of process variables in the complex batch sequence of waste pulper. A single batch comprises a sequence of

1. filling water,
2. filling waste,
3. pulping,
4. withdrawing pulp,
5. removing contaminants.

In each step the basic controlled process variables are time, speed of the impeller and liquid or pulp level inside the waste pulper. The governing stochastic element, of course, is the waste stream itself. Usually, the objective of the optimation is to improve and balance throughput and selectivity, or product quality.

Throughput maximization is rather straightforward: Identify the constraints that limit throughput along a processing sequence and 1. Minimize their idle time, 2. Increase their capacity. The success of such improvements is easily measured by the time it takes to do a days work, or, more scientifically, in terms of a Waste Loading Rate.

Improving product quality and selectivity often counteracts throughput maximization. In addition, selectivity is much more difficult to quantify because it involves extensive handsorting efforts and mass balancing. Improving selectivity requires much more guesswork and trial-and-error to find out the influential factors. These may be process variables and control algorithms as well as critical mechanical components that need to be redesigned or adjusted. And the results of such modifications can reliably be measured only over at least a week or two of operation.

Optimizing a WMP-installation certainly cannot be done at one instant, but takes at least a year. It involves frequent site visits as well as careful and unbiased evaluation of operator recordings and experimental data. It needs hands-on-action and digging-in-the-dirt, communication skills and determination, scientific methodology and daring to be crazy, creative thinking and trust into one´s own common sense. The gradual improvements might have been deducted scientifically by analyzing the process, but the real breakthroughts were rather art than science.

SUMMARY AND OUTLOOK

Wet Mechanical Processing of organic wastes provides a highly flexible interface between contaminated wastestreams in a wide range of water contents. It is capable of turning organic materials poor in structure into feedstocks well suited for low-tech as well as high-tech and high-performance biological processes. The process itself is a completely automated batch operation. The new generation of BTA Waste Pulpers offers the option of completely enclosing the equipment, so that from the tipping floor on, all emissions are easily captured in small volumes of air to be filtered. Hence, emissions and occupational biohazards are minimized.

Despite full automation and encapsulation, the process itself offers high throughput, almost complete contaminant removal and high yield of organic matter for biological utilitzation. Organic matter from biowaste is processed and purified to a point where it can be turned into high-quality products meeting the most demanding standards.

The use of water implies combining wet pretreatment with anaerobic digestion. In any case, the yield of biogas will provide enough primary energy to at least compensate electricity demand and generate a considerable surplus of heat.

With its high flexibility, the front-end approach to reducing emissions and the potential yield of renewable energy from waste, WMP qualifies as a technology suitable to support the transition towards and sustainable biomass management.

REFERENCES

1. Kübler, H.; Schertler, C.; Schnell, R.; Wild, M.: „Versuchsanlage zur anaeroben Vergärung von Nassmüll der Gemeinde Garching", Final Report Verbundvorhaben „Verwertung von Biomüll -Nassmüll" Teilvorhaben I, BMFT Förderkennzeichen 1430352 / 0; BTA GmbH&Co. KG, Munich 1991
2. Hochrein, P.: „Verwertung organischen Gewerbeabfalls und nassen Biomülls mit dem BTA -Verfahren". Research Report; BTA GmbH&Co. KG, Munich 1992
3. Hochrein, P.; Niefnecker U. et al.: Various Research Reports. BTA GmbH & Co. KG, Munich 1991 to 1995
4. Hochrein P.: „Improvements for the Pulping Stage at NSB" and „Manual Test of Improved Pulping Cycle". Unpublished reports, BTA GmbH & Co. KG, Munich 1992/93
5. Carl Bro A/S, personal communication: Compiled process data recorded from pulping stage at NSB 1991 to 1994
6. Personal Communication with Jeppe Obtrup, Carl Bro A/S, May 1994
7. Hochrein, P.: „Begutachtung der BTA-Bioabfallaufbereitung der Stadt Baden-Baden", Expertise for the City of Baden-Baden. Unpublished,;REA GmbH, Munich 1995
8. Bartsch, U.: „Untersuchung der verfahrensspezifischen Emissionen aus einer Anlage zur kalten Behandlung von Bioabfällen (BTA-Anlage der Stadt Baden-Baden)". MPU - Meß- und Prüfstelle Technischer Umweltschutz GmbH; Nürnberg 1995
9. Kuchta, K.: „Emissionsarten, Emissionsquellen und Ursachen ihrer Entstehung am Beispiel der Bioabfallkompostierung - Geruchsemissionen -" in Bibliothek des Instituts WAR (Ed.): „Umweltbeeinflussung durch biologische Abfallbehandlungsverfahren"; Bibliothek des Instituts WAR, TH Darmstadt, Darmstadt 1994
10. Vogel, F.: „Kompostierung von Grünrückständen und vorbehandelten Bioabfällen der Stadt Baden-Baden" in [11]
11. Arbeitskreis zur Nutzbarmachung von Siedlungsabfällen (ANS): „Anaerobe Bioabfallbehandlung in der Praxis". Schriftenreihe des ANS Band 30, Stuttgart/Mettmann 1995
12. Hochrein, P.: „Emissionen der Bioabfallverwertung Baden-Baden" in [11]
13. Hochrein, P.: „Aufbereitung und Vergärung von Industrie-, Gewerbe. und Bioabfällen - Erfahrungsberichte" in „RHINO-Fachkongreß Bioabfallmanagement '94". Rheinisches Institut für Ökologie, Recklinghausen/Köln 1994
14. Hochrein, P. and Outerbridge T.: „Anaerobic digestion for soil amendment and energy" BioCycle 33 (1992), No. 6 pp 63, JG Press, Emmaus, PA.

Anaerobic Fermentation of Biowaste at High Total Solids Content - Experiences with ATF-System

Norbert Rilling, Martin Arndt, Rainer Stegmann, Technical University of Hamburg-Harburg, Harburger Schloßstraße 37, D-21079 Hamburg, Germany

ABSTRACT

At the Technical University of Hamburg-Harburg a new one-stage anaerobic dry fermentation process (ATF-process) operating at high total solids content has been developed. The ATF-process is a quasicontinuously fed plug flow operated reactor system. The basic investigations regarding various factors influencing the anaerobic fermentation of biowaste (water content, material-mixing, pretreatment, inoculum, water recirculation, pH, alkalinity etc.) were carried out in bench-scale reactors (volume 290 L) under mesophilic and thermophilic conditions. As a second step a half-technical scale plant (reactor volume 10 m^3) had been designed and constructed. This plant has been operated successfully since 1989. As a further step the ATF-process will be investigated under full scale conditions (reactor volume 100 m^3) during a research project financed by the Stadtreinigung Hamburg and the Deutsche Bundesstiftung Umwelt. In 1994 the full scale treatment-plant located in Hamburg-Bergedorf started operation.

KEYWORDS

Anaerobic digestion, dry fermentation, ATF-process, biowaste

INTRODUCTION

The separate collection and composting of biowaste is an important aim in waste management. In the Federal Republic of Germany, the biobin will be introduced country-wide to the households within the next years. Moreover, organic trade waste will be separately collected to an increasing extent. The government has accelerated this process by passing the „TA Siedlungsabfall" (Technical Directive Municipal Waste) planning largely separate collection of the waste fractions, which is in part also prescribed by law in some Federal States.

The potential of waste that can be treated biologically does not only comprise the mostly regarded yard and kitchen waste separately collected in the households but also a great number of residues, as for example from large-scale catering establishments and canteens, market waste and other low polluted organic residues e.g. from food industry. When separate collection of these kinds of waste is possible, as yet the common disposal practice has been composting or reuse as fodder. Anaerobic fermentation of the organic material in a closed system under air exclusion suggests an alternative. As final product an anaerobically stabilized substrate is produced as well as biogas (about 60% methane and 40% carbon dioxide) which can be used as energy source. In addition it seems that the odour problems can be solved more easily. When comparing composting as only treatment process with anaerobic fermentation combined with composting as postprocessing step the latter can offer the following advantages:

- less space requirements
- smaller constructional volume (interior space)
- minor odour problems
- surplus energy (instead of energy consumption)
- shorter process time

THE ATF-PROCESS

The ATF-process for anaerobic dry fermentation (see Figure 1) is a single stage process working with high solids contents and can alternatively be operated under mesophilic or thermophilic conditions. The process is characterized by a simple system technique. The processes of hydrolysis, acid and methane formation proceed simultaneously in the reactor. The final product of anaerobic degradation is a relatively roughly structured biomaterial of a water content of about 60%, which can be composted in windrows immediately after fermentation.

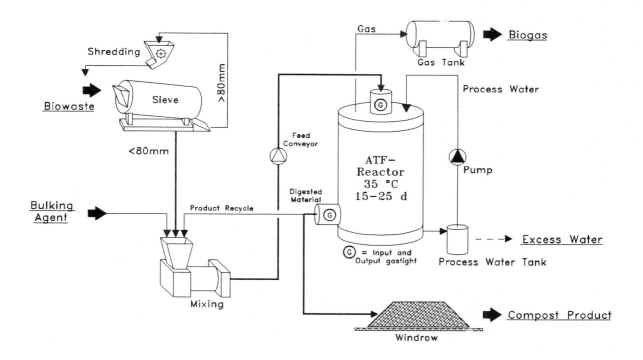

Fig. 1: Scheme of the ATF-process [1]

The most significant difference between the ATF-process and other methods of anaerobic fermentation of solid biowaste is that it works without any external water addition at the beginning of the process, without agitation of the substrates in the fermenter and without costly dewatering of the fermented material. The ATF-process is promising because it is relatively simple in handling, secure in operation and cost effective.

FUNDAMENTALS OF ANAEROBIC SOLIDS FERMENTATION

To be able to better describe the processes taking place in landfills, Stegmann (1981) developed a laboratory test procedure to simulate the biochemical decomposition processes under anaerobic conditions [2]. Subsequently, research and development work was conducted to enhance the anaerobic degradation processes in municipal solid waste (MSW) landfills [3].

On the basis of these and additional investigations as well as experiences made with large-scale landfills, a new process for anaerobic fermentation of biowaste was developed at the Technical University of Hamburg-Harburg (TUHH). The following objectives should be reached within the scope of a research and development programme:

- simple procedure of the process
- single-stage process conduction
- fermentation at high solids content (dry fermentation)
- low construction costs by using common unit elements
- low energy consumption
- no agitation of the material in the reactor

At first, new laboratory reactors have been developed adapted to the particular requirements of dry biowaste fermentation. Figure 2 shows the general design of such a bench-scale reactor of 290 L volume. The reactors are constructed as double-walled cylindric metal vessels with removable cover. They are heated by thermostat-controlled water recirculation and can be operated within arbitrary temperature ranges (mesophilic at about 35°C or thermophilic at about 55°C). The vessels are fed from above and are gas-tightly sealed.

The leachate is collected in a pump sump and recycled into the reactor via a distribution device installed at the cover. The substrate in the reactor is not agitated. The biogas produced is continuously released through an opening in the cover and the gas quantity is measured by gas meters. The gas composition is analyzed by gas chromatograph (GC) with heat conductivity detector (HCD) for CH_4, CO_2, H_2, N_2, O_2. The leachate is regularly analyzed (e.g. for pH, conductivity, BOD_5, COD, Kjeldahl-Nitrogen, ammonia, organic acids). The leachate quantity removed is replaced by the same quantity of fresh water to not change the total amount of water in the reactors. When the reactors are fed and discharged, solids samples are taken from the input or output materials for analyzation (e.g. for dry solids, H_2O, organic dry solids, C_{total}, N_{total}).

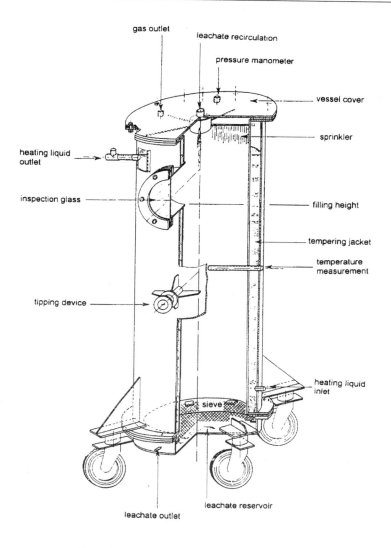

Fig. 2: Design of a 290-L-reactor [4]

The objective of anaerobic dry fermentation is to degrade the waste at lowest possible water content so that mechanical dewatering of the fermented material is not necessary. A sufficient moisture content of the waste and the transport of the nutrients dissolved in the water is achieved by leachate recirculation.

The investigations of fermentation of biowaste were conducted as batch-load tests, keeping the reactors in the mesophilic temperature range at 35°C. The biowaste was not ground. The water content was adjusted at 65-70%. The process water was discontinuously recirculated to achieve distribution of the hydrolyzed substances and organic acids. Depending on the individual feed substrate, gas production rates of up to 190 m^3 biogas per ton dry solids with a methane content of 60-65% were measured [5]. The essential gas production indicating degradation of the material lasted over a time interval of 20-30 days.

In 1989, the knowledge and experiences from these successful investigations with biowaste fermentation in bench-scale reactors was applied to a first technical-scale plant (reactor volume 10 m^3), which was constructed and operated within a joint project of the Department of Waste Management, and the Foundation „Jugend in Arbeit Hamburg e.V.". The reactor consists of a vessel with openings at the top to fill in the waste and at the bottom to discharge the substrate. It is heated at mesophilic or thermophilic temperatures. The reactor is insulated to maintain the temperatures. The process water is collected at the bottom of the reactor, discontinuously pumped upwards in the reactor and trickled over the material. By this the moisture content of the biowaste is adjusted. Via this pathway, also water addition is possible when the feed substrate is very dry. The biogas produced escapes the vessel by natural pressure and is continuously monitored as to quantity and composition. When the methane concentration is sufficient, the biogas is stored in a tank after cooling to lead off the condensed water. This gas can be used to heat the reactor. In a pretreatment step, all waste material is screened to 100 mm and the sieved material is blended with fermented substrate for inoculation prior to placing into the reactor. Since no water is added to the material before feeding into the reactor it cannot be pumped. After a three-week fermentation time the substrate is composted in a postprocessing step lasting 4-6 weeks. During this phase, further conversion and degradation processes take place leading to a significant rise in temperature and thus to hygienization of the material. After a 10-mm screening, a compost is obtained having a degree of maturity of IV-V according to the German standards of the Bundesgütegemeinschaft Kompost [6].

ANAEROBIC DRY FERMENTATION OF BIOWASTES

The 290-L-reactors and the 10-m^3-ATF-plant were used to investigate anaerobic dry fermentation of various organic wastes. In addtion to kitchen and yard waste, the fermentability of a great number of specific organic wastes, such as wastes from slaughterhouse and central market, lecithin sludges and bleaching earth, etc. was studied [7,8].

Figure 3 gives a comparison of the cumulative biogas production rates during anaerobic dry fermentation in experiments with 290 L volume reactors and the 10-m^3 ATF-plant. For the tests biowaste with a very high proportion of yard waste was used. The biowaste was screened (screen size d<80 mm) and blended at a ratio of 2:1 with anaerobic inoculum and fed into the reactors. The water content of the mixed material was adjusted at 52.5% H_2O in the 10-m^3-plant and at 55.6% H_2O in the laboratory reactors. The gas production from the biowaste was relatively low. In total, 230 L biogas/kg dry organic matter was produced during 64 days, the methane production was 130 L/kg dry organic matter. This is above all due to the relatively high proportion of yard waste.

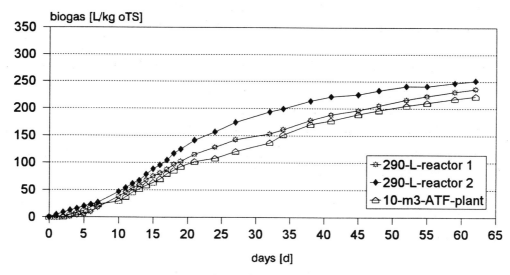

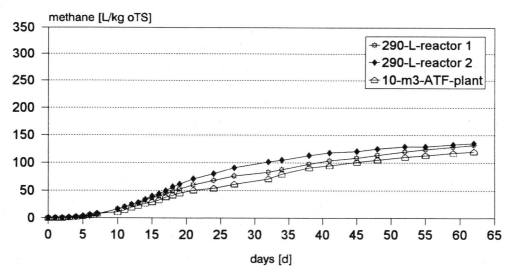

Fig. 3: Cumulative biogas and methane production from biowaste [1]

The yard waste fraction is hardly degraded under anaerobic conditions and thus the gas potential is low. The main gas production involved with the essential phase of organic degradation took place over a period of 20 days between the 9th and 29th testing day. During this time interval 65 L CH_4/kg dry organic matter was produced at an average, which is about 50% of the total methane potential. The average specific methane production rate during this phase amounted to about 3.3 L CH_4/kg dry organic matter and day. Biogas production of the laboratory reactor was largely consistent with that of the ATF-testing plant.

ENGINEERING TRANSLATION OF THE ATF-PROCESS TO PILOT-SCALE

The experiments conducted with the laboratory reactors and with the 10-m^3 testing plant have shown that the ATF-process can be operated steady even with very dry biowaste (solids content >45% TS) without agitating the material in the reactor. The positive results of the investigations finally led to the advancement of the ATF-process and its translation to a full-scale plant. The following objectives shall be reached:

- integration of the plant in the practice of daily biowaste collection
- full-scale pretreatment (screening)
- quasicontinuous feeding and discharge
- integration of curing phase (composting) in the overall process
- extensive automatization
- demonstration of the plant as state of the art
- keeping all standards for approval by the authorities

The research project was approved in the spring of 1993. It is funded by the „Stadtreinigung Hamburg" and the „Deutsche Bundesstiftung Umwelt, Osnabrück". After a year of planning, approval and construction, the plant started operation in August 1994.

As basic substrate for fermentation separately collected biowaste was used. The principle of the ATF-process by the TUHH for a full-scale plant is presented in Figure 1. Prior to feeding into the ATF-reactor, the biowaste is conditioned and blended with material already anaerobically fermented at a mass ratio of 2:1 to 5:1. The 100-m^3-reactor is filled from above and the temperature is maintained at 35°C (mesophilic) or 55°C (thermophilic). The material is tipped by a front-end truck into the feed funnel of a drag chain conveyor and is gas-tightly fed into the reactor after having passed a double lock. According to the type of substrate, the material stays in the reactor for 15-25 days. During this time, it succesively moves downwards the reactor and is discharged at the bottom. During the fermentation process the material is neither agitated nor otherwise turned but only moistened in intervals with process water. The water percolating through the substrate provides the microorganisms with the nutrients necessary for the degradation process.

The fermented material is discharged via a sliding floor in combination with a worm conveyor system. By applying a sliding floor the discharge according to the 'first in - first out' principle is guaranteed and defined retention times can be assured. The system can be seen as a quasicontinuously fed plug flow operated reactor.

The complete engineering design has been worked out in cooperation with two enterprises of the region. Haase Energietechnik GmbH, Neumünster, was responsible for the gas and security technique, Saxlund GmbH, Soltau, for the conveyor technique to charge and discharge the material. The excellent know-how of these two firms guaranteed the smooth completion of the pilot plant. Although the process is simple in principle, for the full-scale development a series of engineering problems, such as gas-tightness, explosion protection, charge and discharge of the material, etc. had to be solved. During operation of the plant, a great number of further questions will have to be investigated. Because of its uncomplicated engineering concept the ATF-fermenter is working safely and reliably. The process is not susceptible to operating trouble and can be charged with miscellaneous organic wastes. In combination with a postprocessing composting step, the plant is expected to operate at relatively low cost. After the running-in period and intensive runs to test the individual aggregates, the plant is now being optimized under operation conditions. The ATF-reactor is fed with about 5 m^3 of biowaste per day thus having an annual capacity of 1,000 tons.

The ATF-process is also suited for the treatment of various organic wastes of different origin, e.g.:

- fruit and vegetable waste from central markets
- slaughterhouse waste (paunch manure)
- wastes from food processing industry
- food waste from hotels, restaurants, canteens
- wastes from fish processing industry
- bleaching earths
- algae, seaweed washed ashore
- agricultural wastes
- draff

CONCLUSION

The anaerobic dry fermentation of organic wastes with the ATF-process was studied in bench-scale (reactor volume 290 L) and in half-technical scale (reactor volume 10 m^3). The studies showed that anaerobic biological treatment of biowaste is possible even with high contents of solid material of about 45 %. After a detention time of about 20 days readily and medium degradable organics are converted to biogas. After a postprocessing composting step compost is obtained. As the scale up from bench-scale to half-technical scale causes only little technical problems it is expected that the ATF-process can be operated under full-scale conditions (reactor volume 100 m^3) successfully, too.

REFERENCES

1. Rilling, N., Stegmann, R. (1994). Neue Techniken der anaeroben Trockenfermentation organischer Abfälle, TU Hamburg-Harburg
2. Stegmann, R. (1981). Beschreibung eines Verfahrens zur Untersuchung anaerober Umsetzungsprozesse von festen Abfallstoffen im Labormaßstab, Müll und Abfall, Heft 2
3. Stegmann, R. (1983). New Aspects on Enhancing Biological Processes in Sanitary Landfill, Waste Management and Research,
4. Stegmann, R. (1989). Das Forschungsprojekt der TU Hamburg-Harburg, Schriftenreihe des Arbeitskreises für die Nutzbarmachung von Siedlungsabfällen, Heft 16
5. Spendlin, H.-H., Stegmann, R. (1987). Versuche zum anaeroben Abbau (Faulung) der vegetabilischen Abfälle des Hausmülls, Zwischenbericht, Hamburg, (unpublished)
6. Anonymus (1992). Kompost-Gütesicherung RAL-GZ 251, Deutsches Institut für Gütesicherung und Kennzeichnung e.V., Sankt-Augustin
7. Rilling, N. (1994). Untersuchungen zur Vergärung organischer Sonderabfälle, in: Anaerobe Behandlung von festen und flüssigen Rückständen, Dechema-Monographien Band 130
8. Rilling, N., Stegmann, R. (1992). High Solid Content Anaerobic Digestion of Biowaste, Proceedings ISWA'92, Madrid

The Paques Anaerobic Digestion Process: A Feasible and Flexible Treatment for Solid Organic Waste

Drs. J. Brinkman, Ir. P.J.F.M. Hack, PAQUES Solid Waste Systems B.V., P.O. Box 52, 8560 AB BALK, The Netherlands

ABSTRACT

PAQUES Solid Waste Systems B.V. has developed an anaerobic digestion process for different kinds of solid waste. The process has the following general characteristics:
- It is a wet mesophilic digestion process. This means a dry matter contents in the reactor of 5-12% and a process temperature of 30-40°C.
- The process can be carried out in one reactor (BIOLAYER® process) for relatively constant and slowly degrading waste streams, such as mixed household waste and biowaste, or in two reactors (PRETHANE - BIOPAQ® process) for variable loads with rapidly degrading waste, such as unsold fruits and vegetables.

Since 1987 a full scale PRETHANE -BIOPAQ digestion plant treats 7,000 to 15,000 tonnes annually of unsold fruits and vegetables from the Auction in Breda, The Netherlands. Since 1992 a pilot plant has been running, in which the pretreatment and digestion of Mixed Municipal Solid Waste, industrial organic waste and biowaste has been tested. Also extended research on post treatment, off gas treatment and effluent treatment was conducted. Based on the information from the pilot plant, designs and feasibility studies have been made for the treatment of 10,000 to 30,000 tonnes of biowaste annually with the BIOLAYER process. The results of the feasibility study for 20,000 tonnes of biowaste are presented here. The main conclusions are:
- Investment and operation costs are relatively low compared to other digestion processes. The investment costs are 900 - 1050 DM/tonne input/year and the operational costs 175 - 210 DM/tonne, for complete plants, including power generation and according to the highest German standards.
- Organic Matter Conversion during the digestion is comparable to other digestion processes, namely 45 to 60%, depending on characteristics of the biowaste.
- The biogas production is 80 to 120 m^3 per tonne of biowaste, or 300-450 m^3 per tonne of Organic Matter. The methane contents of the biogas is 55 to 65 volume percent.
- When using the biogas in a heat power generator, there is a clear surplus for electricity and thermal energy. The electricity surplus varies between 50 and 140 KWh electrical/tonne input, the thermal energy surplus between 150 and 295 KWh thermal/tonne input. The wide variation is due to the great influence of local factors and/or specific demands of the customer.

PRESENT MARKETS FOR THE PAQUES PROCESS

There are 3 main areas for anaerobic solid waste treatment, in which the PAQUES Solid Waste Systems technology can be applied:

Household waste in Europe

Anaerobic digestion is becoming a treatment method for household waste, because of following:

- the environmental problems of the most common alternatives, incineration and disposal on landfills;

- related to the environmental problems the treatment costs of household waste are getting higher. In The Netherlands 200 guilders per tonne is a normal price for landfill and incineration. Prices are still going up;
- source separation of biowaste for compost production.

PAQUES Solid Waste Systems B.V. has carried out the engineering of a digestion plant in Leiden, The Netherlands, for 75,000 tonnes of household waste and 25,000 tonnes of source separated biowaste from households, with a project value of 60-70 million guilders. Also, the engineering of Biowaste (BiomÅll) digestion plants for several projects in Germany is taking place. The average capacity of these plants is approximately 20,000 tonnes of biowaste per year.

Agro industrial waste in Europe
In the food and the agro industry large amounts of waste slurries and solid waste are produced.

The market for solid waste treatment in this field is developing, because of following reasons:

- The acceptance of food industry waste as cattle feed is decreasing. More and more farmers want to control the food of their animals thoroughly. Less material is accepted as cattle feed, due to lower feed quality, and the occurrence of pollutants (1);
- Transport and landfill disposal prices for this waste are going up, due to government legislations. Mass reduction becomes interesting for industries. In a few years landfilling of organic wastes will not be allowed anymore in Germany and The Netherlands (1);
- In some countries in Europe electricity production from alternative energy sources, such as sun, wind and biogas is very well paid. In Switzerland one can receive Dfl. 0.20 - 0.30 per KWh produced from biogas, in Germany Dfl. 0.15 - 0.20 per KWh produced from biogas (1).

The digestion process of PAQUES Solid Waste Systems B.V. is already applied since 1987 for treating unsold fruits and vegetables from the Auction RBT Breda in the Netherlands.

Biogas production from organic waste in developing countries
In developing countries the treatment costs for waste are almost zero at the moment.
   On the other hand labor and investment costs are far lower when compared to Western countries, and the energy price is as high as in these countries. The production of an energy source, biogas and a fertilizer, compost, is the reason to build solid waste digestion plants.

A pilot plant in India has been put into operation in March 1995. It is running on mechanically pretreated household waste. A 150,000 tonnes/year demonstration plant for the city of Pune is being engineered.

PROCESS DESCRIPTION

In Figure 1 a general block scheme of the PAQUES Solid Waste Systems B.V. digestion process is presented.

In the mechanical pretreatment the waste is classified and homogenized. Coarse non-biodegradable components (plastics, textiles) and Ferro components are separated. For biowaste the following pretreatment concept has been developed:

In a slowly rotating drum sieve the biowaste is being mixed and classified. By recycling process water from the digestion the waste is brought in suspension. The fraction smaller then 55 mm is now pumped directly into the anaerobic digestion process with a total solids concentration of 8 to 12%.

The fraction bigger than 55 mm is brought into a shredder, which consists of three slowly rotating screws. Here a selective refinement of the waste takes place: thin, flat components, e.g. plastics and textiles pass the screws and are not refined. The fraction from the shredder can be recycled to the drum sieve for extraction of biodegradable components. In case it mainly exists of plastics and textiles, it is disposed off.

Ferro components can be removed from the waste streams by a magnetic belt.

The anaerobic process is a wet, mesophilic process. It can be carried out in a one step process (the BIOLAYER process), or a two step process, in which hydrolysis/acidification of the solids is separated from the methanization process (the PRETHANE -BIOPAQ process).

The one step process is being used for relatively slowly degrading materials and relatively constant streams, such as biowaste from households or mixed household waste. The two step process is being used for rapidly degrading materials, such as fruits and vegetables from auction and other agricultural wastes and large variations in load.

Under this conditions PAQUES Solid Waste Systems B.V. has experienced, that the one step digestion process becomes unstable. pH drops in combination with decrease of biogas production.

The fraction smaller than 55 mm is brought into the first anaerobic reactor, the PRETHANE reactor, in case of two step digestion, or BIOLAYER reactor in case of one step digestion.

In this reactor the biological degradation of the waste takes place, and inorganic components like glass, sand and stones are separated by sedimentation. Upgrading of this inorganic fraction to recyclable products (support material for buildings and roads) takes place by a washing step outside the reactor.

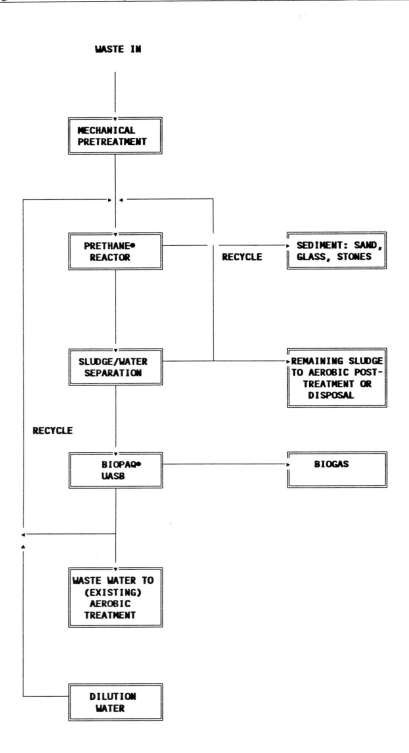

Fig. 1: The PAQUES Solid Waste digestion process - Technological concept

In Figure 2 a design scheme for the PRETHANE or BIOLAYER reactor is given. The reactor consists of four zones:

In the intensively mixed zone the mechanically pretreated waste is brought into the reactor. Mixing takes place by recycling biogas. The reactor temperature is kept constant between 35 and 40°C by heat exchangers. The heavy inorganic fraction sedimentates at the conic bottom and is removed from there.

The coarse fibrous material in waste flotates spontaneously and forms a floating layer with a thickness of 1-2 meter and a Total Solids Concentration of 10-20%. In the floating layer hydrolysis and acidification takes place. The biological process in the floating layer is controlled by percolation of recycled process water through the floating layer. The fatty acids produced are transported to the methane zone. The floating layer is transported to the floating layer removal mechanically. The retention time of the coarse solids in the floating layer is 4 days for rapidly degrading materials and 1-2 weeks for slowly degrading materials.

In the methane zone the fatty acids produced in the floating layer are converted into biogas. The total solids concentration is 4 to 7%. This zone is mixed by injection of recycled biogas.

In case of one step digestion the conversion to methane is complete. The retention time for the very fine solids is 3-4 weeks.

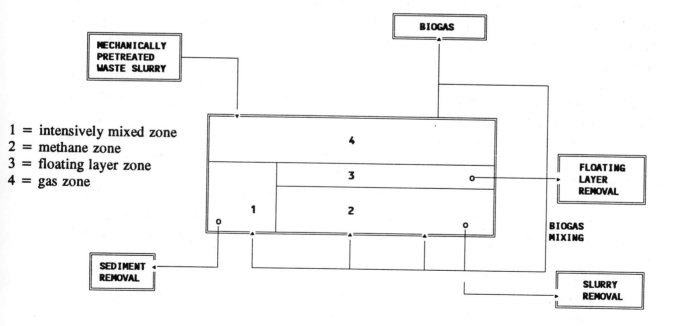

1 = intensively mixed zone
2 = methane zone
3 = floating layer zone
4 = gas zone

Fig. 2: The PRETHANE or BIOLAYER reactor

In case of two step digestion a separate treatment of the process water in a separate, second reactor (BIOPAQ) is necessary. The biogas is collected in the gas zone, above the floating layer. The advantages of this reactor type, when compared to other digestion concepts, are:

- Separation of inorganic fraction reduces wear of equipment, which reduces maintenance costs;
- When compared to completely mixed systems, the required mixing energy is smaller, because there is no need to prevent formation of the floating layer;
- There is no mechanical mixing. Pollution of mechanical stirrers with fibrous materials will cause serious technical problems. The use of recycled biogas as a mixing agent does not have such problems.

For biowaste, the sludge/water separation exists of a slowly rotating drum sieve, a screw press and a decanter centrifuge.

The slurry from the methane zone is first dewatered in a drum sieve. The solid material is dewatered to 50% Total Solids in the screw press, together with the removed floating layer. The solid material from the screw press is mixed with the fine solid material, removed from the decanter centrifuge.

The water streams from the drum sieve and the screw press are collected. A part of this water is used in the pretreatment drum sieve and in the reactor as recycle water, and a part is pumped to the decanter, where fine solid particles are separated and dewatered till 20 to 30% Total Solids.

In case of a two step process the filtrate from the decanter is brought into the BIOPAQ UASB (Upflow Anaerobic Sludge Bed) reactor [Figure 3]. Here the remaining dissolved fatty acids are converted to biogas. The BIOPAQ UASB reactor is applied since 1983 in the treatment of waste waters from pulp and paper industry, beer and beverages industry, potato processing industry and leachate treatment from landfills. Worldwide, the number of full scale references of the BIOPAQ UASB is 245 (January 1996).

The effluent of the BIOPAQ UASB is discharged or recycled to the PRETHANE reactor or discharged to the sewer or post treated in an aerobic activated sludge process.

In case of a one step BIOLAYER process the filtrate from the decanter is directly discharged or post treated.

The solid output of the digestion process has a moisture contents of 50-60%. It can be converted to compost in a conventional aerobic composting process. A retention time of 2 to 4 weeks is sufficient to produce a stable compost (Rottegrad V).

The biogas can be desulphurized in a scrubber. It can be used in a heat power generator, a boiler, or it can be upgraded to natural gas quality.

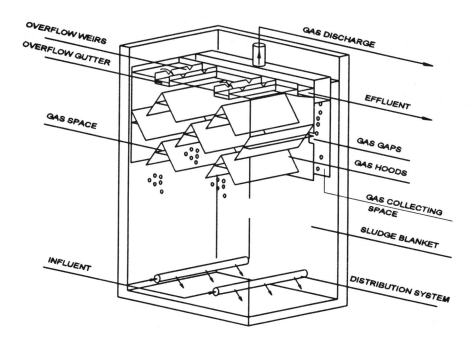

Fig. 3: The BIOPAQ reactor

RESULTS

Results of the PAQUES digestion process are given for two examples: the two step PRETHANE -BIOPAQ digestion of unsold fruits and vegetables at RBT Breda and a one step BIOLAYER process for the treatment of 20,000 tonnes/year source separated biowaste.

PAQUES also has several years of semi technical scale experience with the digestion of Mixed Municipal Solid Waste in The Netherlands and in India. These results will be published elsewhere.

Application on unsold fruits and vegetables, Auction Breda

Since 1987 at Auction in Breda (The Netherlands) the unsold fruits and vegetables are digested in a two-step PRETHANE -BIOPAQ anaerobic digestion system.

The digestion plant is consisting of a 50 m^3 volume mixing and shreddering unit, a 350 m^3 PRETHANE reactor and a 280 m^3 BIOPAQ UASB reactor and a 12 m^3 aeration tank for effluent treatment. The biogas is converted to heat and electricity in a 2 x 85 KW heat power generator. The odorous off gases from the equipment are scrubbed in the aeration tank. The off gas from the aeration tank is treated in a compost filter.

The main features of the digestion plant, based on the experience of the latest 6 years, are given in table 1 below. A mass balance of the process is given in figure 4. The mass balance is presented on dry weight and on wet weight basis. Because the waste consists of water for Ò 90% on wet weight basis, the biggest part goes to waste water. The solid residue is dewatered to 5-10% Total Solids. It is used as a fertilizer.

On dry weight basis, it becomes clear the biggest part of the solids goes to biogas. This is due to the fact, that the inert fraction is negligible and the material is well degradable.

Table 1: Main features of PRETHANE -BIOPAQ digestion plant, auction breda

Incoming waste				
Quantity (during 6-8 months)	t/year	7,000	-	15,000
	t/week	200	-	700
	t/day	0	-	200
Dry matter contents	%	4	-	15
Process Conditions				
Process temperature	°C	30	-	35
pH PRETHANE reactor		4.5	-	6.5
pH BIOPAQ reactor		7.0	-	8.0
Load PRETHANE reactor	kg VS/m^3 reactor/day	0	-	20
Load BIOPAQ reactor	kg COD/m^3 reactor/day	0	-	20
Process Performance				
Biogas Production	m^3/m^3 reactor/day	0	-	3.0
	m^3/tonne vegetables/fruits	8	-	40 *
	m^3/tonne Volatile Solids	250	-	400
Org. Matter Conversion	%	55	-	75
Methane contents Biogas	volume %	70	-	80
Wet mass reduction	% of incoming waste	60	-	70
Economic aspects				
Total Investment	Dfl		4.5 Million	
Operational costs	Dfl/tonne input waste	40	-	50 **

* Biogas production per tonne of vegetables/fruits is mainly depending on dry matter contents
** Operational costs are exclusive capital costs but inclusive all other operational costs, such as operating personnel, chemical use, costs for waste water discharge, disposal of solids, maintenance.

Following characteristics of the plant should be mentioned:

- The table shows that the digestion plant handles extreme variations of the waste load. Despite of this fact process stability is high
- Because the waste water treatment is integrated in the wet digestion process, discharge costs for waste water are low.

Application on source separated biowaste
During the winter and spring of 1992, the PRETHANE -BIOPAQ anaerobic digestion of the Auction in Breda could be used for treating source separated biowaste on a scale of 80-100 tonnes/week.

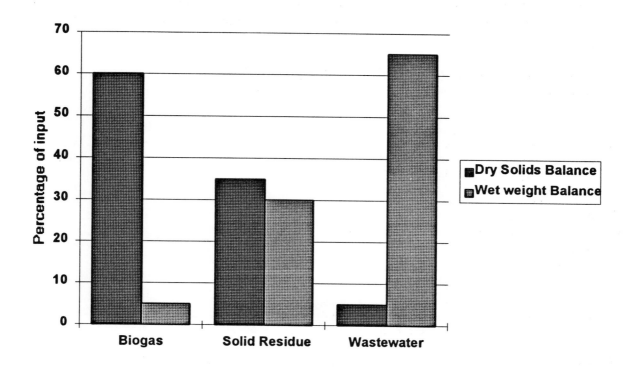

Fig. 4 : Mass Balance (Dry Solids) of PRETHANE -BIOPAQ Digestion, Auction Breda

From May 1994 till April 1995 the pilot plant of PAQUES Solid Waste Systems B.V. has been operational on biowaste on a scale of 800 tonnes/year. Both the one step BIOLAYER and the two step PRETHANE -BIOPAQ process have been tested. Based on the pilot research a choice has been made for the BIOLAYER process.

Further engineering and feasibility studies have been carried out for BIOLAYER plants with capacities of 10,000 to 30,000 tonnes of Biowaste per year, for the German situation.

The figures in table 2 below are based on a scale of 20,000 tonnes of Biowaste per year. An aerobic composting post treatment, an off gas treatment to deal with odour, a waste water treatment, a power plant, as well as the civil costs are included.

An example of a mass balance is given in figure 5. It should be mentioned, that the mass balance can vary with the input composition of the biowaste. Parameters like amount of pollutants and amount of garden waste influence biodegradability of the waste and separation of inorganics, and therefore mass balance.

The mass balance is presented on wet weight and on dry weight. On wet weight the biggest part goes to waste water. This is due to the fact, that the biowaste has a moisture contents of 60-75%.

Table 2: Main features of 20.000 tonnes/year BIOLAYER digestion plant, biowaste

Incoming waste				
Quantity	t/year	20,000		
	t/day	60	-	105
Dry matter contents	%	25	-	40
Amount of pollutants	Mass percentage	2	-	5
Amount of garden waste	Mass percentage	40	-	75
Process Conditions				
Process temperature	°C	35	-	40
pH PRETHANE reactor	7.0	7.0	-	7.7
Load PRETHANE reactor	kg VS/m^3 reactor/day	6	-	12
Process Performance				
Biogas Production	m^3/m^3 reactor/day	3	-	4
	m^3/tonne Biowaste	80	-	120
	m^3/tonne Volatile Solids	300	-	450
Org. Matter Conversion	% anaerobic conversion	45	-	60
	% total conversion	60	-	70*
Methane contents Biogas	volume %	55	-	65
Wet mass reduction	% of incoming waste	70	-	80
Economic aspects				
Total Investment	Dfl.	18	-	23 Mill**
	Dfl./tonne input waste/year	900	-	1050
Operational costs	Dfl./tonne input waste	175	-	210***
Energy Balance				
Energy Production	KWh electrical/tonne input	150	-	210
	KWh thermal/tonne input	220	-	340
Energy Use	KWh electrical/tonne input	70	-	100
	KWh thermal/tonne input	45	-	70
Energy Surplus	KWh electrical/tonne input	50	-	140
	KWh thermal/tonne input	150	-	295

Notes:
* The total conversion of organic matter is based on the anaerobic digestion and the aerobic post composting process together.
** Exact figures depending on local circumstances and demands. The investment figures are based on a turn key installation, including civil works and biological cleaning post treatments for the waste water and for the off gases (odour reduction). Also a heat power generator is included.
*** Operational costs are inclusive capital costs and all other operational costs, such as operating personal, chemical use, costs for waste water discharge, disposal of solid residue, maintenance, insurances.

During the digestion process the remaining material is dewatered to 50-60% moisture. Pollutants are separated during the pretreatment, inorganics during the digestion process by sedimentation. The inorganics are washed and dewatered and if necessary, they can be separated in a sand fraction and a glass, stones fraction.

The off gases are mainly produced during the aerobic composting of the solid end products of digestion. The off gases consist of degraded organic material (= mainly CO_2), but for the biggest part of evaporized water. On dry weight basis the remaining compost and the biogas are the most important fractions.

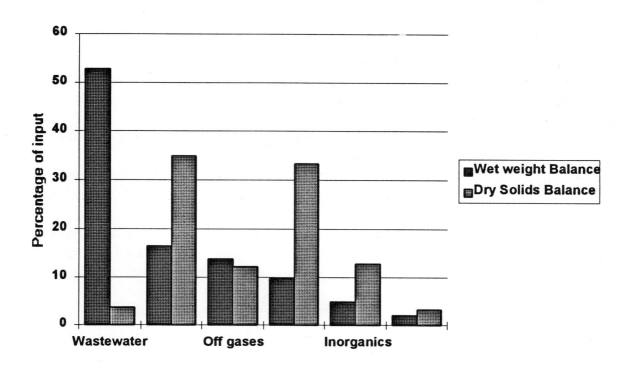

Fig. 5 : Mass Balance (Wet weight) of BIOLAYER Digestion of Biowaste

With regard to the solid end products following has been observed during the pilot research:

- The sand fraction meets the Dutch standards for heavy metals and poly cyclic aromatic hydrocarbons for use as a support material in roads and buildings. Also the contents of Volatile Solids in the sand is lower than 1.0% of the Total Solids. Based on this information the material is suitable for use as a support material [2].
- The compost fraction has a high stability after aerobic post treatment (German Rottegrad V) and meets all German standards of the "GÅtesicherung RAL-GZ-251" for compost [3]. The organic matter contents is higher when compared to aerobic compost: 50-65% of total solids, due to the separation of sand. The salt contents is relatively low .

Oetjen-Dehne and Ries [4] mention investment prices of 1100 to 1400 Dfl./tonne Biowaste at a size of 20,000 tonnes Biowaste/year for one step digestion processes. According to this reference the operational costs are 225 to 300 Dfl./tonne Biowaste.

With regard to organic matter biodegradation, biogas production per tonne Volatile Solids input, and organic waste load and specific biogas production other wet digestion systems claim values in the same range [5].

CONCLUSION

The BIOLAYER process for anaerobic digestion is an economically feasible and proven treatment method for the treatment of biowaste.

When compared to other one step digestion systems the BIOLAYER process comes out as a cost effective process for the treatment of biowaste.

The process is flexible: both the mechanical pretreatment and the BIOLAYER digestion process can handle a varying composition of biowaste.

At the Auction Breda the PRETHANE -BIOPAQ two step digestion process has operated since 1987. Varying loads with unsold fruits and vegetables have been treated successfully.

It can be concluded, that PAQUES Solid Waste Systems B.V. has a wide experience with the anaerobic treatment of different kinds of solid organic waste. Based on this, tailor made and flexible solutions can be offered for treatment of solid organic waste.

REFERENCES

1. PAQUES Solid Waste Systems B.V. Communications with different waste treatment organizations and government institutions in The Netherlands, Germany and Suisse, 1994.
2. Dutch Government. Appendix 1a on the : Decision on building materials with regard to soil and surface water protection. 1989 (In Dutch).
3. Deutsches Institut fÅr GÅtesicherung und Kennzeichnung. E.V. Kompost. GÅtesicherung RAL-GZ-251. 1992 (in German).
4. Oetjen-Dehne R., Ries G. What does the biological waste treatment cost ? In : "Biologische Abfallbehandlung". K.J. ThomÇ-Kosmiensky (Ed). E.F. Verlag, Berlin, 1995. ISBN 3-924511-72-1. Page 168-177. (In German)
5. Scherer P.A. Anaerobic Digestion Processes. In : "Biologische Abfallbehandlung". K.J. ThomÇ-Kosmiensky (Ed). E.F. Verlag, Berlin, 1995. ISBN 3-924511-72-1. Page 373-403. (In German)

Anaerobic Digestion of Biowaste in Full-Scale Plants in Brecht, Belgium and Salzburg, Austria by Means of the Dranco Process

Winfried Six and Anja Hofman, Organic Waste Systems, Dok Noord 4, B-9000 Gent, Belgium

ABSTRACT

During the last years the interest in anaerobic treatment of biowaste, as well as the application of this technology, has clearly been increasing. The DRANCO process has been developed for the anaerobic treatment of biodegradable organic substrates like the organic fraction coming from household refuse, source separated garbage, restaurant wastes, industrial organic wastes, etc. Two full-scale DRANCO plants are in operation and more plants are in the planning phase. This paper reports on the results of the 10,000 tpy (ton per year) plant in Brecht, Belgium and the 20,000 tpy plant in Salzburg, Austria. Both plants are treating biowaste. The total solids content in the digester varies between 15 and 35%. The volumetric loading rate ranges from 6.3 to 8.4 kg VS/m^3 day. The biogas production is ranging from 100 to 140 Nm^3 per ton of biowaste, depending on the substrate.

KEYWORDS

Biowaste, source-separated collection, DRANCO-technology, anaerobic digestion, biogas.

INTRODUCTION

During the last years the interest in anaerobic treatment of biowaste, as well as the application of this technology, has clearly been increasing. Aerobic composting plants are facing difficulties due to the increasing amounts of wet and compact kitchen and restaurant waste in the biowaste. Especially biowaste coming from city areas contains a relatively high amount of such kitchen waste and only little structure material. As a result considerable amounts of bulking material must be added to the biowaste to allow aeration of the waste and keep the odor emissions under control. Anaerobic digestion plants do not face these problems and have furthermore some additional advantages such as compact size, short retention times and the production of alternative energy which makes them more than selfsustaining concerning energy demands.

The DRANCO process has been developed for the anaerobic treatment of biodegradable organic substrates. These substrates can be the organic fraction coming from household refuse, source separated garbage, thickened aerobic sludges, restaurant wastes, solid or semi-solid industrial organic wastes, agricultural surpluses, etc. They are simultaneously converted into a stabilized and hygienically safe compost, called Humotex, and energy in the form of biogas by means of an anaerobic high-solids digestion followed by an aerobic composting phase.

For the moment 2 full-scale DRANCO plants are in operation, more plants are in the planning phase. This paper reports on the results of the 10,000 tpy (ton per year) plant in Brecht, Belgium and the 20,000 tpy plant in Salzburg, Austria. An overview is given of the collection and composition of the biowaste, the start-up of the plant and the operational parameters.

THE DRANCO PLANT IN BRECHT

Since June 1992 a full-scale DRANCO plant in Brecht, Belgium treats about 10,000 ton of biowaste per year. The waste is coming from a more rural area and is composed of 15% kitchen waste, 75% garden waste and 10% paper waste. The average characteristics of the incoming biowaste are : a total solids content of 40%, a volatile solids content of 55% and a C/N ratio of 20. The schematic overview of the installation is given in Fig. 1.

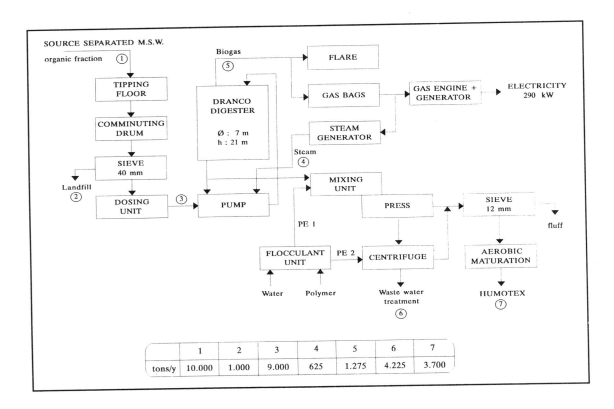

Fig.1. Mass balance of the DRANCO plant in Brecht

The biowaste is first comminuted in a homogenizing drum and sieved afterwards over a 40mm screen. The oversize of the sieve is landfilled and the fraction less than 40mm is intensively mixed with digested residue from the digester, heated with steam and pumped into the digester. The digester has a volume of 808 m^3 with a diameter of 7 m and a height of 21 m.

After ca. 20 days of thermophilic digestion the residue is dewatered to a solids concentration of ca. 50% by means of a screw press. The press water is treated by a centrifuge, before being sent to the municipal wastewater treatment plant. The press cakes are refined with a vibrating sieve and composted aerobically for a duration of about 10 days prior to selling as a high-quality soil amendment. The biogas is stored in a gas bag and partially used to produce steam needed for process heating. The rest of the biogas is transformed into electricity by means of a 290 kW gas engine. The electricity is used for operating the installation and is partially sold to the grid. The biological start-up of the installation was very successful. In less than 2 months the plant was working at full capacity. The total solids content in the fermentor is about 34%. The average volumetric loading rate is 8.4 kg VS/$m^3_{reactor}$.day. The amount of methane per ton VS is ranging from 250 to 300 Nm^3, which means a biodegradability of 60 to 65%. The biogas production rate is 4.1 $Nm^3/m^3_{reactor}$.day. Figure 2 shows the amount of incoming biowaste and the biogas production in the second half of 1994.

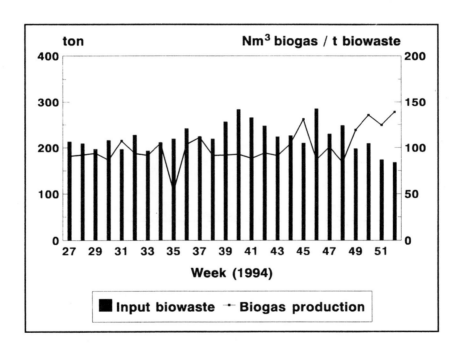

Fig.2. Results DRANCO plant Brecht.

THE DRANCO PLANT IN SALZBURG

Since December 1993 a plant for the treatment of 20,000 tons of biowaste per year is operating in Salzburg (Austria). The Salzburger Landesregierung started already in 1988 to analyze the amount and the composition of the waste produced in the Salzburger area. About

30% of the total waste was considered to be biowaste. It was expected, that the biowaste could be collected separately with an efficiency of 80%. Depending on the season this fraction consisted of about 60-90% kitchen waste and 10-40% garden waste. On the base of these data a DRANCO plant with a capacity up to 20,000 tpy was ordered. The start-up of the plant in Salzburg was done together with the introduction of the separate collection in a part of the city. During the next 6 months the collection area was gradually enlarged until it covered the whole area. A flow-sheet of the installation is given in Fig. 3. The pretreatment in the plant consists of a manual sorting, a shredder, a 40 mm sieve and a magnet and removes about 10 % of the incoming waste stream. The fermentor has a size of 1800m^3. The biogas is collected in a gas storage of 2500 m^3, where it is mixed with desulphurized biogas coming from a landfill. The gas is transformed into electricity by gas engines with a total installed power of 1.6 MW. The digested residue from the fermentor is dewatered in a screw press before it is sent to an aerobic posttreatment, being a tunnel composting unit, for about 14 days to obtain a stable and high quality compost. Table 1 summarizes the technical data of the plant.

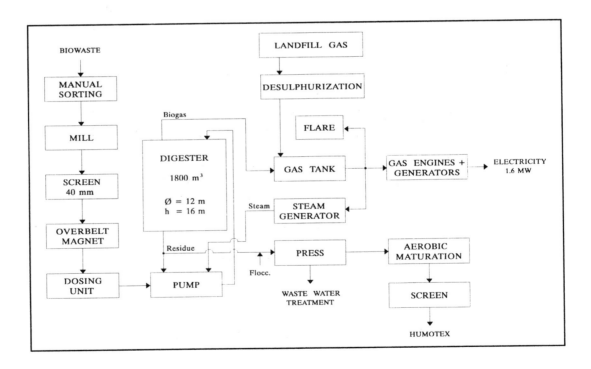

Fig.3. Flow-sheet of the DRANCO plant in Salzburg.

Table 1. Technical Data of the DRANCO plant in Salzburg

Capacity of the plant:	average	15.000 tons/y
	maximum	20.000 tons/y
Amounts to be processed:	average	60 tons/d
	maximum	90 tons/d
Working hours:	250 d/y; 5 d/w; 8 h/d	
Operating staff:	5 people	
Amount of compost/ton of input:	220 kg	
Level of compost maturity:	"Rottegrad" V	

Start-up phase of the DRANCO plant in Salzburg

The biological start-up of the plant began in December 1993. At the same time, the SAB corporation (Salzburger Abfallbeseitigung Gesellschaft), which is responsible for the collection of the waste and after the commissioning also for the operation of the DRANCO plant, started with the separate collection of biowaste in some districts of Salzburg and enlarged those areas during the next months. During the first two weeks of the start-up phase, 200 tons of inoculum from the DRANCO plant in Brecht were put in the reactor and heated up to 50-55°C. In a next step, biowaste was added. In Figure 4, the increase of the digester content and the simultaneous increase of biogas production during the first six weeks of operation is shown. In this time the digester content could be doubled while a steady biogas production could be observed. During the first months of the start-up, the plant was operated at a total solids concentration of 32% in the reactor. Due to the relatively low TS-content of the incoming biowaste (Table 2) the total solids concentration in the reactor went down to 18% during the first months of operation. This lower TS-content does not have any negative effect on the mechanical or biological operation of the plant.

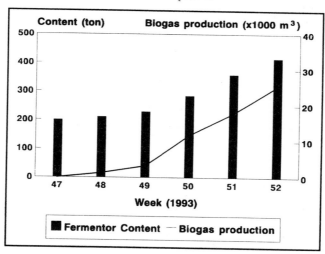

Fig. 4. Increase of digester content and simultaneous increase of biogas production during the first six weeks of operation

Week 16 of the year 1994 was chosen as the test week to control the performance of the plant. In other words five months after biological start-up and less than two years after the order for the plant, the commissioning of the plant took place. During this commissioning week, 200 tons of waste, which represented all of the available waste collected separately during that week, were fed to the fermentor. Table 2 shows the composition of the waste fed to the digester during this week.

Table 2. Composition of the waste fed to the digester of the DRANCO plant Salzburg during the commissioning week

Total input of organic waste (tons)	% TS	% VS	C/N
200	32	71	18

During the test the plant met all guaranteed values. The average biogas production per ton of biowaste in this week was 157 Nm^3 with an average methane content of 61.3%. Figure 5 shows the daily amount of waste fed to the digester and the biogas production per hour during this week.

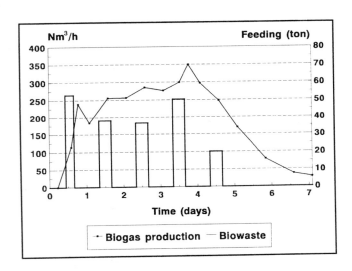

Fig.5. Daily amount of waste fed to the digester and the biogas production per hour during the commissioning week.

The biogas production starts immediately after feeding biowaste to the digester and decreases quite soon when feeding is interrupted (e.g. during the weekend). This reflects that the biowaste in Salzburg is highly and readily degradable. With this highly degradable waste acidification of the reactor content can not be observed. This shows that the DRANCO technology can treat such waste without any problems.

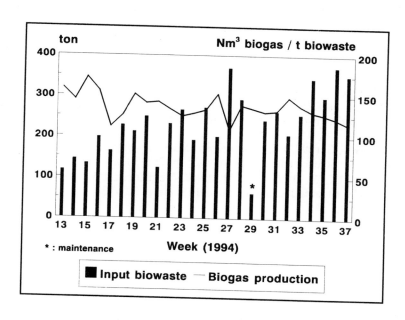

Fig.6. Results of the start-up of the DRANCO plant in Salzburg

Performance of the DRANCO plant in Salzburg after the start-up phase

The total solid content in the fermentor is about 18%. The volumetric loading rate is 6.3 kg $VS/m^3_{reactor}$.day. The amount of methane produced per ton VS is around 350 Nm^3, which means a biodegradability of ca. 80%. The biogas production rate is 3.7 $Nm^3/m^3_{reactor}$.day. An electrical production of 260 kWh per ton of incoming biowaste is achieved. Figure 6 shows the amount of incoming biowaste and the biogas production in spring and summer 1994.

CONCLUSION

The DRANCO process has a high degree of flexibility towards the composition of biowaste. In Brecht the incoming biowaste contains a lot of garden waste and non-recyclable paper, resulting in a TS-content in the fermentor of 34% In Salzburg the biowaste contains more kitchen and restaurant waste resulting in a much lower TS content. The results of the two plants described, show that the DRANCO process allows operation by a total solids content in the digester ranging from 15 to 40%. The figures show that the design concept is not only very flexible regarding changes in feedstock but also can counteract seasonal fluctuations. The data presented show that the DRANCO plant in Brecht is treating the designed capacity of biowaste and that the plant in Salzburg is started up very well, without any significant operational problem. Anaerobic composting of biowaste can be considered as a proven technology.

REFERENCES

1. BOELENS, J., DE BAERE, L. and SIX, W. (1993). Methanization of biowaste in the full-scale DRANCO plant in Brecht, Belgium, a EEC supported project. In : Proceedings, Thermie Workshop on "Energy efficient technologies for the treatment of municipal solid waste in Greece and other Balkan countries", Athens, 1993.
2. SIX, W. and DE BAERE, L. (1992). Dry anaerobic conversion of municipal solid waste by means of the DRANCO process. Water Science and Technology, 25, p. 295-300.
3. KAENDLER, CH. and SIX, W. (1995). Das DRANCO-Verfahren zur Vergärung organischer Abfälle. In : ANS, Heft 30 : Anaerobe Bioabfallbehandlung in der Praxis, p. 327-333.

Biological Waste Treatment through Biogas Digesters in Rural Nepal

Amrit B. Karki, Krishna M. Gautam, Upendra Gautam, Consolidated Management Services (CMS) Nepal (P). Ltd. Lazimpat, P. O.Box # 10872, Kathmandu, Nepal

ABSTRACT

Poor sanitation experienced due to shortage of energy has been a serious constraint to the achievement of sustainable development particularly in the developing countries like Nepal. Predominant (95%) dependence upon traditional energy source (95 percent) such as fuelwood, agricultural residues and animal wastes characterize the energy scenario of Nepal. This has adversely affected the environment, health and hygiene of the inhabitants. Poorly managed forests have to shoulder this immense burden in order to meet the increasing demand for energy caused by the rising population and the lack of development of alternative energy resources. Although Nepal has the largest hydropower potential in the world less than one% of this potential has been exploited so far because of its high initial investment requirements. Thus, it is of paramount importance to develop various alternative options to meet the energy needs in rural communities of Nepal.

Thus, taking into account of (a) the time involved in collecting fuelwood, (b) smoke generated in the poorly ventilated kitchen, (c) diseases spread by the haphazardly scattered wastes, (d) decreasing forage availability for animal population in rural areas, (e) poor sanitation, (f) poor economic condition of the people and (g) pressing energy needs for household and other purposes, the rural communities in Nepal have to be given due attention to combat these problems. Biogas technology was first introduced in Nepal in 1955. Since then, the technology is getting increasingly popular for its potential to ameliorate the aforesaid problems. Nepal's experience in this area should be relevant for other developing countries in the world.

KEY WORDS

Biogas technology, biogas plants, waste, rural, Agriculture Development Bank of Nepal (ADB/N), His Majesty's Government of Nepal (HMG/N).

INTRODUCTION

Nepal, with 20 million population, is predominantly an agriculture country where more than 90% people live in the rural areas in traditional way. Low income generation, illiteracy, lack of proper transportation facilities and infrastructures, etc are among the principal constraints for the development of the communities. Being ignorant, they have very little knowledge about sanitation, health and hygiene. Human and animal wastes scattered haphazardly beside their houses and at open places pollute the environment. People of the rural areas have the habit of defecating in open places for they are not habituated to use latrines or they are not willing to construct them. Due to this, water-borne pathogenic disease are posing serious threat to the lives of people. Also, contamination of surface and ground water is a common phenomenon in rural areas of Nepal. The use of human feces in biogas plants is promoted not only to enhance gas production but more so to improve the sanitation as most of the harmful pathogens are destroyed in the process of anaerobic digestion. Because of this value of biogas technology in treating human feces, more than 30 percent of household biogas plants are attached to family latrines.

People are utilizing animal wastes in traditional ways to cater their energy needs. Burning of cattle dung cakes and agricultural residue for cooking purposes is very common, particularly in rural areas of Nepal. In an agricultural country like Nepal, this is a tremendous loss in terms of the manurial value of dung and other agricultural wastes. Such practice has adversely affected the environment, health and hygiene of the people. Women, who are mainly responsible for entire household management, are constantly exposed to smoke in an environment of poorly ventilated kitchens. Because of this, they have been suffering from eye, lungs and heart diseases.

Since people have been using wood for energy purpose, deforestation has become a serious problem which has resulted in soil erosion and poor soil quality leading to low productivity. Rivers have also become unpredictable with dangerous flow patterns, especially during the rainy season. This has forced the villagers to migrate from rural areas to urban places in search of better life.

Realizing this, Government Organizations (GOs), Non-Government Organizations (NGOs), Community Based Organizations (CBOs) etc. started involving themselves to combat these problems.

Popularization of biogas technology in rural community in Nepal has been found useful for localized production of household energy along with other benefits such as conservation of forest and improvement in health, particularly that of women.

HISTORY

Father B.R. Saubolle S. J installed the first historic biogas plant in 1955 at St. Xavier's School, Godavari on his personal initiative. His pioneering demonstration on production of methane from cow dung helped create some awareness of this technology among the local people [2].

On the inauguration of the Agriculture Year in 1974/75, His Majesty's Government of Nepal (HMG/N) launched special program to promote biogas technology by establishing 250 units of family-size biogas plants in different parts of the country. HMG/N in its seventh Five Year Plan (1987-1992), was able to establish 4,000 biogas plants in the country. Agricultural Development Bank of Nepal (ADB/N) provided finances to the farmers willing to install biogas plants. During the Eighth Five Year Plan (1992-1997), the National Planning Commission has adopted a policy to promote the private sectors' participation in various development programs including the biogas technology. Accordingly, a target has been set forth to install 30,000 units of biogas plants with the involvement of private biogas companies. If this target is realized, the country will benefit from an additional 72 MW of energy, primarily for household consumption. At present, the rate of installation of biogas plants is about 10,000 plants per annum [2,3].

A plant size of 8 to 10 m3 is the most common and found to be sufficient to meet cooking and lighting requirements of an average household. These plants of about 20 years of life are found to be economically and financially viable in the present context of ever increasing prices of conventional fuels [3,5]. One of the merits of this technology is its simplicity. However, training is necessary to ensure proper operation, maintenance and repair of biodigester and its appliances and to familiarize periodically the users with the technology development in the field of biogas.

PRESENT RURAL ENERGY STATUS

The annual average per capita rural household fuelwood consumption in Nepal is about 708 kg in the mountain and 689 kg in the plain regions [3]. It has been reported that 62,000 metric tons of dry weight dung cakes are collected throughout the country annually for domestic cooking. If this amount of dung were fed into biogas digesters instead of burning, apart from the energy obtained from gas production, it would result in a national saving of approximately 4,600 metric tons of nitrogen and 3,200 metric tons of phosphorus annually which could be used for fertilizing the crops [2].

In 1988/89, the total demand for energy in Nepal was estimated to be about 252 million gigajoules (GJ). Out of this, 95% was used in the domestic sector, mainly for cooking. To meet the demand in this sector, firewood (79%), agricultural residues (11%), dung cakes (9%), petroleum products (1%) were burnt and electricity was used (0.3%). Compared with 1980/81, the burning of agricultural wastes and dung cakes increased 15 times. This has obvious disadvantages as organic matter needed to improve soil fertility is no longer available. The loss of nutrients and organic matter decreases the soil fertility and consequently reduces crop yield[6].

As biogas technology has proved to be beneficial to women, government agencies as well as donor countries are taking keen interest in promoting and implementing this program on a massive scale in Nepal. At present, more than 16 private companies have joined the business of biodigester construction. CMS has been supporting them in building an inter-company group which can conduct dialogue with HMG/N on the policy aspect on energy development in the country With the involvement of all these organizations, it is expected that the installation of 30,000 biogas plants as targeted by the government will be met easily even much before the end of the proposed Five Year Plan period. This achievement will substantially increase the availability of organic manure. Also, at least these 30,000 households will not be compelled to harvest nearby forest for their energy requirements.

GOVERNMENT SUBSIDY IN BIOGAS DEVELOPMENT

At present, a massive campaign on the dissemination of biogas technology is on its way. This has helped in gaining popularity of this technology at the household level. HMG/N has planned to extend biogas program in all the potential areas of the country. Until now, more than 20,000 biodigesters have been installed in 67 districts out of 75 districts of Nepal. The country has potential of about 1.3 million biogas plants but until now only 1.5 percent of it has been exploited. People seem to have prejudice against using the gas if a latrine is connected to the cow dung plant. However, at present, the rate of the installation of latrine connected biogas plants is increasing slowly and gradually resulting into an improved sanitation in the rural communities. If all this potential is exploited, it is expected that about half of the country's population will be benefitted from this technology [1].

Realizing the need to implement this program effectively, government is providing subsidy to the people willing to install plants both in hill and plain regions. The present subsidies for each plant constructed in the hill and plain regions are NRs 10,000.00 (US$ 200.00) and NRs 7,000.00 (US$ 140.00) respectively. This has encouraged the beneficiaries in commissioning the plants. Presently, the Netherlands Development Organization (SNV/Nepal) has been assisting HMG/N in the promotion and dissemination of biogas technology [1]. With the biogas inter-company working group's inputs, it is expected that government policy towards biogas development in the country will be more promotion oriented and sustainable.

LATRINE ATTACHED PLANTS: A CASE STUDY

About 70 to 75% of the installed biodigesters are of the integrated type (i.e. latrine connected to cow dung plant). Hence, self-participation is already gaining ground in Nepal for biological waste treatment process. However, besides concerted efforts from various sectors such as GOs, NGOs, CBOs etc, the need for active involvement of academic and research institutions is also felt to promote the R & D activities [7].

In 1989, Ram Prasad Kafle, a resident of Swauli Bharatpur Municipality, installed a 10 m3 fixed dome design biogas plant with the technical assistance from the Gobar Gas and Agricultural Development Company and financing from the ADB/N (Agricultural Development Bank/Nepal).

Mr. Kafle who had seven members in his family owned two buffalos. Prior to the establishment of the biogas plant, he had been using fuelwood for cooking and electricity for lighting. He used to buy fuelwood from street vendors who collected firewood from the nearby forests. He estimated that his annual fuelwood requirement was equivalent to two full-grown salwood trees (Shorea robusta). The nearby forest, as the time passed, receded farther and farther away.

Mr. Kafle feeds his digester with about 30 kg of dung every day. He has stall fed his buffalo so as not to lose the dung. He strongly recommends biogas plant owners to stall feed their animals which could help them operate a biogas plant even with few animal heads.

Mr. Kafle attached his latrine to the biogas digester pit to enhance biogas production. However, he is hesitant to speak openly about this to many of his orthodox friends and relatives.

After installing the biogas plant, the gas produced from the plant became suffficient for cooking and lighting which has ultimately made him self sufficient in energy for household purpose. Moreover, he is now satisfied with the increased production of the crop in the land where he utilized digested slurry as manure. He is also happy that he does not have to spray synthetic fertilizers in his crop fields. As a highly satisfied biogas plant owner, Mr. Kafle strongly recommends his other farmer friends to install biogas plants.

He also advocates the attachment of latrines to the digester pit. By doing so, one can save the cost of constructing soak pit and septic tank separately. Besides obtaining an increased amount of gas and high quality slurry as fertilizer, he reports an improvement in sanitation and health of the family members. Furthermore, he wants to draw attention of the authority to establish an appropriate R & D program [5]. Based on this identified need to biogas plant owners, CMS through its affiliation organ Rauka Impex, has started supplying the biogas appliances which are considered efficient and maintainable on the basis of biogas plant owners/users' feedback.

CONCLUSIONS

Biogas has proven to be the most popular technology not only for localized generation of household energy but also to process bio-degradable waste both in the rural and urban areas of Nepal. The technology has increasingly been socially more acceptable; is economically viable, technically feasible and has proved to be sustainably manageable in Nepal, which is one of the poorest country of the world with a literacy rate of about 40 percent only. The technology has made a significant contribution towards improving the quality of life, particularly that of women by reducing their work load related to firewood collection, incidence of eye, lung and heart diseases. The government policy on subsidy has been initially conducive to biogas development. On the merits of its high economic and social benefits, biogas technology has also been effective in changing people's attitude towards dealing with human faeces, forest conservation and sanitation.

The private sector biogas construction companies have been capable to make use of the development opportunity created by the government subsidy for establishment of the biogas plants. Now, the official party has to be broadened to include technology and sustainable promotion aspects of bioenergy systems.

There are various possibilities to derive benefits from the biodegradable wastes. Integrated system of biogas plants, simple technology, collaboration of different actors and agencies (households, municipality, wards, CBOs, NGOs, Mayors, relevant government department) are required to make the program successful. These possibilities also need to be properly reflected in the up-coming alternative energy development policy of government.

REFERENCES

1. Biogas and Natural Resources Management (BNRM), Issue, 48, April 1995.

2. Karki A. B., Kunda Dixit (1984) Biogas Fieldbook.
3. Karki A. B., Gautam K.M., Karki A. (lg94) Biogas for Sustainable Development in Nepal. Paper presented at 'Second Conference on Science and Technology for Poverty Alleviation' organized by Royal Nepal Academy for Science and Technology (RONAST) from 8-11 June, 1994.
4. Karki A. B., Gautam R. and Gautam U. (1995) Municipal Solid Waste in Kathmandu Valley: A Review. Consolidated Management Services Nepal (P) Ltd.
5. Pokhrel, R.K., Yadav, R.P (1991) Application of Biogas in Nepal: Problems and Prospects.
6. Wim J. Van Nes (1994) The Biogas Support Program in Nepal. Paper presented at ECOTECH 1994 Computer Conference and IUFRO Workshop.
7. Urban Management Program for Asia and Pacific (UMPAP/UNDP) National Workshop on Kathmandu Valley's Municipal Solid Waste and Role of Urban Community Based Organizations (CBOs). 1-3 December 1994, Kathmandu.

Section 6

Biodegradability

Delignification of Wheat Straw by Wet Oxidation Resulting in Bioconvertible Cellulose and Hemicellulose

A.B. Bjerre, T. Fernqvist, A. Plöger and A.S. Schmidt, Environmental Science and Technology Department, Risø National Laboratory, P.O. Box 49, DK-4000 Roskilde, Denmark

ABSTRACT

The wet oxidation process (water, oxygen, elevated temperature, alkaline conditions) was optimised as a pre-treatment of wheat straw in order to solubilize the hemicellulose, degrade the lignin and open the solid crystalline cellulose structure. The effects of temperature, oxygen pressure, reaction time and concentration of wheat straw were investigated. The higher temperature the higher degree of delignification and hemicellulose solubilization could be obtained. The purity of the solid cellulose fraction also improved with higher temperature. In general, most of the conversions occurred during the first 10-15 minutes. Optimum wet oxidation conditions for obtaining enzymatic convertible cellulose (90%) was 10 minutes at 170°C using high carbonate addition. By applying a high straw density and lower carbonate addition sufficient hemicellulose was solubilized (measured as monosaccharides) to act as the sole carbohydrate source for ethanol producing micro-organisms. Furfurals, which are known inhibitors of microbial growth in other pre-treatment systems, could not be identified in significant amounts after wet oxidation treatment.

KEYWORDS

Lignocellulosic biomass, fractionation, pre-treatment, biowaste, enzymatic hydrolysis.

INTRODUCTION

The conversion of renewable lignocellulosic biomass, such as wood and agricultural crop residues, into fuel and chemicals has great environmental potential. In Denmark, biowaste in the form of wheat straw can be used as a source of fermentable sugars. Lignocellulose has a rigid structure and in order to fractionate its three major constituents (cellulose, hemicellulose and lignin) pre-treatment is necessary [7].

Various high temperature pre-treatment methods can be applied [1,6]. When biomass is treated with water or steam alone or with small amounts of acid, the pre-treatments are called autohydrolysis, steaming or steam explosion. These processes are only effective for fraction purposes at high temperature (>200°C) [1] and will, due to the severe treatment, generate many degradation products (from lignin and carbohydrates) e.g. furfural and 5-hydroxymethyl-furfural which are potential inhibitors of micro-organisms in the subsequent fermentation [4]. When high pressure of oxygen or air is present, the process is known as wet oxidation and has several advantages over the other pre-treatment processes. Wet oxidation is the most effective way of separating lignocellulosic materials into a cellulose and a hemicellulose rich fraction [1,6,7] at lower temperature (150-200°C) without generating many potential inhibitors.

The requirement of oxygen and alkaline conditions makes the wet oxidation process more expensive than e.g. steam explosion. However, due to the absent inhibitors after wet oxidation, the high cost detoxification step, needed after steam explosion [4, 12], can be bypassed making wet oxidation a promising pre-treatment process. The wet oxidation normally takes place in a reactor requiring long heating and cooling times, consequently, the process is very time consuming (harsh conditions). This paper describes an investigation of the wet oxidation processing of wheat straw biowaste using a reactor which allows shorter reaction times [2, 11], and hence, milder process conditions. The process parameters were optimised with respect to either bioconvertible cellulose or solubilized hemicellulose. The possible precipitation of dissolved hemicellulose was also examined.

MATERIALS AND METHODS

Wet oxidation

The wheat straw was grown at Risø National Laboratory in 1990 and 1993. The wet oxidation was carried out in a specially designed loop-autoclave constructed at Risø National Laboratory with very short heating and cooling times [2, 11] (about 2 minutes) suitable also for kinetic studies of suspensions. Wheat straw was mixed with Na_2CO_3 and water; and the oxygen pressure was applied before heating. After the reaction, the suspension was filtered to separate the solid cellulose-rich fraction (with some insoluble lignin) from the liquid hemicellulose-rich fraction (with some lignin degradation products).

Fibre analysis

The gravimetric method of Goering & van Soest [5] was used to determined the different fibres: hemicellulose, cellulose and lignin in the solid fractions.

Enzymatic convertibility of cellulose

The convertibility of the cellulose in the filter cake to glucose was determined by applying a mixture of two enzyme products: Celluclast (13.9 NCU/mL) and Novozym 188 (0.46 CBU/mL) kindly donated by Novo Nordisk A/S, Bagsværd, Denmark. The sample was suspended in 0.2 M acetate buffer (pH = 4.8) with the enzymes and rotated for 24 hours at 50°C [3]. The concentration of glucose in the filtrate was determined by HPLC.

Monosaccharide analysis

The hemicellulose in the liquid fractions from wet oxidation was hydrolysed by 4%w/v H_2SO_4 at 121°C for 10 minutes. Interfering ions were removed by a combination of precipitation and ion exchange [3]. The monosaccharides were then quantified by HPLC (Biorad Aminex HPX-87H column) with 0.004 M H_2SO_4 as eluent.

Analysis of furfurals
The presence of furfural and 5-hydroxymethyl-furfural was examined by HPLC (Nucleosil 5C-18 column) with a linear eluent gradient of methanol (10-90%) at pH 3 [3].

Precipitation of hemicellulose
The soluble hemicellulose fraction was precipitated by acetone addition (1:1 volume) during mixing for 30 minutes. After centrifugation (1500 rpm, 5 minutes) the concentrated hemicellulose suspension was dialysed against water for 24 hours and freeze-dried.

RESULTS AND DISCUSSION

The two types of raw material used in this study differ in their compositions (Table 1). The straw from 1993 was greyish whereas the one from 1990 was yellowish containing much less non-cell wall material (NCWM). A larger content of carbohydrate (cellulose and hemicellulose) was present in the wheat straw from 1990 compared to the 1993 straw probably caused by the different weather conditions during the ripening and harvesting period in those two years. The year 1993 had a very wet summer which might have led to contamination of the fields by fungi giving more NCWM and lower carbohydrate content in the straw.

Table 1: Fibre composition of the wheat straws (based on dry weight)

Straw	NCWM[a] [%w/w]	Hemicellulose [%w/w]	Cellulose [%w/w]	Lignin [%w/w]	Ash [%w/w]
1990	12.0	35.4	40.7	10.5	1.4
1993	18.8	32.8	38.1	8.9	1.4

a: NCWM = Non-Cell Wall Materials

The effects of the process parameters: temperature, oxygen pressure, reaction time and straw concentration on the fractionation of the wheat straw were investigated. The temperature and time effects are shown in Figures 1-3. At 150°C and 170°C, the content of lignin in the solid fraction decreased rapidly to its maximum degree of delignification (Figure 1), whereas at 130°C only 20% lignin was degraded after 10 minutes. The content of hemicellulose in the solid fraction decreased even faster (Figure 2) than the lignin content. Already after 2 minutes 50% of the hemicellulose was solubilized at 170°C. The relative cellulose content of the solid fraction increased rapidly (Figure 3) as both the lignin and the hemicellulose content decreased, whereby also the cellulose structure became more exposed [7]. Cellulose was found to be more resistant to the treatment than hemicellulose, which may be explained by the more compact and crystalline nature of cellulose [9]. Figures 1-3 all show that a higher reaction temperature gave a better and faster fractionation of the three main constituents in wheat straw. Furthermore, by opening the solid crystalline cellulose structure a better enzymatic convertibility of the present cellulose to glucose could be obtained (Figure 4) [7]. Clearly, the optimum wet oxidation conditions for bioconvertible cellulose was 10 minutes and 170°C giving a 90% conversion of cellulose to fermentable glucose. The oxygen pressure had a profound effect on the filterability of the suspension after wet oxidation [3], but did not significantly affect the degree of fractionation [10]. Although, oxygen might influence the inhibitor production, this was not investigated in detail in this preliminary study.

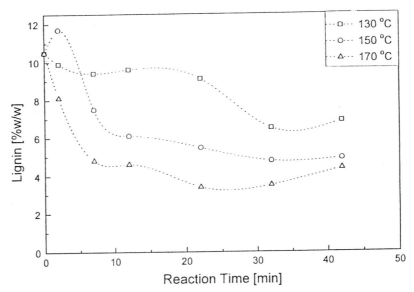

Fig. 1: *The effect of wet oxidation reaction time and temperature on the delignification of the solid fraction (20 g wheat straw (1990), 10 bar O_2, 10 g Na_2CO_3).*

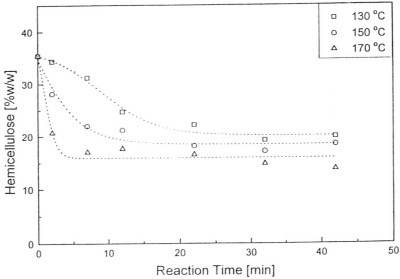

Fig. 2: *The effect of wet oxidation reaction time and temperature on the solubilization of hemicellulose from the solid fraction (20 g wheat straw (1990), 10 bar O_2, 10 g Na_2CO_3).*

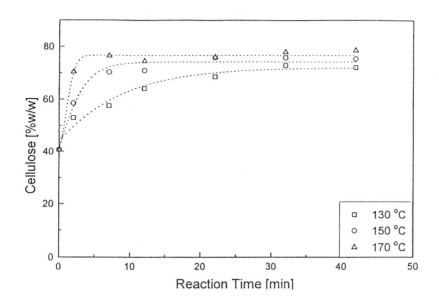

Fig. 3: The effect of wet oxidation reaction time and temperature on the cellulose content in the solid fraction (20 g wheat straw (1990), 10 bar O_2, 10 g Na_2CO_3).

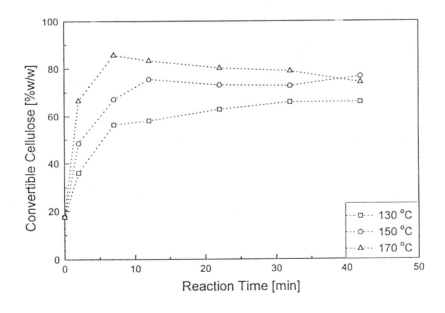

Fig. 4: The effect of wet oxidation reaction time and temperature on the enzymatic convertibility of the cellulose into glucose (20 g wheat straw (1990), 10 bar O_2, 10 g Na_2CO_3).

By solubilization the hemicellulose became available for either enzymatic or acid hydrolysis to fermentable pentoses [3]. As the wet oxidation at low wheat straw density (20 g/L) used in Figures 1-4 gave too low a concentration of soluble hemicellulose for it to act as the sole carbohydrate source for ethanol producing micro-organisms, hence, a higher starting concentration of wheat straw (60 g/L) was employed. Despite the use of less carbonate, the fractionation at high straw density also took place within the first 10-15 minutes (Figure 5), but the fractionation was less complete. However, the accessibility of the cellulose increased with increasing reaction time in accordance with others [9] giving a 67% enzymatic conversion after a residence time of 15 minutes. Using the higher concentration of wheat straw and less severe conditions (less chemicals), the concentration of solubilized hemicellulose increased to more than 9.5 g/L (measured as monosaccharides after acid hydrolysis) (Figure 6) sufficient for ethanol production. This was achieved due to the fact that the amount of available hemicellulose is a balance between the solubilization and degradation rate. At longer residence times the rate of degradation exceeded the rate of solubilization, hence, an optimum residence time of 15 minutes was observed at 185°C.

The hemicellulose content determined as monosaccharides after acid hydrolysis only accounted for 62% of the theoretical value. Even though, some hemicellulose degrades to carboxylic acids, CO_2 and H_2O both during the wet oxidation [7] and during acid hydrolysis [8], this hardly explains the 38% apparent loss of hemicellulose. Accordingly, the hemicellulose was also quantified by precipitation with acetone determining 20-30% more hemicellulose (Table 2), which still accounted for only 60% of the theoretical hemicellulose in solution. By acetone addition, the lower molecular weight hemicellulose units are not precipitated but remains in solution. Furthermore, some larger molecular weight lignin degradation products might have been co-precipitated with the hemicellulose. Presently, the hydrolysis is being investigated in more detail as the analytical acid hydrolysis is a compromise between incomplete hydrolysis and sugar degradation to *e.g.* furfurals [8] and some loss in materials must be expected due to oxidation.

Table 2: Comparison of precipitation and acid hydrolysis for quantification of solubilized hemicellulose

Tempera-ture	O_2-pressure	Time	Hemicellulose precipitated	Total sugar	Hemicellulose by sugar	Difference
[°C]	[bar]	[min]	[g/L]	[g/L]	[g/L]	[%w/w]
150	6	5	3.3	2.8	2.5	36
150	6	15	4.6	4.0	3.5	30
185	6	5	7.7	7.5	6.6	16
185	12	5	8.0	7.5	6.6	21

Furfurals, which are known inhibitors of microbial growth in other pre-treatment systems, could not be identified after the wet oxidation at low wheat straw density [3]. By treatment of higher concentration of wheat straw a furfural-like compound was detected (3 ppm), which compared to furfural production by steam explosion must be regarded as very low. This low concentration did not affect microbial ethanol production by thermophilic anaerobic bacteria (data not shown) or enzyme production by *Aspergillus niger* [3]. Currently, also the presence of other inhibitors and their effect on the fermentability of the hemicellulose-rich fraction is being examined.

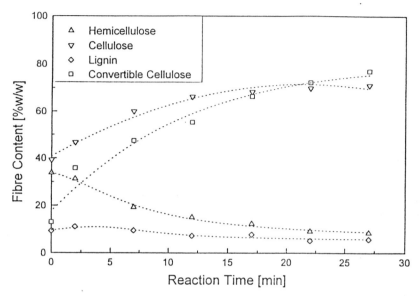

Fig. 5: The effect of wet oxidation reaction time on the fractionation of lignin, hemicellulose and cellulose in the solid fraction and the enzymatic convertibility of the cellulose into glucose (60 g wheat straw (1993), 185°C, 12 bar O_2, 6.5 g Na_2CO_3).

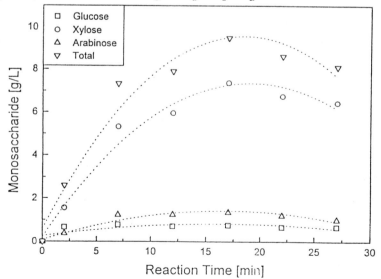

Fig. 6: The effect of wet oxidation reaction time on solubilized hemicellulose measured as monosaccharides after acid hydrolysis (60 g wheat straw (1993), 185°C, 12 bar O_2, 6.5 g Na_2CO_3).

CONCLUSIONS

This preliminary study of wet oxidation of wheat straw has shown some interesting results:

- Wet oxidation was found to be an efficient process for fractionating lignocellulose to convertible cellulose and hemicellulose using condition of relatively low temperature and short reaction time.
- Straw density and chemical addition were interdependent parameters giving different optimal temperature and reaction time for different purposes:
    1. Optimal enzymatic conversion of the cellulose fraction was achieved by using a low straw density with a high chemical addition at 170°C and 10 minutes giving 90% conversion.
    2. Optimal recovery of hemicellulose was achieved by using a high straw density and low chemical addition at 185°C and 15 minutes giving up to 9.5 g/L solubilized hemicellulose.
- The solubilized hemicellulose fraction was sufficient as sole carbohydrate source for microorganisms without any inhibitory effects.
- In addition, methods for quantification of the solubilized hemicellulose fraction must be continued investigated in order to improve the mass balance.

ACKNOWLEDGEMENTS

Funding was provided by the Danish Energy Ministry in the project "Development of chemical and biological processes for bioethanol production" (EFP1383/94-0003).

REFERENCES

1. Biermann, C.J. (1983) The development of a new pre-treatment method (rapid steam hydrolysis) and the comparison of rapid steaming, steam explosion, autohydrolysis, and wet oxidation as pre-treatment processes for biomass conversion of southern hardwoods. Ph.D. Thesis, Mississippi State University, USA.
2. Bjerre, A.B. and E. Sørensen (1992) Thermal decomposition of dilute aqueous formic acid solutions. *Ind. Eng. Chem. Res.* 31, 1574-1577.
3. Bjerre, A.B., A.B. Olesen, T. Fernqvist, A. Plöger and A.S. Schmidt (1995) Pre-treatment of wheat straw using combined wet oxidation and alkaline hydrolysis resulting in convertible cellulose and hemicellulose. *Biotechnol. Bioeng.* (in press)
4. Buchert, J. (1990) Biotechnical oxidation of D-xylose and hemicellulose hydrolyzates by *Gluconobacter oxydans*. Ph.D. Thesis, Technical Research Center of Finland.
5. Goering, H.K. and P.J. van Soest (1970) Forage fiber analyses. Apparatus, reagents, procedures, and some applications. In *Agricultural Handbook*, Vol. 379, Agricultural Research Service, United States Department of Agriculture, Washington DC. pp. 1-20.
6. Hörmeyer, H.F., W. Schwald, G. Bonn and O. Bobleter (1988) Hydrothermolysis of birch wood as pre-treatment for enzymatic saccharification. *Holzforschung* 42, 95-98.
7. McGinnis, G.D., W.W. Wilson and C.E. Mullen (1983) Biomass pre-treatment with water and high-pressure oxygen. The wet oxidation process. Ind. Eng. Chem. Res. 22, 352-357.
8. Puls, J. (1993) Substrate analysis of forest and agricultural wastes. In *Bioconversion of forest and agricultural plant residues* (J.N. Saddler (ed.)), CAB International, London, pp. 13-32.

9. Saddler, J.N., L.P. Ramos and C. Breuil (1993) Steam pre-treatment of lignocellulosic residues. In *Bioconversion of forest and agricultural plant residues* (J.N. Saddler (ed.)), CAB International, London, pp. 73-91.
10. Schmidt, A.S. and A.B. Bjerre (1995) Optimization and kinetic studies of wet oxidation pre-treatment of wheat straw for bioconversion. *Bioresource Technol.* (submitted)
11. Sørensen, E. and A.B. Bjerre (1990) Soil recovery by wet oxidation. *Environ. Technol. Lett.* 11, 429-434.
12. Von Sivers, M., G. Zacchi, L. Olsson and B. Hahn-Hägerdal (1994) Cost analysis of ethanol production from willow using recombinant *Escherichia coli*. *Biotechnol. Progr.* 10, 555-560.

C/N Ratio Effect on Degradation of Cellulose in Composting of Food Waste and Paper

Hang-Sik Shin, Yeon-Koo Jeong,, Eung-Ju Hwang, Department of Civil Engineering, Korea Advanced Institute of Science and Technology, 373-1, Kusong-dong, Yusong-gu, Taejon 305-701, Korea

ABSTRACT

The degradation of paper fraction was investigated in the composting of source separated food waste and paper mixture with change of C/N ratios. In low C/N ratio of 14.0, the cellulose in paper fraction was degraded in maturing stage concurrent with peak cellulose activity, temperature rise, and sub peak of CO2 evolution rate. This result might be ascribed to the initial low pH and high ammonia content caused by the abundance of readily degradable organic matters. However, the cellulose was degraded mainly at the thermophilic stage in C/N ratios of 20.4 and 29.9, indicating that these ranges of C/N ratio might be also appropriate for cellulose degradation. The cellulose activities increased abruptly during the thermophilic stage and reached maximum value at cooling stage in composting of appropriate C/N ratios. The cellulose was degraded significantly during the increase of cellulose activity.

KEYWORDS

Source separated food waste, composting, C/N ratio, cellulose degradation, cellulose activity.

INTRODUCTION

The source separated collection and composting of food waste encounters a number of problems such as liquid leaking from storage containers and transportation vehicles, odour problems as well as large requirements of bulking agents, while it has obvious merits over the composting of mixed wastes. In order to reduce these problems, it has been tried to wrap kitchen garbage with waste paper or to place paper on the bottom of the container [1], which enables the source separated collection and composting of food waste to be a viable alternative. And the compost of better quality could be obtained with the addition of waste paper because the high salt concentration and organic content in compost could be controlled. The source separated composting program for household organic waste is becoming well established in Europe, with pilot and full scale biowaste collection programs in at least eight countries in 1992 [2]. In Korea, the similar scheme for food waste management is under consideration in recent years.

The waste paper could also be a good substitute for wood-chip and bark as an amendment in composting process. Actually, it was also tried to use the waste paper scrap as bulking agent in composting of sewage sludge in USA[3]. In addition, utilisation of non-recyclable waste paper for this purpose could extend the life of landfill [1]. Therefore, it was attempted to compost the source separated food wastes and waste paper mixture without any other amendment.

Although many researches on the degradation of cellulose in composting of various wastes were made [4-9], a few were devoted to understanding the degradation of cellulose especially in composting of source separated food waste and paper mixture. And while the effects of C/N ratios were investigated in relatively higher range above 35 with cellulosic wastes such as cotton wastes and bark [4,9], it was hardly examined in lower ranges. The degradation of paper fraction in composting of food waste and paper mixture is also of importance in that the undegraded paper may be hard to separate after composting. This work was undertaken in order to understand the degradation of cellulose in relatively low C/N ratios comprehensively resulted from the different mixing ratios of food wastes and paper and to elucidate the quantity of waste paper required for the well-balanced composting reaction. The degradation of paper fraction was investigated by analysing the cellulose activity, cellulose content as well as overall composting parameters.

MATERIALS AND METHODS

Wastes and Experimental Conditions

Food waste collected from a cafeteria was dried in open air and stocked to get uniform feed material. Waste paper consisting of equal amount of newspaper and office paper was shredded to less than 1 cm in width prior to mixing with food waste. The physical and chemical characteristics of feed materials were described in Table 1. Composting of food waste and paper mixtures was conducted under various mixing in different C/N ratios, as shown in Table 2. About 4.0 kg (dry weight) of waste mixture including 1.0 kg of compost was placed in the reactor. The seed compost was obtained from the preliminary composting operation conducted with the same wastes. Initial water content was adjusted to about 60% by adding tap water. Whenever the excessive moisture loss was observed in mixing operation, water was supplied to maintain the water content above 55%.

Table 1: Physical and chemical characteristics of wastes

Wastes	VS(%)	FS(%)	Cellulose content(%)	NH3-N (mg/kg ds)	Org.-N (g/kg ds)
Food waste	86.3	13.7	4.6	387.3	47.6
Newspaper	95.6	4.4	33.7	-	-
Office paper	85.7	14.3	43.4	-	-

(% of dry matter)

Composting Reactor

For the cylindrical composting reactor, an acrylic column (300 mm in diameter, 500 mm in depth) with 28 l of working volume was used. At the bottom of the reactor, a perforated plate was installed to distribute air evenly. The reactor was insulated with glass wool in order to reduce the loss of reaction heat. Two sensors for temperature monitor and control were placed in the centre of the reactor. The reactor temperature was controlled so as not to exceed 55°C by regulating the air flow rate in the range of 0.8 to 8.0 l/min. Exhaust gas from the reactor passed to a condensate trap, lN H2SO4 solution, and SN KOH solution in series to capture ammonia and carbon dioxide, respectively. The compost was mixed every other day in early stage, and every three to five days in the following stage. About 50g of composite sample was taken from the reactor after every complete mixing.

Table 2: Experimental conditions (dry weight)

	Run A	Run B	Run C
Food waste: paper(kg)	2.0: 1.0	1.5: 1.5	1.0: 2.0
C/N ratio	14.0	20.4	29.9
Cellulose content (% of dry matter)	15.2	20.0	24.2

Seed compost: 1.0 kg, water content: 60%, carbon content: volatile matter/1.8, nitrogen content: TKN

Physical and Chemical Analysis

Water content was determined by drying the sample in an oven at 105°C for 24 h. Determination of volatile solid was made by burning the pre-dried sample at 550°C for 3 h. These parameters were determined in triplicates. For pH measurement, test solution was made by blending 95 ml distilled water with 5 g compost sample for 30 min. The ammonia content in sulphuric acid solution and compost were determined by distillation method, and carbon dioxide by titration with 0.1 N H_2SO_4 standard solution [10]. The condensate generated mainly during the thermophilic period was also analysed due to its large quantity and high content of ammonia. Cellulose content was determined by the modified Updegraff's method [11]. Prior to the measurement, the compost sample was dried at 60°C for 48 h and crushed by a commercial crusher.

Enzyme Assay

The cellulolytic enzyme activity was determined by using 0.1 M citric acid/0.2 M Na_2HPO_4 buffer of pH 6.0 [12]. The wet compost of 5 g was mixed with 95 ml distilled water for 30 min and centrifuged at 4000 rpm for 5 min. The supernatant was used as an enzyme solution for cellulose activity. Enzyme reaction was conducted by mixing 0.5 ml compost extract and 0.5 ml buffered substrate solution at 50°C for 30 min. The standard substrate used was 1.0% buffered CMC solution. Sodium azide of 0.015% was added to inhibit microbial growth. Enzyme reaction was stopped by adding 3 ml of DNS reagent and boiled for 5 min to get colour development. Samples were then allowed to cool down to room temperature, and the optical density was read at 550 nm. The cellulolytic enzyme activity was expressed as the amount of reducing sugars released from 1 g dry compost for 30 min.

RESULTS AND DISCUSSION

General Composting Reaction

The general composting reactions in Run A were described in Figure 1 in terms of pH, temperature, CO_2 evolution rate, and ammonia evolution rate. The initial pH drop below 5.0 caused by acidification of readily degradable matters in food waste discouraged the composting reaction. As long as pH was low, temperature did not increase above 50 °C. The temperature showed another peak on the 30th day, being ascribed to the degradation of cellulose which was not attacked in the previous stage. Similar increases were also observed in CO_2 evolution rate and cellulose activity at this time, confirming the degradation of cellulose.

The pH drop was not observed in Run B and C, which promoted the rapid degradation of raw materials (Figure 2 & 3). As a result, the temperature increased to 55°C within a day and remained over 10 days in both runs, which meant well-balanced composting reaction. The CO_2 production rates in Run B and C reached the maximums on the 4th day at the early thermophilic period. In all cases, the peak CO_2 production rate was observed at the initial thermophilic stage, indicating the active degradation of organic materials, as reported by other researchers [13,14]. And the peak values decreased with the increase of C/N ratio.

In Run A, during the initial low pH, ammonia loss was not observed. But as the pH increased, it suddenly increased from the 9th day to reach a peak in thermophilic period. In Run B and C, significant ammonia escaped after two days lag, and peak losses were also observed in the thermophilic period. The cumulative quantities of ammonia volatilized in Run A, B, and C during the experiment were 30.6 g, 23.8 g, and 9.9 g, respectively, implying that 22.9%, 22.7%, and 15.5% of the initial nitrogen escaped as ammonia gas.

Owing to the low C/N ratio of Run A, high ammonia content above 8 g per one kg compost was accumulated and lasted over a week as shown in Figure 4. The high content of ammonia probably delayed the degradation of cellulose in paper fraction. However, such a high level of ammonia was not accumulated in Run B and C because of appropriate C/N ratios.

In Run A, the ammonia content dropped abruptly on the 25th day, being exactly coincided with the change of cellulose (Figure 4). These changes of ammonia content could not be explained by the escape of ammonia. Taking the reduction of ammonia to nitrite and nitrate was inhibited at high ammonia concentrations [15] and the probably low microbial immobilisation of inorganic nitrogen at this period into accounts, another mechanisms of ammonia transformations as an initial stage of humification facilitated by the degradation of cellulose might be involved.

According to Stevenson [16], ammonia could be fixed chemically with lignin and organic matter in soil, which was associated with oxygen uptake and promoted in alkaline condition. As expected, ammonia fixation increased not only with pH increase, but also with increasing amount of ammonia in Run A. As the cellulose was degraded, the lignin may be brought to surface and transformed, which encouraged the fixation of ammonia, as reported by Morisaki et al. [17]. As the ammonia adsorbed to the bark, organic nitrogen increased concurrently.

However, they stated that this did not mean ammonia to be changed to organic forms but rather ammonia bound tightly with component of bark.

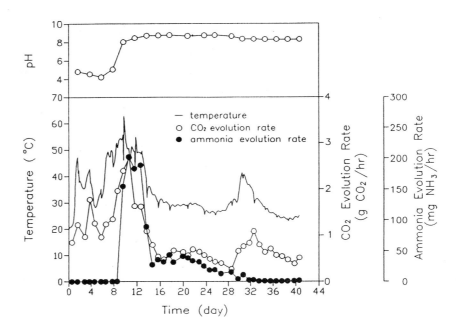

Fig. 1: Changes of temperature, pH, CO2 evolution rate, and ammonia evolution rate in Run A

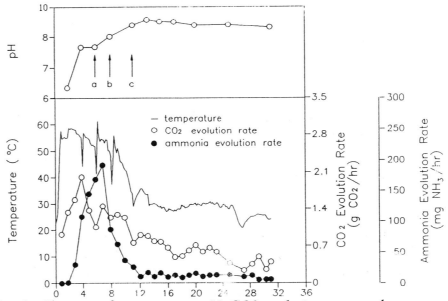

Fig. 2: Changes of temperature, pH, CO2 evolution rate, and ammonia rate in Run B. Water added for the water content over 55%: a = 800 ml, b = 280 ml, c = 75 ml

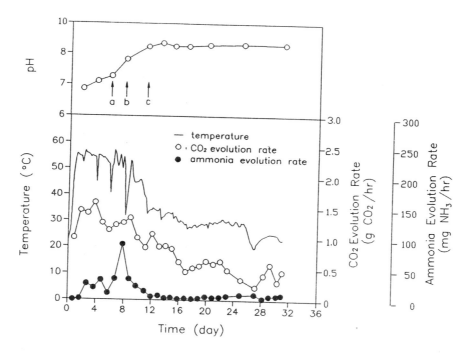

Fig. 3. Changes of temperature, pH, CO2 evolution rate, and ammonia evolution rate in Run C Water added for the water content over 55%: a=800 ml, b=230 ml, c=70 ml.

These phenomena were not observed apparently in Run B and C due probably to the earlier degradation of cellulose before large accumulation of ammonia and the immobilisation of greater portion of nitrogen into biomass. The cellulose degradation indicated by the increase of cellulose activity and the volatilisation of ammonia occurred simultaneously in Run B, while the increase of cellulose actively in Run C preceded the volatilisation of ammonia. The microbial immobilisation of inorganic nitrogen was also responsible for these results because of the relatively higher C/N ratios in both runs. Therefore less ammonia might be fixed chemically which was not observed clearly.

Cellulose Activity and Cellulose Degradation

The evolution of cellulose activity in three runs were illustrated in Figure 5. After two weeks delay, the cellulose activity in Run A increased gradually and reached the maximum on the 35th day. However, the cellulose activity in Run B and C increased steadily during the thermophilic stage, and peaked at the cooling stage. The delay in Run A might ascribed to the abundance of readily available organic matters which gave rise to the adverse conditions such low pH and high ammonia content. And it also seemed to affect the cellulose degradation due to greater availability than cellulosic substances. As long as readily degradable matters were available, polymerised material like cellulose was reported to be slightly attacked [4].

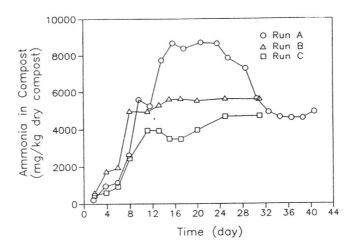

Fig. 4: Changes of ammonia content in compost in three runs

The peak cellulose activities in Run A, B, and C were attained on the days of 35, 13, and 11, respectively. As the C/N ratio increased, the time required for peak activity decreased. This result might ascribed to the relative abundance of cellulose which had significant effects on cellulose induction. The cellulose activity in Run C decreased abruptly after the maximum activity, indicating that the available cellulose in paper was depleted rapidly. Similar results have been reported by Ashbolt & Line [9]. They stated that the activity of cellulose was greater in the higher C/N ratio during the mesophilic stage, but reverse was the case at thermophilic stage

The characteristic single peak of cellulose activity was observed, as reported by Bono et al. [5], which is contrary to the results of others [6, 9]. Ashbolt & Line [9] reported that at least two regions of peak activity were observed, in the mesophilic and the thermophilic stage of composting. Similarly, the cellulose activity increased during the mesophilic period, followed by decline in thermophilic period and increased again in the late cooling period [6].

The cellulose content in Run A increased in the beginning 12 days and showed gradual decrease afterward until the 29th day, followed by rapid decrease by the 33rd day. The increase of initial cellulose content seemed to be offset by rapid decomposition of readily decomposable and low molecular weight substrates. However, the cellulose was steadily decomposed during the experiment in Run B and C without significant increase of cellulose content. This result indicated that cellulose was attacked from the start and the C/N ratio of 20.4 and 29.9 were appropriate for cellulose degradation. The cellulose degradation from the start implied that the cellulose in waste paper was readily attackable if the composting conditions were proper and initial amount of cellulose was relatively high.

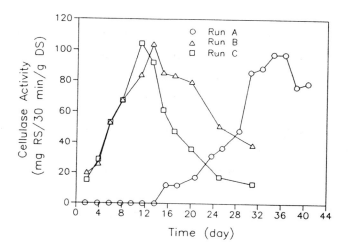

Fig. 5: Changes of cellulose activity in three runs

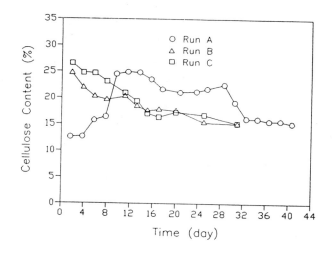

Fig. 6: Changes of cellulose content in three runs

The cellulose content decreased drastically during the increase of cellulose activity at the thermophilic period and slowly decline after the peak in three runs. This result might be explained by the cellulose binding to the cellulose and modification of cellulose in a way that reduce the binding constant with the degradation of cellulose. While the cellulose was bound in cellulose during the active degradation, it desorbed from the cellulose as the cellulose was degraded and exhibited the maximum actively at the cooling stage [24,25]. However, Godden et al.[6] stated that the maximal rate of cellulose degradation was observed during the cooling phase, when the cellulose activity reached the maximum.

CONCLUSIONS

The waste paper fraction was degraded substantially in the thermophilic periods concurrent with the increase of cellulose activity, when the C/N ratios were in the range of 20 to 30. However, the C/N ratio of 14.0 delayed the degradation of paper fraction to the maturing stage concurrent with peak cellulose activity, temperature rise, and sub peak of CO_2 evolution rate. This delay might be ascribed to the factors such as initial low pH and high ammonia content in compost caused by the abundance of readily degradable matters. Sufficient amounts of paper should be added in collection stage and/or composting stage in order to rapid degradation of paper fraction. These results also indicated that the food waste and paper mixture could be composted successfully without adding amending material. And it was not necessary to separate the undegraded paper fraction after composting.

REFERENCES

1. De Baere, L., Six, W., Tillinger, R. and Verstraete, W. (1992). Wastepaper improves biowaste composting, *Biocycle*, Sep., 70-71.
2. Spencer, R.L. (1993). European collection programs for source separated organics. *Biocycle*, Jun., S6 59.
3. Goldstein, N. (1992). Adding paper to the mix, *Biocycle*, Aug., 54-58.
4. Baca, M.T., Fornasier, F. and de Nobili, M. (1992). Mineralization and humification pathways in two composting processes applied to cotton wastes, *J. Ferment. Bioeng.*, 74, 17g-184.
5. Bono, J.J., Chalaux, N. and Chabbert, B. (1992). Bench-scale composting of two agricultural wastes. *Biores. Technol.*, 40, 119-124.
6. Godden, B., Penninckx, M., Pierard, A. and Lannoye, R. (1983). Evolution of enzyme activities and microbiol populations during composting of cattle manure, *Eur. J. Appl. Microbiol. Biolech.*, 17, 306 310.
7. Godden, B. and Penninckx, M.J. (1984). Identification and evolution of the cellulolytic microflora present during composting of cattle manure: on the role of actinomycetes sp., *Ann. Microbiol. (Inst. Pasteur)*, 135 B, 69-78.
8. Stutzenberger, F.J., Kaufman, A.J. and Lossin, R.D. (1970). Cellulolytic activity in municipal solid waste composting, *Can. J. Microbiol.*, 16, 553-560.
9. Ashbolt, N.J. and Line, M.A. (1982). A bench-scale system to study the composting of organic wastes, *J. Environ. Qual.*, 11 (3), 405-408.
10. APHA, AWWA and WPCF. (1985). Standard methods for the examination of water and wastewater, 16th edition.
11. Updegraff, D.M. (1969). Semimicro determination of cellulose in biological materials, *Anal. Chem.*, 32, 420-424.
12. Stewart, B.J. and Leatherwood, J.M. (1976). Derepressed synthesis of cellulose by Cellulomonas, *J. Bacteriol., 128(2)*, 609-615.
13. Nakasaki, K., Yakuchi, H., Sasaki, Y. and Kubota, H. (1992). Effects of C/N ratio on thermophilic composting of garbage, *J. Ferment. Bioeng.*, 73(1), 43-45.
14. Nakasaki, K., Yaguchi, H., Sasaki, Y. and Kubota, H. (1990). Effects of oxygen concentration on composting of garbage, *J. Ferment. Bioeng.*, 70(6), 431-433.
15. Focht, D.D. and Chang, A.C. (1975). Nitrification and denitrification processes related to waste water treatment, *Adv. in Appl. Microbiol.*, 19, 153-186.
16. Stevenson, F.J. (1982). Humus chemistry; genesis, composition, reactions, John Wiley and Sons, New York.

17. Morisaki, N., Phae, C.K., Nakasaki, K., Shoda, M. and Kubota, H (1989). Nitrogen transformation during thermophilic composting, J. *Ferment. Bioeng.,* 67(1),57-61.
18. Beguin, P., Eisen, H. and Roupas, A. (1977). Free and cellulose-bound cellulases in a Cellulomonas species, J. *Gen. Microbiol.,* 101, 191-196.
19. Langsford, M., Gilkes, N.R., Wakarchuk, W.W., Kilburn, D.G., Miller, R.C., Jr. and Warren, R.A.J. (1984). The cellulose system of Cellulomonas fimi, J. *Gen. Microbiol.,* 130, 1367-1376.

Options for a Common Treatment of Biodegradable Plastics and Biowaste

K. Hoppenheidt and J. Tränkler, Bavarian Institute for Waste Research, Am Mittleren Moos 46 B, D-86167 Augsburg, Germany

ABSTRACT

This study demonstrates the necessity of biodegradability and compostability tests for different compost raw materials. The German waste legislation demands such a testing in order to ensure a production of high quality composts with low content of pollutants and impurities.

Biodegradability of selected polymeric materials was studied under different test conditions (aquatic and compost environment). The extent of mineralization observed varied from < 5% to > 95% depending on the test conditions and the raw material of the test substance. Polymeric materials made from natural raw materials as well as selected mineral oil based polymers demonstrated the required biodegradability for 'suitable compost raw materials'.

KEYWORDS

Biodegradable plastics, biowaste, composting process, common treatment, biodegradability testing.

INTRODUCTION

The change of waste legislation has promoted the composting of organic residues in Germany. In the near future there will be about 300 composting plants with a total capacity of 2.8 million metric tons per year [1].
Waste legislation and different guidelines limit the input of the composting process, so far [2-4]. A catalogue of demands for compost raw materials should ensure the production of high quality composts. Compost raw materials have to be collected source-separated. They have to have a low content of pollutants and impurities and a sufficient biodegradability and compostability [4]. Therefore, a classification of organic residues concerning their suitability to the composting process was proposed (Table 1) [4]. Native, biogenic organic residues like biowaste and yard waste are supposed to be suitable compost raw materials without any restriction. High contents of pollutants, non-degradable organic components or hygienic doubts are reasons for the refusal of some organic residues as compost raw materials.

Some organic residues have to demonstrate their suitability as compost raw material e.g. bioplastics and other polymeric materials. Bioplastics are new materials from different and in the most cases renewable resources [5, 6]. Their properties are supposed to be similar to conventional plastics but bioplastics have to be biodegradable. Bioplastics are mainly developed for short-term or one way applications like packaging materials, one-way dishes, diapers, etc.

In Germany bioplastics and other polymeric materials have to demonstrate their suitability as compost raw material in a multi-stage test scheme (Figure 1) [4]; standard methods for the test procedures are in discussion [7]. Only materials which pass all test stages get the permission for a common treatment with biowaste.

Table 1: German classification of organic residues with regard to the suitability as compost raw material

Suitable compost raw materials	Suitability testing required	Unsuitable materials
Biowaste	Biogenic commercial waste	Organic residues with high contents of pollutants and impurities
Yard waste	Biogenic residues from centre strips	Products with non-degradable components
Organic residues from the preservation of the countryside	Products made from completely biodegradable materials, **Bioplastics**	Diapers, faeces, textile, leather
	Paper, cardboard, ...	

MATERIALS AND METHODS

Two different respirometric methods were used for biodegradation studies. The BODIS-test is an aquatic biodegradation test for the determination of the biological oxygen demand for insoluble substances [8]. The biodegradation in the compost environment was studied in a closed system by gaschromatographic analysis of the oxygen and carbon dioxide content in the headspace of the reactor vessel [9].

Comparisons of the observed oxygen demand and/or carbon dioxide evolution with the theoretical values for a total degradation of the test substances were used for the calculation of results (% mineralization). Normally powders as well as fine fragments of raw materials and selected products were used as specimen.

RESULTS AND DISCUSSION

Necessity of Biodegradation Studies

The attribute biodegradability is used for marketing purposes. But so far there is no generally accepted definition of the term biodegradation for polymeric materials [10]. The term biodegradation describe simple biotransformations as well as the total degradation, termed mineralization. Meanwhile a working group of the German Institute for Standardisation (DIN) proposed that polymeric materials should only be attributed biodegradable, if all organic components undergo a total biodegradation (mineralization) [7]. Considering this definition the German composting guideline demands a proof of the total biodegradability [4]. Materials which only undergo a biocorrosion (biodeterioration, biodisintegration) instead of a mineralization shall no longer be attributed to be biodegradable.

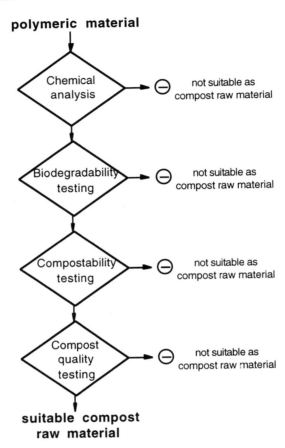

Fig. 1: Proposed test scheme for the suitability testing of compost raw materials

Figure 2 shows some results of biodegradation studies with so-called biodegradable plastics marketed in Germany. Only two of four tested polymeric materials showed a significant biodegradation.

Biodegradation Results are Specific to the Used Test System

Biodegradation studies in aquatic test systems are very convenient and they give the opportunity for a detailed balance of the carbon mass flow. On the other hand it is known, that the compost environment offers better conditions for biodegradation (higher temperature, different microbial populations, etc.). Therefore the biodegradation of selected polymeric materials was additionally tested in a compost environment (compost made from biowaste; incubated by 55 °C). The biodegradation results from the BODIS-test and the compost test are compared (Figure 3).

Materials with high extent of mineralization in the BODIS-test exhibited comparable results in the compost environment; in most cases the extent of mineralization in the compost environment was slightly higher. On the other hand one polymeric material with insufficient biodegradation in the BODIS test showed a high extent of mineralization in the compost environment and has to be assessed as biodegradable material too. But neither the aquatic nor the compost environment enabled a sufficient biodegradation for another test material which seems to be non-degradable according to the above mentioned definition. These results correspond to the high content of LDPE, which is known to be only slightly biodegradable [11, 12].

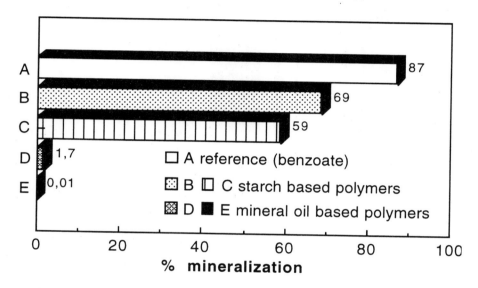

(B: loose-fill chip; C: foil fragments from MATER-BI; D: granules from BIONOLLE; E: foil fragments from ECOSTAR)

Fig. 2: Extent of biodegradation of selected polymeric materials in the BODIS-test after 6 weeks.

Biodegradability of Products based on the Same Raw Material

Different products can origin from the same basic raw material. An important question is, whether it is possible to deduce from biodegradation results of the raw material or one product to the biodegradability of all products made from the same raw material. Figure 4 summarises the results of biodegradation tests with cellulose based materials.

Two of the tested products, a dust cloth and a biobag, showed nearly the same biodegradation as cellulose powder. But a copy paper gave evidence of an incomplete biodegradation despite its high cellulose content.

These results demonstrate that biodegradation results should only be transferred to other products if they are indubitable made from the same raw material and if modifications of the polymer during the manufacturing process are out of question.

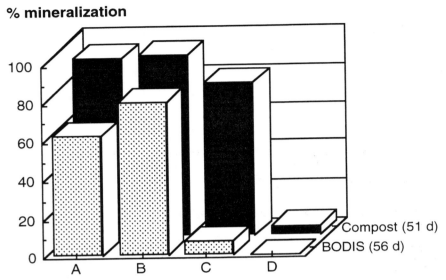

(A: foil fragments from MATER-BI; B: Cellulose powder; C: granules from BIONOLLE; D: foil fragments from ECOSTAR)

Fig. 3: Comparison of biodegradation results in different test systems.

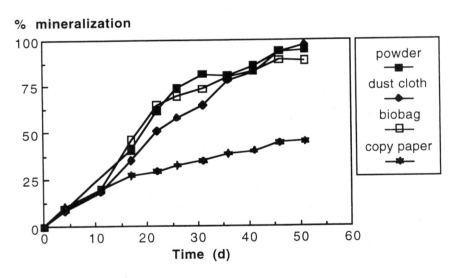

Fig. 4: Comparison of the biodegradation of different cellulose based products (compost environment, 55 °C).

Assessment of the Biodegradability of Selected Biobags

Figure 5 gives another example for the necessity of biodegradation studies. Different types of biobags are sold in Germany for the collection of biowaste from households. The four tested biobags are marketed as biodegradable products. The test results confirm the biodegradability only for two biobags. The remaining biobags have to be classified as incomplete or non-degradable.

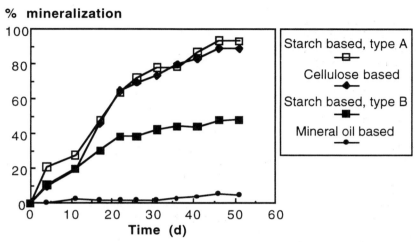

(Starch based: type A = MATER-BI; type B = 4P; mineral oil based: ECOSTAR)

Fig. 5: Comparison of the biodegradation of four types of biobags

CONCLUSION

With high priority the German Cycle Economy and Waste Act promotes the production of high quality composts. Therefore, materials should be excluded from the composting process if they contain non-degradable organic components, high contents of pollutants or impurities or other quality interfering constituents.

The biodegradation testing is an important step in the suitability testing of organic residues as compost raw materials. Standard methods for the suitability testing are in discussion [6]. Development and waste management of really biodegradable and compostable polymeric materials require generally accepted test procedures.

But results of the proposed test scheme are only applicable for the aerobic biodegradation and treatment. Provided that the anaerobic treatment of biowaste will become more important, it is necessary to develop a test scheme for the assessment of the anaerobic biodegradability and treatability of innovative materials like bioplastics.

ACKNOWLEDGEMENTS

Biodegradation tests were performed by Mrs S. Lerch and we thank for her assistance. The authors thank Mr R. Koch, Bayer AG Leverkusen, and M. Jahnke, IBR Walsrode, for the preparation of test procedures for the compost- and BODIS-test. Montedison Deutschland GmbH, Showa Denko (Europe) GmbH, 4P and Ecostar have supported the study with their products.

REFERENCES

1. Bundesgütegemeinschaft Kompost e. V (1995). Personal communication, May 1995.
2. Gesetz zur Förderung der Kreislaufwirtschaft und Sicherung der umweltverträglichen Beseitigung von Abfällen (Kreislaufwirtschafts- und Abfallgesetz - KrW-/AbfG) vom 27. 09.1994 (German Cycle Economy and Waste Act), Bundesgesetzblatt BGBl. I S. 2705.
3. Dritte Allgemeine Verwaltungsvorschrift zum Abfallgesetz (TA Siedlungsabfall) - Technische Anleitung zur Verwertung, Behandlung und sonstigen Entsorgung von Siedlungsabfällen (Technical guideline for the utilization, treatment and other disposal of domestic refuse). Bundesanzeiger Nr. 99a, G 1990 A, vom 29.05.1993.
4. Länderarbeitsgemeinschaft Abfall (1995). LAGA Merkblatt M10: Qualitäts-kriterien und Anwendungsempfehlungen für Kompost (LAGA guideline M10: Quality criteria and recommended applications of compost) - Stand: 15.2.1995.
5. Brune, D., Utz, H. and Korn, M. (1991). Untersuchung zum Einsatz bioabbaubarer Kunststoffe im Verpackungsbereich (Study on practical applications of biodegradable plastics for packaging purposes). Forschungsbericht Nr. 01-ZV 8904, Bundesministerium für Forschung und Technologie, Bonn.
6. Huang, J.-C. et al. (1990). Biodegradable Plastics: A Review. Advances in Polymer Technology, Vol. 10, No. 1, p. 23-30.
7. DIN Deutsches Institut für Normung e. V., Normenausschuß Kunststoffe (FNK), Arbeitsausschuß 103.3 „Bioabbaubare Kunststoffe". Chairman: M. Pantke, Bundesanstalt für Materialforschung (BAM), Unter den Eichen 87, 12205 Berlin (unpublished).
8. Jahnke, M. (1993). BOD Test für schwer- und unlösliche Prüfsubstanzen (biol. Abbautest) (Guideline for the determination of the biological oxygen demand of insoluble substances). Draft guideline for BODIS ring trail (IBRG, DIN FNK 103.2). Hannover (unpublished).
9. Koch R. (1994). Bestimmung der mikrobiellen Abbaubarkeit von Kunststoffen in Kompost (Determination of the microbial degradation of polymers in the compost environment). Bayer AG, Leverkusen (unpublished).
10. Swift, G. (1992). Biodegradability of polymers in the environment: complexities and significance of definitions and measurements. FEMS Microbiology Reviews, 103, p. 339-346.
11. Albertson, A. C., Karlson, S. (1990). Polyethylene Degradation and Degradation Products. In: Glass, J. E., Swift, G. (ed.), *Agricultural and Synthetic Polymers - Biodegradability and Utilisation.* ACS-Symposium Series 433, American Chemical Society, p 60-64.
12. Johnson, K. E. et al. (1993). Degradation of Degradable Starch-Polyethylene Plastics in a Compost Environment. Appl. Environ. Microbiol., Vol. 59, No. 4, p. 1155-1161.
13. Krupp, L. R, et al. (1992). Biodegradability of Modified Plastic Films in Controlled Biological Environments. Environmental Science and Technology, 26, No. 1, p. 193-198.

Section 7

Aerobic & Other Processing

A Steady-State Model of the Compodan Composting Process

Lars Krogsgaard Nielsen, Krüger A/S, Gladsaxevej 363, DK-2860 Soeborg, Denmark

ABSTRACT

The aim of the paper is to present a steady-state computer model of Krüger's composting process Compodan. The model calculates the air flow, water and energy content of the process air based on set parameters of retention time, degradation of volatile solids, relative humidity and process temperature of the process air. The purpose of the model is to have a tool to calculate air injection for plants incorporating the Compodan process.

The model has been validated on results from the start-up phase of the first full-scale plant using the process. There is a good correlation between the air flow calculated by the model and the air flow measured on-line on the plant during a period with a stabilised process.

The potential of the model is to predict the performance of the composting process and the compost quality dependent on process temperature and degradation of volatile solids. A further potential will be to optimise a quick start-up of the process and a well functioning biofilter, which are the characteristic features of the Compodan process.

KEYWORDS

Compost, source-separated waste, Compodan process, steady-state computer model, air parameters, air injection, enthalpy in air, water in air, mass balance, energy balance.

INTRODUCTION

Aeration is the key factor to achieve a well-functioning composting process. The aeration ensures a certain oxygen level for the microorganisms that degrade the waste and aeration removes generated energy in order to regulate the process temperature to an optimal level. Temperature is one of the major single factors influencing the degradation rate.

It is well known that process control based on keeping an optimal temperature determines the air requirements. Also the Compodan process is a product of this knowledge.

The Compodan process is Krüger's patent-pending design for aeration during composting in a reactor. The process has been used in 2 newly built plants for the composting of source-separated household waste in Aarhus, Denmark 1995 (20 tons/day) and in Creusot Montceau, France 1996 (80 tons/day).

A standard reactor is 20 m wide, consisting of 10 sections of 2 m. The reactor length is adapted to the plant capacity. Ideally, the sections can be connected to separate blowers sucking the process air down through the material. In Aarhus the reactor is divided into 4 zones for which reason there are common blowers (3 in all) for several sections, see figure 1. The air is sucked through the material in zones 2, 3 and 4, whereas it is blown up through zone 1. The air chamber over the compost - called zone 5 - contains a mixture of process air from zones 1-3 and outdoor air.

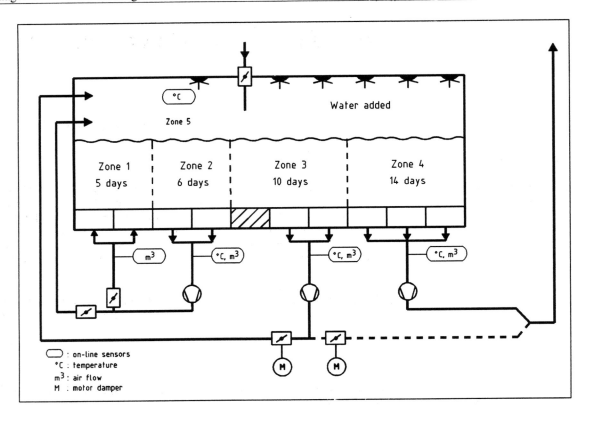

Fig. 1: Composting reactor with Compodan process

Waste is fed into zone 1 and compost is extracted from zone 4. The retention time in the reactor is approx. 35 days. Screw conveyors suspended from a traversing bridge transport the compost through the zones.

The air supply is controlled on the basis of on-line temperature measurements taken in the air ducts so that the temperature is maintained at a level ensuring the highest degradation rate. When the temperature exceeds the set-point (e.g. 55°C), the air flow is increased and vice versa. The process is also controlled through addition of water to avoid drying-up of the compost which reduces the degradation rate.

One section is not aerated to obtain a temperature increase which ensures hygienisation of the compost. In Denmark, the hygienisation requirement is minimum one hour at 70°C.

Hot process air from the very active zone 2 is injected into zone 1 and all exhaust air passes through the normally least active zone 4. This makes the composting process set off quickly as the waste is subjected to the temperature at which the degradation rate is highest, and zone 4 gets the function of a built-in biofilter. Finally, by re-circulation of process air through zones 2 and 3, uniform process conditions are obtained throughout the approx. 2 m thick compost layer as the temperature and oxygen concentration differences are minimised.

MATERIALS AND METHODS

Total solids (TS), volatile solids (VS) and chemical oxygen demand (COD) have been determined on the basis of Standard Methods 2540B, 2540E and 5220B [2].

The steady-state computer model Comp-air for the Compodan process is based on formulas and concepts from [1].

Calculation procedure and formula expression for Comp-air:

Zone 5 is given estimated start values for temperature (t) and relative humidity (ϕ) after which the calculation is iterated until the calculated final values of specific humidity (x) and enthalpy (I) for zone 5 are equal to the start values (x and I) in the latest iteration. The iteratively calculations are done on the basis of the formula for the air's x and I when the saturation point is not exceeded. If saturation occurs, the enthalpy expression cannot be inverted directly to a temperature expression, but must be solved iteratively [1].

The start value for x and I in zone 5 are calculated with a presumed temperature (t):

$x = 0.62198 * p_d/(p-p_d)$
$I = 1.006t + x(2501 + 1.8t)$

In the subsequent iteration, the x and I in zone 5 are calculated as the sum of x and I added from zones 1-3 and outdoor air, with reduction of an estimated heat loss through the structure.

Entalpy (I) in zones 1-4 is calculated as the sum from zone 5 and from the respective zone, and specific humidity (x) in zones 1-4 is calculated with presumed temperature (t):

$x = (I-1.006t)/(2501 + 1.8t)$

The relative humidity (ϕ) in zones 1-4 are presumed, where it in zone 5 is given estimated start values and subsequently calculated by iterations:

$\phi = p_d/p_{dm}$
$p_d = x*p*(x+0.62198)$
$p_{dm} = e^{(23.5771-(4042.9/(T-37.58)))}$, applies to 0-80°C

Comp-air calculates the air flows through the zones so that the energy balance in the respective zones is correct.

Notation:
- ϕ: relative humidity
- p: total pressure (Pa)
- p_d: water vapour partial pressure (Pa)
- p_{dm}: saturated water vapour pressure (Pa)
- $p_d(in)$: water vapour partial pressure in incoming air (Pa)
- x: specific humidity (kg H_2O/kg dry air)
- I: enthalpy, kJ/kg dry air
- t: temperature, °C
- T: temperature in Kelvin

The model calculates the air flows on the basis of the system's energy balance. The energy generated must be removed from the process while the desired temperatures are maintained in the respective zones. The set parameters of the model are:

- Amount of incoming waste,
- retention time in the zones,
- waste and compost total solids (TS),
- waste volatile solids (VS),

- volatile solids degradation (%) in the zones,
- energy generated from VS degradation,
- water produced at VS degradation,
- outdoor air temperature and relative humidity (RH%),
- process air temperature and relative humidity in zones 1, 2, 3 and 4,
- specific heat for VS, ash and water.

RESULTS AND DISCUSSION

The first composting plant with the Compodan process was put into operation in Aarhus, Denmark. The content of total solids (TS), volatile solids (VS) and chemical oxygen demand (COD) in source-separated waste including straw (bulking agent) and the compost from the plant in Aarhus is stated in table 1.

Table 1: Characterisation of waste and compost in Aarhus (standard deviation in brackets).

	TS %	VS % of TS	COD g/kg VS
Waste	37.3 (3.1)	76.4 (6.1)	1400 (150)
Compost	65.0 (11)	50.2 (5.8)	-

A constant for energy produced is 14 MJ/kg COD released [3]. With 1.4 kg COD/kg VS, a total of 19.6 MJ/kg VS degraded is released. Another constant for water generated is 0.7 kg H_2O/kg VS degraded [4].

The supply of waste during the commissioning of the plant has constituted approx. 55% of the design capacity for which reason the retention time is increased from approx. 35 to 65 days. The long retention time has resulted in a considerable VS degradation totalling approx. 70% throughout the 4 zones.

On-line temperature measurements taken in a stable operating period show 50, 52, 51 and 48°C in zones 2, 3, 4 and 5, respectively, see figure 2. On-line measurements of the air flow show 6.5, 11.0 and 2.6 m_/(h*metric ton of added waste) in zones 2, 3 and 4, respectively, in the same operating period, see figure 3. During the commissioning period practically all process air from zone 2 is re-circulated to zone 1.

As set values in the model, Aarhus data are used concerning waste composition, measured process temperatures in zones 1-4, VS degradation and retention time in the zones and presumed relative humidity in zones 1-4. The calculated air flows in zones 2-4 and temperature in zone 5 by the model are compared with measurements taken at the plant in a period with stable operation during the commissioning, see table 2 and 3. Set values (input) are shown in bold.

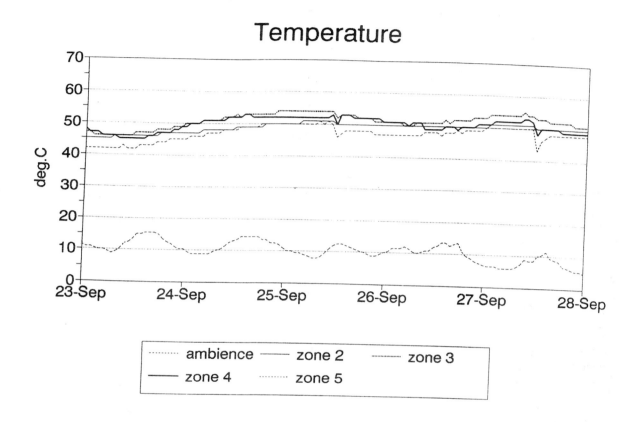

Fig. 2: On-line measurements of temperature from the plant in Aarhus with the Compodan composting process.

When comparing the model calculated air flows and the measured air flows, the deviations are considerable within each zone. However, the model's total air flow is only 8% higher than the measured flow. Note that the air flow through zone 1 is a part of the air flow from zone 2. Likewise, zone 5's temperature is calculated to be approx. 8% lower than the measured temperature. These results lie within the uncertainty margins connected with e.g. the determination of the VS degradation.

In the commissioning period, the temperature in the air duct from zone 4 was generally 5-10°C below that of the compost.

As a result, water has been condensed in the air ducts from zone 4 under the reactor floor. The condensation involves an energy loss which reduces the air requirement for removal of energy from the reactor. This energy loss due to condensation is not included in the model. This is part of the explanation for the reason why the calculated air flow is higher than the air flow measured.

Table 2: Mass and energy balances of comp-air, example 1 metric ton of Aarhus waste.

MASS BALANCE		
Waste	kg/65 days	**1000**
Waste, TS	%	**37**
Waste, VS	% of TS	**76**
VS, degradation	%	**70**
Compost	kg/65 days	267
Compost, TS	%	**65**
Generated H_2O	kg/kg VS decomp.	**0.70**
Generated H_2O	kg/65 days	140
Evaporated H_2O	kg/65 days	1114
Added H_2O	kg/65 days	437
Compost H_2O	kg/65 days	93
ENERGY BALANCE		
Generated energy	MJ/kg VS decomp.	**19.6**
Generated energy	MJ/65 days	3910
Temperature, outdoor	°C	**10**
rh, outdoor	%	**60**
Water, outdoor	kg/kg	0.0045
Enthalpy, outdoor	MJ/65 days	276
Heat capacity, H_2O	kJ/(kg*K)	**4.19**
Heat capacity, VS	kJ/(kg*K)	**1.29**
Heat capacity, ash	kJ/(kg*K)	**1.67**
Heat capacity, waste	kJ/(kg*K)	3.10
Temperature, waste	°C	**10**
Temperature, compost	°C	**55**
Heating up waste	MJ/65 days	224
Enthalpy, exhaust air	MJ/65 days	3699
Conduction	MJ/65 days	**263**

Table 3: Air parameters of comp-air, example of 1 metric ton of Aarhus waste.

		zone 1	zone 2	zone 3	zone 4	zone 5 model	zone 5 measured
Retention time	days	9	11	19	26	-	-
VS degradation	% of VS start	10	15	25	20	-	-
Relative humidity, air	%	100	100	100	100	100	100
Temperature, air	°C	51	50	52	51	44	48
Air flow, measured	m³/h	6.0	6.5	11.0	2.6	-	-
Air flow, model	m³/h	5.0	5.6	8.3	7.8	-	-
Energy to air	kJ/h	214	537	895	716	-	-
Water, air	kg/kg	0.0912	0.0862	0.0965	0.0912	0.0617	-
Enthalpy, air	kJ/kg	342	292	303	288	202	-

The distribution of the VS degradation between the zones is estimated on the basis of samples of the raw compost in the reactor. Consequently, the model's calculations are unreliable with regard to the distribution of the air flows in the zones. Calculations using the same total VS degradation throughout the reactor, but with a different distribution between the zones, show a considerable variation in the calculated air flows. For instance, the air flows are increased in zones 2 and 3 (unchanged in zone 4) if some of the VS degradation is moved from zone 4 to zone 3. At the same time, the calculations result in a higher temperature in zone 5.

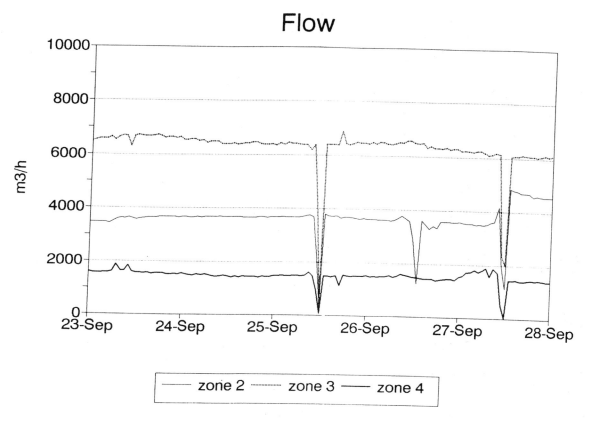

Fig. 3: *On-line measurements of air flow from the plant in Aarhus with the Compodan composting process.*

The next step in the validation of the model on the basis of operating data should be calibration of the water condensation and the VS degradation and their distribution throughout the 4 zones.

CONCLUSIONS

A steady-state computer model of the Compodan composting process has been developed to calculate the air flow, water content and enthalpy of the process air on the basis of the energy balance. The model is verified and theoretically validated on the basis of the same formulas that describe an Ix-diagram. Furthermore, the model is partly validated on the basis of process data during the start-up period of the first composting plant using this process.

The model's calculations of the total air flow in zone 2-4 and zone 5's temperature agree well with the measured values.

The future work will include calibration of the model with exact determination of water condensation and VS degradation in the 4 zones during stable full-scale operating conditions.

One of the future aims is to develop the model from being steady-state to dynamic. A dynamic model will include time-conditional changes of parameters that influence the VS degradation rate e.g. process temperature and water content in the compost and include this effect on the energy balance. This will make it possible to predict the Compodan composting process and to regulate the process in advance in order to optimise different factors such as VS degradation, blower energy consumption and compost quality. A dynamic model will be a useful tool for optimisation of a rapid process start in zone 1 and the biofilter function in zone 4.

ACKNOWLEDGEMENTS

This work forms part of INTEC (INtegrated TEChnologies for a global management of wastes), a comprehensive development project with the participation of companies from 7 European countries. INTEC is a EUREKA project and Krüger's project is supported financially by the Danish Agency for Development of Trade and Industry, which covers 40% of the costs.

The author especially wishes to thank his colleagues Nils Byager and Svend Broholm, Krüger A/S, for technical computer assistance in connection with the development of the model.

REFERENCES

1. Kjerulf-Jensen, P. (1987): Chapter 5 Humidity, In: Hansen, H.E.; Kjerulf-Jensen, P.; Stampe, O.B. (eds.): Basics of Heat and Climate Technique, pp. 109-128 and Appendix A.05.01-A.05.03. DANVAK Publishers (451 pp.), ISBN 87-982652-1-0. In Danish.
2. Standard Methods for the Examination of Water and Wastewater. 18th Edition 1992. Published by: APHA, AWWA and WEF.
3. Kleff, B.H.A. van; Kuenen, J.G.; Heijnen, J.J. (1993): Continuous Measurement of Microbial Heat Production in Laboratory Fermentors. Biotechnology and Bioengineering, Vol. 41, pp. 541-549 (1993).
4. Haug, R.T. (1993): The Practical Handbook of Compost Engineering, p.412. Lewis Publishers (717pp), ISBN 0-87371-373-7.

Accelerated Composting in Tunnels

Charlotta Lindberg, Gicom Composting Systems B.V., Plein 11-13, 8256 AZ Biddinghuizen, The Netherlands

ABSTRACT

Gicom Composting Systems B.V. is developing, designing and manufacturing computer controlled climatisation units and turn key tunnel composting systems. The key parts in the tunnel composting system are the air re-circulation and the computer managed process control, which together give optimised composting climate in a system that is completely closed, resulting in a biodegradation process that is faster than in other systems and that is proceeding under better control. The Gicom tunnel composting system is able to transform a variety of organic waste streams into compost. For biowaste the processing time is 3-4 weeks, for sewage sludge it is 2 weeks.

Since 1991, some 20 full-scale facilities for composting of biowaste or sludge, designed by Gicom, are in operation or under construction. They are located in the Netherlands, Germany, Spain, United Kingdom and the United States. The treatment capacities range from 8.000 tons/year to 100.000 tons/year.

KEYWORDS

Solid waste management, organic waste, biowaste, sludge, aerobic composting, tunnel composting, composting plants, process control

INTRODUCTION

Gicom Composting Systems BV is developing, designing and manufacturing facilities for accelerated composting in tunnels, with a process based on computer controlled climatisation units. The tunnel composting system was originally developed for the mushroom industry, and the treatment of different waste streams are basically new applications of an existing technique.

Mushroom production is highly depending on the quality of the growing substrate, the compost. There is a direct relation between compost quality and the amount of mushrooms growing per unit area, and thus the income from the crop. It is therefore of great importance to create conditions that allow sufficient process control to obtain a stable, high quality compost. This has led to development from open, extensive and marginally controllable ways of composting, to the closed tunnel system. The personal computer has made continuous control feasible. The investments that are required can be justified by improved compost quality, a faster composting process and a higher mushroom production. The introduction of the tunnel prevented infection from harmful bacteria and seeds, with the organic material loaded into the tunnel from one side and the compost unloaded from the other side.

The tunnel composting process is completely closed and the process control is optimised, which results in a very short processing time. The Gicom tunnel composting system is able to transform a variety of organic waste streams into compost, such as:

- Biowaste (source separated domestic waste)
- Municipal solid waste
- Sewage sludge
- Industrial sludges
- Manure
- Anaerobic digestion process residues
- Contaminated soil

Since 1991, some 20 full scale facilities for composting of biowaste or sludge, designed by Gicom, are in operation or under construction. They are located in the Netherlands, Germany, Spain, United Kingdom and the United States. The treatment capacities range from 8.000 tons/year to 100.000 tons/year.

SYSTEM DESCRIPTION

The Gicom tunnel composting system is based on accelerated batch composting in a closed vessel: the tunnel. The tunnel is divided by a patented floor into two parts; the aeration plenum and the process room. Each tunnel is handled as one separate unit, with it's own control and air handling systems.

Biowaste, or source separated household waste, is transformed into compost in two steps of together three to four weeks.

- Precomposting phase   1-2 weeks
    - mass reduction   45%
- Postcomposting phase   2 weeks
    - mass reduction   10%
- Overall mass reduction   60%

After a total process period of a minimum of three weeks the compost produced out of biowaste has the following characteristics:

- Dry matter   70%
- Organic matter   30%
- C/N ratio   12-15
- pH   7.5
- No pathogens   according to Dutch standard tests

Sewage sludge composting is made in one step, normally using wood chips as bulking agent.

Composting period — 2 weeks
- dry matter content in sludge — 18-20%
- dry matter content in compost — 60%
- reduction of organic matter — 60%
- overall mass reduction — 85%

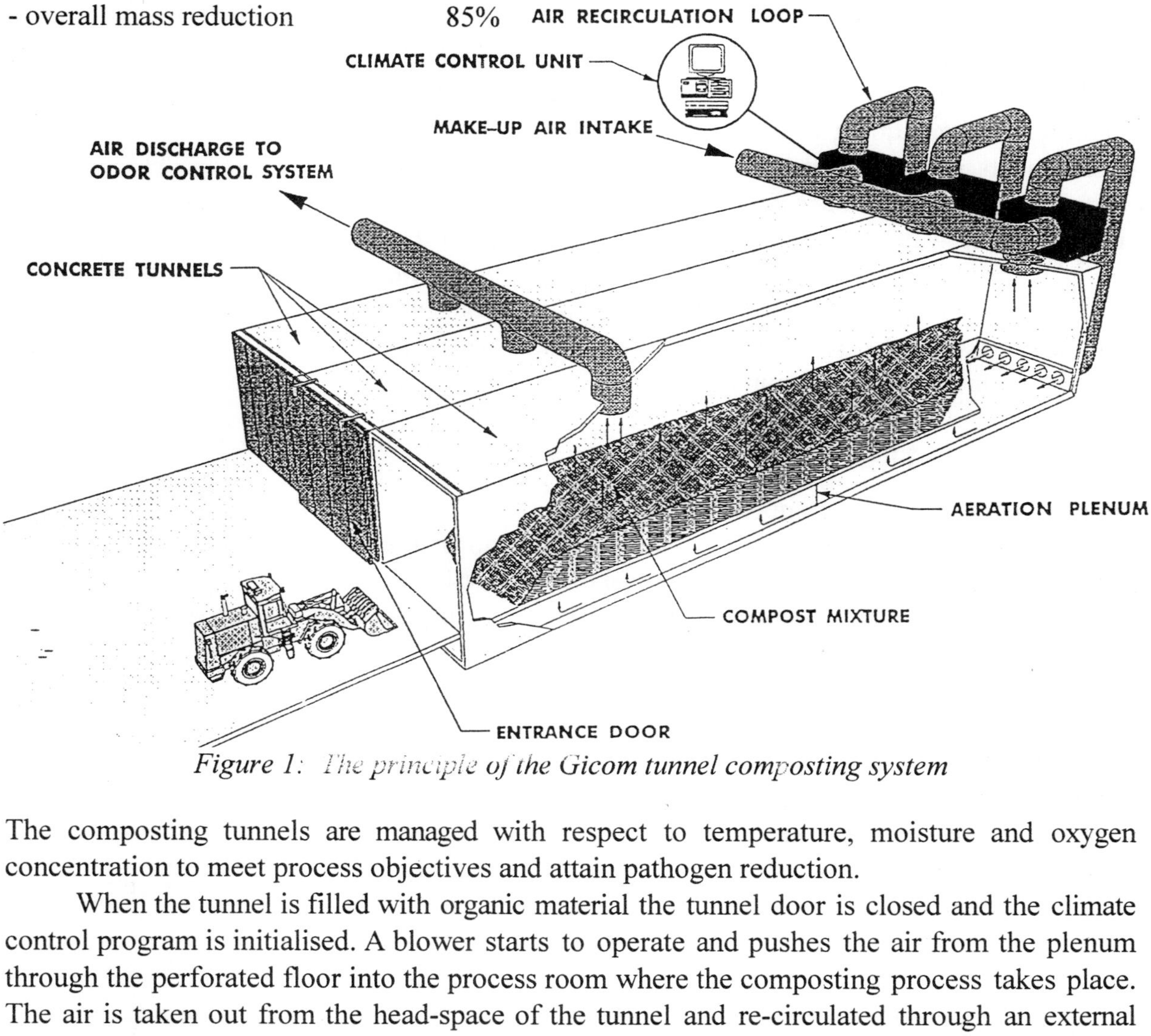

Figure 1: The principle of the Gicom tunnel composting system

The composting tunnels are managed with respect to temperature, moisture and oxygen concentration to meet process objectives and attain pathogen reduction.

When the tunnel is filled with organic material the tunnel door is closed and the climate control program is initialised. A blower starts to operate and pushes the air from the plenum through the perforated floor into the process room where the composting process takes place. The air is taken out from the head-space of the tunnel and re-circulated through an external ductwork back to the blower. If needed, fresh air can be introduced, and at the same time an equal amount of process air is exhausted to the odour control units. Those consist of a scrubber and biofilters.

By re-circulating the air, the aeration capacity per unit floor area in the tunnel is very high, more than twenty times greater than in comparable systems. The high air flow ensures fully aerobic conditions within the tunnel. After one to two weeks the material is taken out of the tunnel and screened. This agitation and a homogenisation means that additional mechanical turning of the compost not is needed. Due to the re-circulation of the air, the need of fresh air input and thus the need of exhausting used process air can be limited, despite the high total airflow. Heat can be taken out by a heat exchanger. The air re-circulation also promotes capture and breakdown of any volatiles produced.

A composting process is creating a very aggressive and corrosive environment, due to high temperature, high humidity and raised levels of CO_2 and NH_3. However, with the tunnel system, these conditions remain within the tunnel, and they do not affect the working conditions or the building structures. Since there are no electrical or mechanical parts installed in the tunnels, the maintenance costs are limited.

The condensate and percolate produced during the composting process are also re-circulated. The water is used both for adding moisture to the compost within the tunnel by use of a sprinkler system, and in the scrubber for moistening the exhaust air prior to the biofilters. Normally when composting biowaste, the only water leaving the plant is evaporated and discharged via the biofilters. The environmental impact of a Gicom tunnel composting facility is thus minimised.

PROCESS CONTROL

To obtain an optimum composting climate, it is necessary to control both temperature, moisture and oxygen concentration. The re-circulation of the process air together with the computer controlled process operation make that possible. The climate in each tunnel is measured and individually regulated by the process control computer. For each tunnel, the following parameters are monitored every two minutes:

- Compost temperature,
- air temperature,
- air humidity,
- oxygen/carbon dioxide concentration,
- fresh air flow,
- re-circulation air flow,
- static pressure.

During the composting process, all measured values can be shown in graphs on the computer screen. It is also possible to show graphs for the following calculated parameters:

- Oxygen consumption   (g/kg; g/h),
- total amount of used oxygen   (ton),
- total amount of water evaporation   (g/kg; g/h),
- total amount of evaporated water   (ton),
- energy content of air flow   (J/kg; J/h),

- total amount of emitted energy (J),
- total amount of circulated air (m^3),
- total amount of fresh air input (m^3),
- water content of composting matter (%).

The user can call for any possible combination of the mentioned parameters.

Normally, the primary control parameter is the compost temperature with moisture and oxygen level next in importance. The results of the measurements are sent to the Gicom process computer, where they are compared with setpoint data. On basis of these comparisons, the process computer controls the climatisation units. By measuring every two minutes, accurate control is possible, ensuring an optimising of the climate and a rapid composting process. An optimised water content in the compost is essential for the process. With the computer control a more accurate figure of the actual water content can be given than with any other system.

Table 1: Characteristics of compost produced at the Bladel facility

Parameter	Compost quality after 3 weeks	Dutch Standard clean compost 1995	Unit
Dry matter	70		%
Organic matter	27.7	>20	% d.m.
Nitrogen (N)	13.3		mg/kg d.m.
Phosphate (P_2O_5)	5.6		mg/kg d.m.
Kalium (K_2O)	12.8		mg/kg d.m.
Magnesium (Mg)	4.9		mg/kg d.m.
Cadmium (Cd)	0.47	0.7	mg/kg d.m.
Chromium (Cr)	11	50	mg/kg d.m.
Copper (Cu)	23	25	mg/kg d.m.
Nickel (Ni)	3	10	mg/kg d.m.
Lead (Pb)	35	65	mg/kg d.m.
Zinc (Zn)	72	75	mg/kg d.m.
Arsenic (As)	2	5	mg/kg d.m.
Mercury (Hg)	0.06	0.2	mg/kg d.m.
Conductivity	2.6	< 4.5	mS/cm 25°C
Cl-value	2,900	< 4,000	mg/kg d.m.
pH-H_2O	7.6		
Stone >5 mm	0.6	< 3	mg/kg d.m.
Glass >16 mm	0.5		
Stability	V	> IV	*
Parasite/pathogen test	passed all tests	meet criteria	*
Weed test	1	< 2	*

* Standardised tests in The Netherlands and Germany

All relevant process data are stored in the computer, and can thus be checked later on.

All computer systems used, both software and hardware, are developed by Gicom. The climatisation unit, which is also used in the mushroom growing industry, is used at more than 180 plants all over the world.

PROCESS FLOW

The waste is passing through a number of different stages during it's transformation into compost. The process flow for biowaste composting is as follows.

1. *Pre-treatment*. When entering the composting plant, the waste is screened in a coarse-meshed (100 mm) rotary screen. The overflow is shredded and mixed again with the incoming waste. No bulking agent need to be added.

2. *Loading*. From the screen, conveyors can bring the waste to an automatic filling cassette. The filling cassette has an telescopic conveyor which is loading the waste into the tunnel in thin, even layers. The filling cassette is equipped with a camera and can be managed from the control room. Loading can also be done by a front end loader.

3. *First stage composting*. The first stage composting takes place in the tunnel during about 10 days. After this period the compost is practically free of odours. During the first phase of the composting, pasteurisation takes place at a pre-set temperature and during the required time. Thereafter the tunnel is cooled down to desired process temperature, normally 48°C. This temperature is kept throughout the composting process. Before opening the tunnel door, the tunnel is cooled down to room temperature.

4. *Unloading/second treatment*. The fresh compost can be removed from the tunnel by use of an automatically operating unloading machine, using a pulling mat, or by a loader. The compost is transported to the treatment area, where it is screened in a 50 mm rotary screen. The oversized materials are mixed with the incoming waste stream. Ferrous metals are removed.

5. *Second stage composting*. The screened compost is brought back to the tunnel, where the second stage composting takes place during two weeks. Normally, water need to be added during this stage.

6. *Final treatment*. The stable compost is undergoing a third screening (12-25 mm) and a second recovery of ferrous metals. It is transported to a storage area or directly delivered to the user.

After 3 weeks of composting, 1000 kg organic waste has been converted into 400 kg compost with the following qualities:

- Organic matter   35%,
- dry matter   70%,

- pH   7.5
- no human pathogens,   according to Dutch standard tests,
- no phytopathogens,   according to Dutch standard tests.

The compost meets the Dutch standards for very clean compost, part from the values of zinc and lead that often are to high.

FEATURES OF THE GICOM TUNNEL COMPOSTING SYSTEM

All factors mentioned in the previous text form the possibility for the user of the Gicom tunnel composting system to run the composting process under optimum (or self-chosen) conditions. Some of the advantages with the system are listed below.

- Excellent process control, which minimises the retention time and thus the size of the facility. The optimised process conditions ensures a fast degradation of easily biodegradable material, which results in a nearly odourless product after only one week of composting. It is a highly sophisticated yet easy to manage process control system.
- Flexibility towards input material and end products and thus towards different markets. Since the tunnels work individually, is it possible to process different types of waste and/or produce different types of compost simultaneously with different feed stocks.
- Sophisticated odour control by use of scrubber and biofilters. Due to the forced aeration with a very high airflow there is a very small risk for local anaerobic conditions that may create smelling H_2S. Re-circulation of air promotes capture and breakdown of volatiles. The Gicom composting plants meet the Dutch standards for odour control, which due to the high density of the population are among the strictest in the world (10 ou/m^3 as 98 percentile).
- For biowaste composting there is no waste water effluent. All percolated water is brought back to the tunnel, and water is leaving the system only in it's evaporated form via the biofilter.
- No devices in the process room - only the tunnel walls are exposed to the aggressive process air. At the same time, this aggressive surrounding is not part of the working environment, and it does not affect the building structure.
- The process handling can be made fully automatically.
- Process energy can if desired be recovered by use of heat exchangers. At Dutch conditions the process energy consumption is about 35-45 kWh/ton input.
- Up-scaling is uncomplicated since a plant is built up by a number of relatively small, individually controlled units.
- The system is suitable for a wide range of capacities.
- Gicom has a world wide experience with full scale plants.

Each Gicom composting facility is individually designed, to suit the needs of the client, including waste stream quantity and quality, the compost quality required and the actual economical situation.

CONCLUSIONS

The first full scale Gicom tunnel composting plants for biowaste and sewage sludge have now been successfully running for a couple of years. Tests with new applications, such as composting manure and digestion residues, are very promising.

The sophisticated process control minimises the time required for achieving a stable compost. The modular system provides the facility with flexibility regarding input materials and required compost qualities, and simplifies upscaling.

The re-circulation of air and water ensures optimum composting conditions within the tunnels, and has together with the odour control system as result that the environmental impact of the facility is minimised.

The completely closed tunnels protect the building structure and the workers from corrosive and foggy air.

REFERENCES

1. Finstein, M.S (1993). Guide to matching composting technology to circumstance, Composting Frontiers, winter 1993
2. Finstein, M.S, and Hogan, J.A (1993). Integration of composting process microbiology, facility structure and decision-making. In H.A.J Hoitink and H.M. Keener (eds.) *Science and Engineering of Composting: Design, Environmental, Microbiological and Utilization Aspects*, Proceedings of an International Composting Research Symposium, Ohio Center, Columbus, Ohio, USA, March 27-29, 1992, Renaissance Publications, pp 1-23.
3. Miller, F.C. (1993). Minimizing odor generation. In H.A.J Hoitink and H.M. Keener (eds.) *Science and Engineering of Composting: Design, Environmental, Microbiological and Utilization Aspects*, Proceedings of an International Composting Research Symposium, Ohio Center, Columbus, Ohio, USA, March 27-29, 1992, Renaissance Publications, pp 219-241.
4. Beordelingsrichtlijn (1993), KIWA, BRL-K256/01, 1993-02-26 (Dutch standards for compost quality, in Dutch)

ATAD - an Effective Technology for Stabilisation and Disinfection of Biosolids

Hans-Gerd Schwinning, ChemEng, Leonhard Fuchs, Principal, Fuchs Gas- und Wassertechnik GmbH, P.O. Box 13 62, D-56703 Mayen, Germany

ABSTRACT

Autoheated thermophilic aerobic digestion (ATAD) of biosolids like sewage sludge or liquid manure is a technology widely applied in Europe, especially in Germany. The process operates typically at a 6 day hydraulic retention time and is capable of achieving a high degree of stabilisation and pathogen reduction. Furthermore ATAD is characterised by high reaction rates which are achieved at thermophilic temperatures ranging from 50 - 60°C. In the completely mixed and aerated environment, thermophilic temperatures are attained and sustained without the addition of supplemental heat by conserving the heat released during biological degradation.

KEYWORDS

Autoheated thermophilic aerobic digestion, ATAD, biosolids, liquid manure, pathogen reduction, sewage sludge, stabilisation.

INTRODUCTION

Increasing amounts of produced biosolids can be expected because of the introduction of more stringent effluent limitations, new wastewater technologies etc. Exploding costs are a result of limited space for disposal as well as required expensive technology before disposal is allowed. For economical as well as ecological reasons it is desirable to return biosolids into the natural cycle. Untreated biosolids like sludge from wastewater treatment plants, manure from animals usually contain putrescible substances as well as pathogenic forms of bacteria, viruses, worm eggs etc. To prevent negative effects on the environment these biosolids have to be treated before disposal or beneficial use. A highly effective technology for achieving stabilisation and disinfection is the autoheated thermophilic aerobic digestion, or ATAD. Such systems have been operated at full-scale in Europe for nearly 20 years.

In 1968 Mr Hubert K.E. Fuchssen., founder of the Fuchs company, observed the strong autothermal conditions during the aeration of liquid swine and poultry manure. This was the beginning of what later was called the ATAD process. Pöpel, Rüprich and Strauch [1] described the use of Circulation Aerators in sewage sludge and liquid manure applications and the attended disinfection performance of the ATAD process in 1970. During the following years several investigations carried out by research institutes and universities were conducted leading to the first commercial size ATAD installation in Vilsbiburg, Germany [2]. The Vilsbiburg ATAD was designed for a population equivalent of 22,000. The Vilsbiburg ATAD has operated successfully for 18 years and remains in service today. This first installation was followed by over 50 other ATAD plants which are mainly located in Germany, with others are located in Austria, Canada, Great Britain, Norway, Switzerland and the US.

PROCESS DESCRIPTION

The biological stabilisation of biosolids from sewage treatment plants is based on the reduction of organic substances contained in the sewage sludge. With the ATAD technology the reduction of these materials is carried out by aerobic microorganisms. As the aerobic energy- metabolic exchange takes place exothermically, the biochemical oxidation of the organic substances releases energy, mainly in the form of heat. Therefore final products are low energy components such as H_2O and CO_2. Because the whole process develops under aerobic conditions, no potential explosive gases are produced. Exhaust gas can be treated effectively in biofilters, which are a combination of scrubbers and a bark filters [3]. Efficient retention of the heat that is released during digestion results in high operating temperatures (> 50°C) which internally results in high degradation rates of volatile solids as well as pathogens. ATAD systems are typically configured with 2 reactors operating in series (figure 1). The system is batch fed once per day after which the reactors are isolated - a practice which has important pathogen control implications.

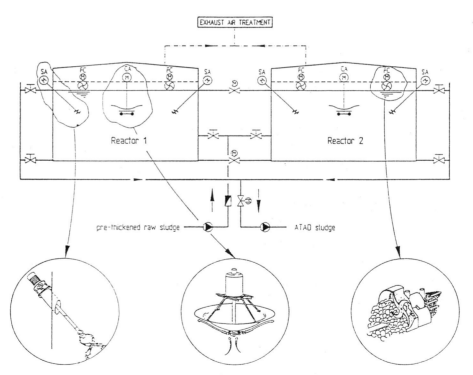

Fig. 1: Typical 2-stage ATAD system with Spiral and Circulation Aerators as well as Foam Controllers

In general biosolids in form of primary as well as waste activated sludge, trickling filter sludge, a mixture of the above or liquid manure are suitable for the application of aerobic thermophilic treatment. Various parameters have to be taken into consideration to utilise the energy which is released during the aerobic degradation of volatile solids for heating up the substrate to thermophilic temperatures ($\geq 50°C$).

The following conditions are necessary to operate the process autothermally at thermophilic temperatures:

- Insulated and covered reactors to minimise heat loss.
- Volatile solids content in the untreated biosolids of at least 2.5 - 3.0% to ensure high degradation rates (if no heat exchanger is used).
- Highly efficient aeration and mixing equipment to ensure an aerobic fully mixed reactor environment and to limit heat loss with the exhaust.
- A foam control system to limit the foam layer in the reactors.

An effective process is obtained with air aeration using self-aspirating aerators (Spiral Aerators, Circulation Aerators) as illustrated in figure 1. These machines have demonstrated their effectiveness during the investigation and development of the ATAD technology. The Spiral Aerator is mounted tangentially through the wall of the reactor and provides vertical as well as horizontal mixing. This aerator has no wetted bearings and a stainless steel shaft to prevent corrosion. The strong mixing results in the rotation of the whole tank contents. Circular reactors above a specified diameter include the addition of a centrally mounted Circulation Aerator to prevent settling in the centre of the tank. This aerator generates a vertical loop which is overlapped by a rotation induced by the side wall aerators, so that the final flow pattern represents a spiral.

Aeration of the substrate results in a rapid development of a dense foam layer which acts as an insulation for the reactor and also leads to an improvement in the utilisation of the oxygen as well as an increase in the biological activity [4]. Foam is delivered to the Foam Controller as a result of the rotation of the underlying sludge and is physically broken up to form a more dense compact layer. The use of special Foam Controllers establishes the formation, depth and to some extent the density of a defined foam layer in the reactors. Therefore the foam controlling system is a substantial aspect of the ATAD technology.

PATHOGEN REDUCTION

The metabolism of aerobic thermophilic microorganisms leads to an inactivation of pathogenic forms of bacteria, viruses, worm eggs etc. On basis of their investigations, Strauch and his colleagues [5, 6] stated that after a treatment of minimum 23 hours at $\geq 50°C$ and a pH-value of above 8 leads to a product which meets typical disinfection requirements for biosolids. Besides this time-temperature-function it is also possible to operate the process at 55°C for

10 hours or at 60°C for 4 hours [7]. Operating the process at one of these time-temperature function results in a product where one gram of sludge contains no salmonella and less than 1,000 enterobacteriaceae.

To prevent short-circuits which could lead to a bleed through of pathogens ATAD systems have to be operated in the batch mode. To maximise the reduction of pathogens ATAD plants typically require a minimum of 2 reactors configured in series (figure 1). In the first stage temperatures are usually in the low thermophilic range (40 - 50°C). Maximum disinfection is achieved in the second stage where temperatures may range between 50 and 60°C. The daily treated sludge is discharged only from the second stage. After discharge has been completed, raw sludge is fed into the first stage while partially treated sludge is displaced into the second reactor. After feeding the reactors remain isolated for 23 hours while thermophilic digestion progresses.

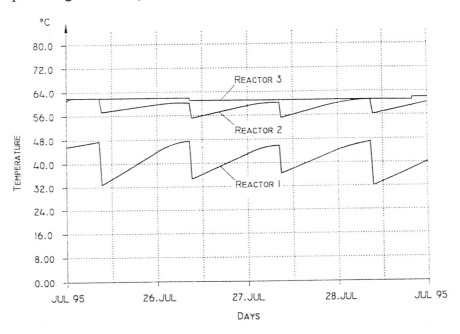

Fig. 2. ATAD Daily Temperature Cycle

Typical long term temperature curves at the two-stage ATAD systems show that the first reactor operates at mesophilic temperatures (35°C-45°C) while the second operates at thermophilic temperatures (50°C-65°C).

During the discharge-feed-cycle the temperature changes in the first reactor. While cold raw sludge is fed to the first reactor (unmixed during feeding) a temperature drop occurs (figure 2). After aeration and mixing have started again, the temperature rises to a maximum value until the next discharge-feed-cycle is initiated. The displacement of sludge from reactor 1 into the following reactors causes only a slight decrease in temperature while in the last reactor nearly constant temperature conditions exist throughout the feed cycle. Wastewater treatment plants operating without primary clarifiers and low f/m ratios usually produce biosolids with a lower fraction of degradable volatile solids. As a result thermophilic temperatures, which are necessary to meet the disinfection requirements, are not always ensured by self-heating without taking additional measures to conserve or recycle heat that is generated during treatment. The lower heat production can be compensated by an internal heat exchange (figure 2). The use of such an internal heat exchanger is applied in a three-stage ATAD design (Europ. Pat. No. 0384162). These in-the-tank integrated heat exchangers are relatively simple to construct and are a very cost-effective application.

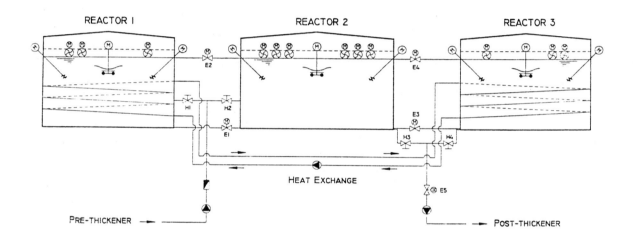

Fig. 3: Typical 3-stage ATAD system with heat recovery

Several studies have been conducted to evaluate the pathogen reduction performance of ATAD [8 - 11]. One investigation carried out at the Ellwangen ATAD facility, Germany, (table 1) was designed to evaluate pathogen reduction at temperatures between 47.5 and 50°C. The minimum temperature requirement for disinfected sludge in Germany is 50°C for a 23 hour treatment period. To limit the temperature to the requested range an integrated heat exchanger in reactor 2 was used for cooling. Subject of the evaluation in the study were the high resistant fecal streptococci, enteroviruses, salmonella and enterobacteriaceae amongst others. The content of fecal streptococci were reduced to a non-detectible level although the temperatures were not higher than 50°C. Regarding the reduction of salmonella all samples of the ATAD sludge were reported as 'non-detect'. Even samples taken from reactor 1 were reported 'non-detect' for salmonella on multiple occasions. Enteroviruses was also found to be non-detectible in the ATAD sludge during this study. The concentrations regarding the enterobacteriaceae were down to non-detectable levels on multiple occasions and always below the requirements of 1,000 enterobacteriaceae per gram sludge in the discharged ATAD sludge.

*Table 1: Summary of Fecal Straptococci, Salmonella, Enterovirus and Enterobacteriaceae Removal at Ellwangen, Germany * [8, 9]*

Day	Fecal Streptococci CFU/ml			Salmonella			Enterovirus PFU/ml			Enterobacteriaceae CFU/ml		
	Raw	R 1	R 2	Raw	R 1	R 2	Raw	R 1	R 2	Raw	R 1	R 2
1	4.6×10^4	1.3×10^4	4.9×10^2	pos.**	pos.	N.D.***	1.4×10^4	8.0×10^3	N.D.	1.9×10^5	6.3×10^4	2.5×10^2
2	6.1×10^4	6.0×10^3	2.3×10^2	pos.	pos.	N.D.	1.4×10^4	N.D.	N.D.	5.4×10^5	1.2×10^4	2.0×10^1
3	3.3×10^4	3.4×10^3	N.D.	pos.	N.D.	N.D.	4.6×10^3	2.0×10^3	N.D.	4.1×10^5	9.0×10^3	N.D.
4	6.6×10^4	3.4×10^3	4.0×10^1	pos.	N.D.	N.D.				1.1×10^6	1.0×10^4	2.0×10^1
5	6.1×10^4	2.3×10^3	2.0×10^1	pos.	N.D.	N.D.				1.5×10^6	9.8×10^3	3.0×10^2
6	8.0×10^4	5.5×10^3	1.0×10^1	pos.	N.D.	N.D.				2.0×10^6	9.5×10^1	N.D.
7	1.1×10^5	1.7×10^4	3.0×10^1	pos.	pos.	N.D.				2.1×10^6	1.0×10^4	N.D.
8	6.5×10^4	4.8×10^4	2.0×10^2	pos.	pos.	N.D.				6.6×10^6	6.5×10^5	2.0×10^0
9	1.4×10^5	4.2×10^4	3.6×10^2	pos.	pos.	N.D.				1.1×10^5	1.0×10^4	1.3×10^2
10	2.1×10^5	2.1×10^4	N.D.	pos.	pos.	N.D.				4.8×10^6	8.0×10^4	5.0×10^1

* Reactor 1 temperature range = 35 - 43°C, Reactor 2 temperature range = 47.5 - 50°C
** Pos. = Positive detection of specified parameter
*** N.D = Non-detection of specified parameter

Another investigation was carried out with cow manure at the ATAD plant in Gemmingen, Germany [12]. During the running-in phase of 9 days no manure was fed to the ATAD (figure 4). When temperatures of above 50°C were achieved, the ATAD was operated in the batch feed mode as described above. The retention time was 6 days. During the whole period of investigations temperatures in the second reactor were always above 50°C. Figure 5 shows the content of fecal streptococci. The numbers relating to reactor 1 and 2 are based on samples taken before the discharge-feed-cycle. Salmonella have never been detected in reactor no. 2 and often have not been detected in reactor no.1.

VOLATILE SOLIDS REDUCTION

In general dry solids concentrations of the biosolids fed to the ATAD should be limited to maximum 6%. In case that mechanical pre-thickening facilities in combination with polymers are used, it might be necessary to operate the ATAD with a lower dry solids concentration. Figure 6 illustrates the performance of the ATAD installation located in Fassberg, Germany, which has been in service for over 10 years.

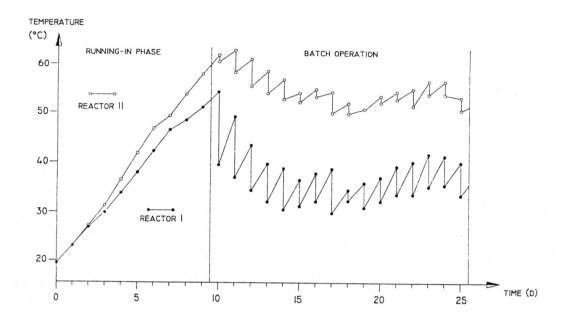

Fig. 4: ATAD reactor temperatures measured at Gemmingen, Germany, during digestion of cow manure [12]

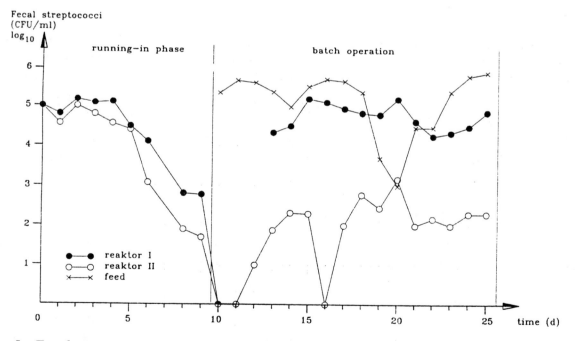

Fig. 5: Fecal streptococci reduction at ATAD Gemmingen, Germany, during digestion of cow manure [12]

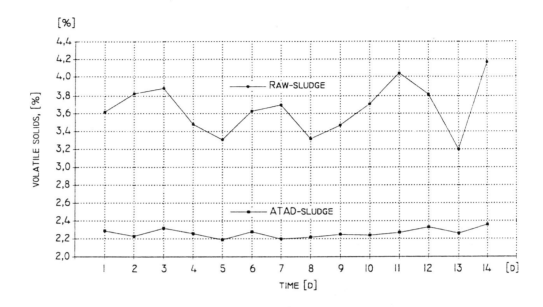

Fig. 6: ATAD volatile solids reduction (Fassberg, Germany)

During the observed period the average dry solids content of the fed sludge was 4.43% while the volatile solids fraction was 82.6%. The digested sludge consists of an average dry solids content of 3.09% while the volatile solids fraction was 73.1%. This data indicates that 38% volatile solids reduction was achieved. The average retention time was 7.3 days and the maximum temperatures registered in the second reactor were between 55 and 63°C. Volatile solids degradation during the digestion of liquid pig and cow manure were reported by Oechsner and Rüprich [13]. They found out that pig manure provides a higher volatile solids degradation than cow manure.

ECONOMIC EFFICIENCY

In general, the economic efficiency of the autoheated thermophilic aerobic digestion process has to be seen in connection of the whole wastewater treatment facility. Especially at small and medium sized facilities ATAD represents the most economic technology regarding total costs (investment and running costs, interest payments). For the economic efficiency also the efficiency of the aeration equipment is important. The total energy input should not exceed 15 kWh/m^3 of sludge throughput (4% dry solids). Besides the low total costs also the trouble-free operation and the high stability of the process speaks for the ATAD technology.

CONCLUSION

The ATAD process has already demonstrated its performance in Europe for more than 15 years. Renewed attention has been focused on ATAD as a process

- capable of achieving simultaneous stabilisation and high rate pathogen reduction,
- which is very stable and
- with low capital costs because of short retention times and providing
- expansion as well as operation flexibility,

which finally leads to an economical biosolids management approach while providing the opportunity to use the final product in beneficial re-use programs.

REFERENCES

1. Pöpel, F., D. Rüprich, W. Strauch, W. Müller and E. Best, E. (1970). Flüssigkompostierung von Flüssigmist und Abwasserschlamm durch Umwälzbelüftung. Landtechnische Forschung 18, No. 5.
2. Strauch, D., H. Wassen, H. Schwab, R. Schaffert, R. Nagel, O.C. Straub (1975). Das Umwälzbelüftungsverfahren (System FUCHS) zur Behandlung von flüssigen tierischen und kommunalen Abfällen. 1.-6. Mitteilung Berl. Münch. Tierärztl. Wschr. 88, pp. 456-460.
3. Bubinger, H., H.-G. Schwinning (1992). Grundlagen und Anwendungsbeispiele der Biofiltertechnologie, 5, pp. 60-70.
4. Wolinski, W.K., A.M. Bruce (1984). Thermophilic oxidative sludge digestion.
5. A critical assessment. Presented at the IFAT conference, Munich 1984.
6. Strauch, D., R. Böhm (1979). Hygienische Probleme der Abwasseraufbereitung und ein neuer Weg zu ihrer Lösung. Forum Mikrobiologie 2, 3, pp. 121-126.
7. Rüprich, W., D. Strauch (1984). Technologische und hygienische Aspekte der aerob-thermophilen Schlammstabilisierung. Korrespondenz Abwasser 31, 11, pp. 946-952.
8. Arbeitsbericht der ATV/VKS-Arbeitsgruppe (1988). Entseuchung von Klärschlamm. Korrespondenz Abwasser 35, 1, pp. 71-74
9. Jakob, J., H.-J. Roos and K. Siekmann (1987). Aerob-thermophile Schlammstabilisierung unter dem Aspekt der Entseuchung. Ein Bericht zum 5. Bochumer Workshop 'Entseuchung von Klärschlamm'. Schriftenreihe Siwawi, Bd. 10, Universität Bochum, pp. 119-149.
10. Strauch, D. (1988). Aerob-thermophile Schlammstabilisierung - Hygiene -. In Bericht des 2. Hohenheimer Seminars 'Entseuchung von Klärschlamm'. Deutsche Veterinär-medizinische Gesellschaft (Hrsg.), pp. 103-133.
11. Braun, H.-J., E. Zingler (1986). Experiences with a Full-Scale Aerobic-Thermophilic Sludge Treatment Plant. Published at 'International Recycling Congress Berlin Oct. 29-31, 1986'.
12. Strauch, D. (1980). Weitere Untersuchungen an einem Verfahren zur aerob-thermophilen Schlammstabilisierung. 1.-8. Mitteilung, gwf-Sonderdruck, 121, pp. 24-27.
13. Rückert, V. (1991). Mikrobiologische Untersuchungen zur aeroben und anaeroben Flüssigmistbehandlung. Dissertation im Fachbereich Tierhygiene, Universität Hohenheim.
14. Oechsner, H., W. Rüprich (1989). Aerob-thermophile Behandlung von Flüssigmist. Landtechnik, 44, pp. 328 - 330.

New Concepts for Biowaste Composting - Dutch Current Practice

Thijs Oorthuys and Heijo Scharff, GRONTMIJ Consulting Engineers, P.O.Box 203, 3730 AE De Bilt, The Netherlands

ABSTRACT

Dutch biowaste is separately collected and converted into recyclable compost. Most plants apply aerobic composting systems, either hangar composting or tunnel composting. Based on operational data the conclusion is drawn that aerobic composting is a reliable and flexible treatment method to generate high quality compost, at acceptable low emissions as well as low costs. A mass reduction of 60% may be achieved.

KEYWORDS

Waste management; hangar composting; tunnel composting; operational aspects; compost quality; compost marketing.

INTRODUCTION

In the Netherlands today about 50 to 70% of household biowaste (vegetable-, fruit-, and garden waste: 'VFG') is collected separately. An estimated amount of 1.3 to 1.4 million tonnes of collected VFG requires treatment. In addition to the separate collection of household biowaste, biowaste also originates from commercial activities, and from public cleansing services, such as park and market wastes. Therefore, in the near future the annual amount of separately collected biowaste from household origin and related activities will be approximately 1.9 million tonnes.

In Figure 1 the seasonal fluctuation of the input to Dutch household biowaste composting plants is shown. The input of the garden waste fraction in the biowaste is the main reason for the seasonal fluctuation. This component drops to nil between November and the end of February. During March the input rapidly increases and may reach its peak during April. Thereafter until October, the load stays at a level of over 120% of the yearly average.

Therefore, the origin of the biowaste in winter time is mainly kitchen waste, which means that the nature can be described as very wet, dense, and hardly containing any structure building material. One may conclude that under Dutch circumstances the operation of composting plants is considerably influenced by seasonal fluctuations, and therefore sufficient flexibility is absolutely necessary.

The treatment aims at the prevention of landfilling or incineration of biowaste, and the production of recyclable compost. In Table 1 an overview of Dutch composting plants, either operational or under construction, is provided.

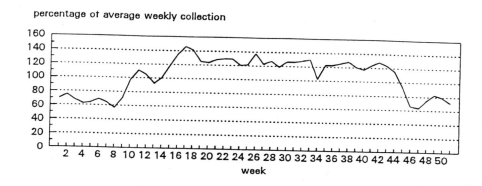

Fig. 1: Typical weekly biowaste load of a Dutch composting plant

Table 1: Composting plants in The Netherlands [1]

Composting systems	Composting capacity (ktpa) (*)			Total number of plants
	At the end of 1994	Realisation after 1994	Total	
Hangar composting	684	240	924	8
Tunnel composting	389	170	559	11
Other aerobic systems	264	11	275	4
Anaerobic fermentation	40	65	105	3
Total	1,377	486	1,863	26

(*) ktpa = ktonne/annum

The ultimate treatment capacity matches the required capacity of approximately 1.9 million tonnes per annum. The first Dutch full scale aerobic composting plant treating household biowaste was a 25 ktpa hangar composting plant that came into operation in December 1990. The present Dutch biowaste composting services are provided at a regional level with an average treatment capacity of the composting plants at about 60 ktpa.

Hangar composting and tunnel composting are the most common methods applied in the Netherlands to produce recyclable compost. From these new concepts, two cases of existing composting plants designed by Grontmij are selected for presentation.

Figure 2 shows the estimated treatment costs per tonne of biowaste for different treatment systems. It may be noted that hangar composting and tunnel composting show a difference in costs in favour of the latter for plants of a capacity less than 50 ktpa.

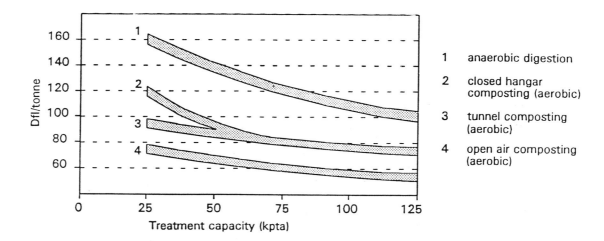

Fig. 2: Annual costs of biowaste treatment in the Netherlands (note: Dfl 100 = DKK 360)

A more detailed specification of investment and operational costs is given in Oorthuys and Scharff [8]. In practice the difference between open air composting and indoor composting tends to become smaller than indicated in Figure 2. A combination of required operational improvements and additional provisions to reduce emissions, have increased the costs.

The reasons for the popularity of the aerobic composting systems are:

- The system is very cost effective;
- the process is simple and easy to control;
- the operation of the aerobic composting systems is very flexible to fluctuations in biowaste input and composition;
- production of polluted wastewater is minimal (if any);
- remaining environmental impact, mainly odour, is limited.

AEROBIC COMPOSTING

In aerobic biowaste composting, aerated biological processes convert organic material into a stable compost. The following main steps in the process flow diagram of a Dutch indoor composting plant are shown in Figure 3:

- Weighing and acceptance of the biowaste delivered at the plant;
- unloading and pre-treatment of the biowaste, if required;
- the biological treatment of the biowaste, i.e. the actual composting process;
- treatment of the process emissions to minimise the environmental impact;
- mechanical treatment to refine the compost;
- normally, residues are used for landfill;

- compost is kept in storage for some time because the ready compost cannot be marketed the entire year round.

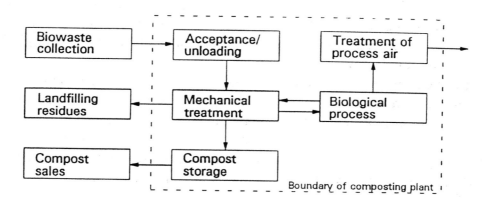

Fig. 3: Process flow diagram aerobic composting

Aerobic composting generally leads to a total mass reduction of about 60% when the process is properly controlled and a sufficient retention time is maintained for generating high quality VFG-compost. This is demonstrated by the evaluation of data obtained from various Dutch composting plants in operation, including both cases presented in this paper. The amount of compost is about 35% of the VFG-input, and the rejects around 5%.

HANGAR COMPOSTING

General

In the Netherlands the largest existing hangar composting plant has a capacity of 300 ktpa. The capacity of the other hangar composting plants ranges now from 45 to about 100 ktpa, with an average around 57 ktpa.

Normally, at a hangar composting plant the following sections, which are fully indoor, can be identified:

- The area for unloading, shredding and sieving of the biowaste;
- the composting hangar, and
- the compost refinement (mainly sieving) and compost storage area.

The closed composting hangar contains an elongated, continuous aerating bed. The aerating bed consists of perforated ducts embedded in a layer of gravel, covered with a layer of wood-chips. After pre-treatment, the material to be composted is placed in piles on top of the wood-chips. Ventilators blow fresh air through the aerating bed into the piles. Once every week, the piles are turned and moistened (if required), and moved further on down the hangar by an automatic turning machine.

Operational experiences of the SOW composting plant [3]

The first case to be presented in this paper is the hangar composting plant from the West Friesland Regional Waste Authority (SOW) situated close to the city of Medemblik in the Province of North-Holland. This plant applies the composting system from the Swiss firm Bühler. An extensive process description can be found in Oorthuys and Koning [2].

The composting plant, with a design capacity of 24.8 ktpa of biowaste, was designed in close co-operation between Grontmij and Bühler. The plant's operation started up in December 1990 and in March 1991 the first compost was produced. Because the operation was very successful and the actual load increased to 40 ktpa in the second year of operation, it was decided at the end of 1992 to enlarge the plant's capacity in order to provide composting services for a larger region. This was done by erecting a second composting hangar, which has been operational since June 1994. Now the net capacity is 60 ktpa of VFG, with a maximum capacity of 80 ktpa.

In Table 3 the cumulative data during the first three years of operation are summarised. The table also shows the relevant data from the second composting plant - the OGAR tunnel composting plant in Groningen - which is presented hereafter.

Table 3: Summary of operational data from sow composting plant and ogar composting plant

		SOW composting plant [3]		OGAR composting plant [6]	
		1991 to 1993		1994 (*)	
		(tonnes)	%	(tonnes)	%
Input:	biowaste	99,153	98.8	20,098	100.0
	bulking material	5,942	5.9	0	
	total input	105,094	104.7	20,098	100.0
	rejects >80 mm	4,687	4.7	0	
	ferrous rejects	28	0.0	0	
	net input	100,380	100.0	20,098	100.0
Output:	compost	22,438	22.4	6,652	33.1
	rejects	17,320	17.3	1,749	8.7
	wastewater	8,770	8.7	0	0.0
	net output	48,688	48.5	8,401	41.8
Mass reduction (**)		51,692		11,697	58.2
Energy consumption:					
	total [MWh]	2,870		752	
	[kWh/ton net input]	28.6		37.4	

(*) input May to December 1994, (**) net input minus net output

Although the impurities in the biowaste of SOW were between 3 and 7%, a considerably larger amount of material needed to be landfilled. It is expected that when the throughput is in accordance with the design capacity and the retention time kept around 10 weeks, more organic material will be degraded. Consequently the amount of material to be landfilled will decrease and the compost production will increase.

Never in previous years was it necessary to find alternative treatment for contracted biowaste. Turning the material to be composted in the hangar can be carried out faster than the standard rate of throughput, thus creating space and time for preventive maintenance. It is also possible to vary the composting time from 7 to 10 weeks, so the treatment capacity is very flexible and may vary within a wide range.

Recently the firm Bühler introduced a further development of its hangar composting system [4]. By optimising the composting process, a reduction of the retention time by 40 to 60% is claimed, so that about 25% of the treatment costs may be saved. In the Netherlands no experience with this new concept exists.

TUNNEL COMPOSTING

General

The tunnel composting technique was developed in the mushroom industry where quality control is essential for obtaining a good and stable mushroom compost, and securing a continuous yield of high quality mushrooms. The application of this technique to treat separately collected VFG, benefits from the long term experience gained in the mushroom industry [5].

The first Dutch full scale tunnel composting plant treating household biowaste came into operation in 1991. The capacity of the individual tunnel composting plants ranges now from 25 to about 75 ktpa, with an average around 40 ktpa. For most tunnel composting plants in the Netherlands the Dutch firm, GICOM Composting Systems, has supplied the essential process equipment and integrated central computer control system.

Dutch tunnel composting facilities always apply a two stage composting process, with a set of so called 'tunnels' as the first step. Each working day the VFG, without being pre-treated, except removal of very bulky or heavy non-compostable components by a front-end loader, is directly loaded into one tunnel while another tunnel is unloaded. Therefore, the retention time in the pre-composting tunnels is approximately one week, which fits in perfectly with the entire operation of the plant. According to the demands of the required treatment capacity, the tunnel dimensions may vary between plants, i.e. widths between 3 and 5 m, a height normally at 4 m, and lengths between 15 and 35 m.

After one week, the material is unloaded and sieved over a 70 mm mesh to remove coarse particles which are shredded. The shredded particles are either mixed with the VFG to be composted again, or are taken with the pre-composted material to the post-composting yard. From this point there are two possibilities:

- Either post-composting takes place in piles in a post-composting yard situated within the building, or
- it takes place in tunnels.

If the post-composting yard is provided with an aerobic pile composting floor, the required residence time is 4 to 6 weeks, depending on the compost quality. Air is sucked through the material for aeration. If the tunnel system is used, the required retention time in the second stage tunnel is 3 to 4 weeks. The size of this tunnel is exactly the same as a pre-composting tunnel. However, the tunnel is filled with the contents of two pre-composting tunnels and the amount of air needed to control the process is relatively small compared to pre-composting.

After the second stage composting, the raw compost is refined by sieving with a 15 mm mesh. Outside the building the final compost is stored for some time in the open air on a platform, and then sold. Residues which are removed from the compost by the sieve, are landfilled.

Process air from the pre-composting tunnels and the post-composting yard is ventilated through a washer to cool the air to a maximum of 35 °C and moisten it using a mixture of percolate and condensate. A side effect of the moistening is the removal of dust and ammonia from the exhaust air before it enters the biofilter. In this way, the climatic and physical conditions of the biofilter can be controlled, ensuring a high efficiency of odour removal, and an extended life of the biofilter.

Operational experiences in the Province of Groningen [6]

The second case in this paper concerns a tunnel composting plant which is not only designed by Grontmij, but also operated by its waste management subsidiary. The plant is located in one of our most northern provinces bordering Germany, namely the Province of Groningen. The plant was designed to treat on average almost 90 tonnes per day of separately collected VFG (average 22,500 tpa; maximum about 25,000 tonnes). The firm, called 'OGAR', owns the plant, and is responsible for treatment and subsequent marketing of the compost.

Full details about the design of the plant and its operation are provided in the proceedings of a symposium on biological waste treatment, recently held in Bochum, Germany [6].

On 9 May 1994, operation of the plant started with acceptance of the first load of VFG. Almost immediately the daily load of the plant, instead of the expected 90 tonnes, reached an average level of 120 tonnes, with a peak load of 150 tonnes per day, which means the actual load is almost 30 ktpa. The high degree of flexibility of the composting system was fully proven, as no failure of the process occurred. All accepted VFG was converted into sellable compost, while no VFG had to be treated elsewhere.

The mass balance of the OGAR tunnel composting plant [6] shows that almost two thirds of the mass reduction is reached in the pre-composting stage, and approximately one third in the post-composting process. The amount of reject material (>70 mm) is less than 2% of the total, indicating that the awareness of the population in the SOZOG-region of the need to discard 'clean' VFG is generally good. It may be expected that by shredding the residue (<70 mm), presently used as cover material on a landfill, the total flow of rejects may be reduced further. The results during the first year of operation are summarised in Table 3.

QUALITY AND MARKETING OF COMPOST

The quality of the compost, with respect to pollutants, is primarily determined by the quality of the biowaste supplied. On has to remember that the Dutch biowaste concerned is a mixture of kitchen and garden waste. Table 4 summarises the Dutch standards for heavy metal content of 'compost' and 'very clean compost'.

Table 4: Compost quality

Dutch compost standards from 1 January 1995 [7]	TS	VS	Heavy metals (mg/kg TS)							
	(%)	(% of TS)	Pb	Cu	Ni	Zn	Cr	Cd	Hg	As
Compost	-	>20	100	60	20	200	50	1	0.3	15
Very clean compost	-	>20	65	25	10	75	50	0.7	0.2	5
Compost on average from:										
- SOW:	73.7	56.1	87	32	12	152	21	0.5	0.2	4.5
- OGAR:	79.6	30	75	21.8	9.3	129	19	0.43	0.14	2.6

In the table it is shown that the VFG compost from SOW and OGAR, as from most Dutch compost plants, meets the standards for 'compost' from 1995 on. Even the limits for 'very clean compost' are met except for some heavy metals, especially lead and zinc. The awareness of the population, together with the separate collection of household chemical waste like batteries, paint, old medicines, which is now implemented in every Dutch community, secures the relative low content of pollutants in the compost. It is expected that, due to the relatively high natural concentrations of heavy metals in vegetable material, it will be very difficult to meet all the requirements for 'very clean compost'.

The spreading of compost is restricted by governmental regulations. Compost at a quality that complies with the standard for 'compost', is allowed to be applied on arable land to a maximum dose of 6 tonnes (TS) per ha per annum. 'Very clean compost' is cleared for unlimited dosage. The nutrients and soil conditioning properties of the compost can increase the efficiency of fertilisers. It can be calculated that the potential demand for compost in the Netherlands is larger than the annual production of compost. Due to the surplus of manure and the relative ease to apply chemical fertilisers in the industrialised agriculture of the Netherlands, in reality there is very little demand for nutrients present in compost. With the exclusion of some sandy soils the same goes for the demand for organic matter or soil conditioner.

Various owners of composting plants allow the municipalities which bring their biowaste to the facility, to take compost for their own purposes. Also, citizens are allowed to take compost at no charge. Besides this, most compost is used for agricultural purposes. Until now there have been no difficulties in marketing the compost. The marketing efforts differ among the various suppliers of compost. The smaller composting plants in general only have costs for analysis of the compost. The larger plants also spend money on promotion or have contracts with distributors. In most cases the distributors collect the compost from the composting plant at no charge. Transport and other costs of the distributor are paid by charging the customer a price, that in general varies between Dfl 10 and 20 per tonne. There are however examples where the composting plant pays the distributor a marketing fee of up to Dfl 20 per tonne compost, resulting in a negative price for the compost to the producer.

CONCLUSIONS

Data obtained from a large number of operational Dutch aerobic composting plants show the following:

- The moisture content of the separately collected biowaste is often around 60%, but shows seasonal fluctuations. The main reason for this is the garden waste fraction in the biowaste, which may drop to nil between November and the end of February;
- with tunnel composting a minimum retention time of 4 to 5 weeks during the process is feasible. With hangar composting a stable compost is produced after 8 to 10 weeks;
- an overall 60 to 65% VS-degradation may be reached;
- due to the VS-degradation and evaporation of moisture contained in the fresh biowaste, a 60% mass reduction may be achieved;
- the amount of reject material (non-compostables) may reach a value under 5%, indicating that in general the population has a good awareness of the need to separate 'clean' biowaste;
- the electricity consumption for aerobic composting is between 30 and 40 kWh per tonne of biowaste, depending on the composting system applied and the retention time adhered to;
- the Dutch treatment costs related to aerobic composting are on average between Dfl 80 and 100 per tonne biowaste;
- most compost is marketed for arable applications. In the recent past it was possible to sell the compost at a profit, but today the compost is marketed at nil profit. The main reason is an increasing flux of compost in the market. Under present conditions, regional differences affect the marketing efforts required to sell the compost;
- based on present experience regarding operational aspects (mainly process control and flexibility) the tunnel composting system tends to be preferred, especially for a required treatment capacity under 50 ktpa.

References

1. VFG-information. Status report VFG treatment in the Netherlands, October 1994 (Dutch).
2. Oorthuys, F.M.L.J., Koning, R.A.: Advanced aerobic composting of biowaste from refuse in the context of Dutch solid waste management; Conference BIOWASTE'92, Herning, Denmark; June 1992.

3. Scharff, H. and Oorthuys, F.M.L.J.: Operating experiences from indoor composting of separately collected household biowaste; European conference on sludge and organic waste; Leeds, Great Britain; April 1994.
4. Kugler, R. and Leisner, R.: Test on temperature controlled table-pile composting - the system concept Wendelin AirTec; Symposium 'Biological waste management - a wasted chance?'; Bochum, Germany; April 1995.
5. Lindberg, C.: Accelerated composting in tunnels; Conference BIOWASTE'95, Aalborg, Denmark; May 1995.
6. Oorthuys, F.M.L.J., Von Deylen, H., and Lokin, P.C.: Engineering, construction, and first operational experiences of the 25,000 tpa tunnel composting plant in Groningen; Symposium 'Biological waste management - a wasted chance?'; Bochum, Germany; April 1995.
7. KIWA. Guideline for the appraisal of VFG compost; BRL-K256\01.
8. Oorthuys, F.M.L.J., Scharff, H.; Operational aspects of aerobic and anaerobic treatment of biowaste in the Netherlands; Invited communication for BIOWASTE '95, Aalborg, Denmark; May 1995.

Operational Aspects of Aerobic and Anaerobic Treatment of Biowaste in The Netherlands

Thijs Oorthuys and Heijo Scharff, GRONTMIJ Consulting Engineers, P.O.Box 203, 3730 AE De Bilt, The Netherlands

GENERAL

For the comparison of operational aspects of aerobic and anaerobic processing of biowaste the properties of the waste and the conditions of processing are of major importance. For the present comparison the mass balances of aerobic and anaerobic processing as shown in Figure 1 (next page) are used. To a large extent these mass balances are determined by the existing situation in the Netherlands. The most important issues are explained below.

The separately collected biowaste in the Netherlands contains vegetable, fruit and garden waste. In general more than half of source separated biowaste is garden waste. Therefore the Dutch biowaste contains a lot of sand and soil particles. The average composition of Dutch biowaste [1] is 40% TS, of which 50% is VS. In general the biowaste contains 3 to 7% impurities, mainly plastics and some ferrous metal. Per inhabitant on average 100 kg biowaste and 300 kg residual waste are collected annually.

The Dutch guidelines with respect to application of compost require a minimum TS content, sufficient maturity and hygienic features. The guidelines were mainly based on the present experience with aerobic processes. In general an aerobic posttreatment (making the system more expensive) will be required for the dewatered anaerobic slurry (digestate) to comply with the guidelines.

STARTING POINTS MASS BALANCES

Composting [2-6]:
- Pretreatment consisting of mild size reduction (worm screw shredder) and sieving;
- mass reduction during aerobic composting is 60%;
- degradation efficiency: 60% of VS supplied to the aerobic process;
- compost production is 30-35% of the input;
- residues to be landfilled or treated otherwise: 5-10%;
- in some plants oversized material from compost refinement is landfilled, in other plants it is shredded and returned to the process to reduce amounts of material to be landfilled, this refining is not incorporated in the mass balance; compost composition: 30% TS of which 30% VS;
- correctly operated composting plants produce no surplus water (excess water is vaporised).

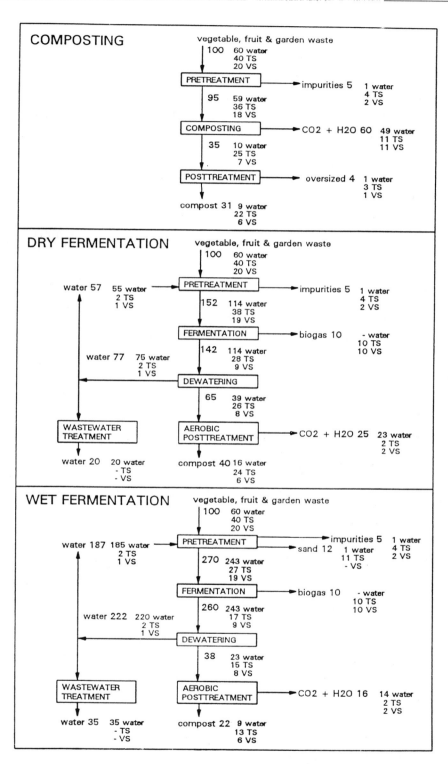

Fig. 1: Mass balances of aerobic and anaerobic waste treatment processes

Dry fermentation [7-8]:
- Pretreatment consisting of mild size reduction (worm screw shredder) and sieving;
- TS content in the reactor is about 25%;
- mesophilic conditions (35-38°C);
- degradation efficiency: 55% of VS supplied to the reactor, corresponding with 85 m^3 STP (standard temperature and pressure: 0°C and 1013 mbar) biogas (55% CH$_4$) per tonne Dutch VFG-waste;
- mechanical dewatering of digestate (residue of the anaerobic process) to 40% TS;
- aerobic posttreatment of digestate;
- compost composition: 60% TS of which 25% VS;
- surplus water to be treated in waste water treatment plant designed for nutrient (N and P) removal.

Wet fermentation [8]:
- Pretreatment consisting of mild size reduction (macerator or pulper) and separation of inerts (e.g. plastics, sand, stones, glass) by flotation and sedimentation;
- TS content in the reactor is about 10%;
- mesophilic conditions (35-38°C);
- degradation efficiency: 55% of VS supplied to the reactor, corresponding with 85 m^3 STP biogas (55% CH$_4$) per tonne Dutch VFG-waste;
- mechanical dewatering of digestate to 40% TS;
- aerobic posttreatment of digestate;
- compost composition: 60% TS of which 45% VS;
- surplus water to be treated in waste water treatment plant designed for nutrient (N and P) removal.

NUTRIENTS

In the Netherlands raising of poultry, pigs and cattle is largely industrialised. In certain areas of the country there is an enormous surplus of manure. Unfortunately in several regions leaching of nitrogen and phosphorous to the groundwater has already occurred and will continue. Therefore regulations have been issued with respect to nitrogen and phosphorous loads on agricultural land. Mature aerobic compost is essentially a soil conditioner and its fertilising value is relatively small. The application of this compost does hardly affect the leaching of nitrogen and phosphorous. Waste water of biowaste processing plants on the other hand contains a lot of ammonia. Application of this waste water as a fertiliser would increase the leaching problems. Therefore in the Netherlands extensive waste water treatment aiming at nutrient removal (N and P) is required for biowaste processing plants. Essentially this means that in the Netherlands a situation has arisen where energy is required to convert ammonia to nitrogen gas instead of saving energy in fertiliser production by applying waste water from biowaste processing.

ENERGY AND CO_2 EMISSION

To support the position of the capital and energy intensive industry of the Netherlands in competition with the rest of the world energy prices for industrial consumers are kept as low as possible. Tariffs are very complicated and depend on the ratio between power made available and power bought (i.e. fluctuations in consumption) and consumption during or outside peak hours. For the average biowaste processing plant the energy price will be approximately Dfl 0.12 to 0.15 per kWh (1 Dfl = 0.6 US$ = 3.6 DKK). Although there are some measures to stimulate sustainable energy production, the price for selling energy at Dfl 0.08 per kWh is relatively low compared to other European countries.

The energy consumption of aerobic processing plants varies from 30 to 40 kWh per tonne VFG-waste [2-5]. With very extensive pretreatment, intermediate treatment and compost refinement (including cyclones, ballistic separators etc.) energy consumption may even rise to 60 kWh per tonne VFG-waste [6].

Using the 85 m^3 STP biogas (55% CH_4) per tonne VFG-waste (generated by anaerobic fermentation of VFG) to fuel a gas engine generator set, approximately 130 kWh per tonne VFG-waste can be produced. The data for energy consumption of fermentation plants [7-8] vary from 20 to 40% of their energy production. These data sometimes do not include pretreatment, aerobic posttreatment and waste water treatment. Therefore it is assumed that 30 to 35% of the energy production (or 40 to 45 kWh per tonne VFG-waste) is consumed in the fermentation plant.

Since the aerobic processing plants consume approximately 40 kWh per tonne VFG-waste and the net electrical power output of anaerobic processing plants is approximately 90 kWh per tonne VFG-waste, approximately 130 kWh per tonne VFG-waste can be saved by applying anaerobic technology. We recognise that an additional saving could made when the nitrogen in the waste water would replace nitrogen from chemical fertiliser. At this moment this is (not yet) an issue in the Netherlands. Therefore we have not quantified the potential energy savings with respect to nitrogen.

The annual CO_2 emission in the Netherlands is approximately 1.9×10^{11} kg CO_2. The fuel mixture that is applied for electricity generation in the Netherlands causes a CO_2 emission of 0.64 kg of CO_2 per kWh [9]. In the near future the annual VFG-waste production in the Netherlands will be approximately 1.9×10^9 kg. In the theoretical event that all VFG-waste would be treated anaerobically, approximately 1.7×10^8 kWh per annum would be produced. This would result in a reduction of the annual CO_2 emission of approximately 1.1×10^8 kg CO_2, or 0.06% of the annual CO_2 emission.

WATERBALANCE

The most conspicuous difference between aerobic and anaerobic processing of biowaste is that in anaerobic processes solar energy stored in the biomass is converted to an energy carrier (biogas) that can be applied elsewhere, whereas in aerobic processes this energy is converted to heat. Aerobic biodegradation of biowaste produces 16 to 18 MJ per kg VS degraded. The evaporation of water requires 2.4 MJ per kg water. Aerobic degradation of half of the VS present in Dutch biowaste is therefore more than enough to evaporate all water present in the biowaste. In practice with optimal operation of the composting process, adding water to the process is required to prevent excessive drying and inhibition of the composting process. Practice shows that in general 60% mass reduction is achieved and 30 to 35% compost (70% TS) is produced [2-6]. This depends on whether oversized material from posttreatment is returned to the process. The latter aspect also determines whether 5 or 10% of the input are residues to be landfilled. Ultimately climatologic conditions and correct design of the biofilter (to reduce odour emission) determine whether all vapour can be emitted. Operational experience of several modern biowaste composting plants in the Netherlands has shown that no surplus water is produced [4-6]. In some cases even a water shortage was reported.

During anaerobic fermentation no water is evaporated. In order to produce a compost the digestate has to be dewatered mechanically and treated aerobically afterwards. In general mechanical dewatering is carried out to approximately 40 to 45% TS. Drying to 60% TS by means of heat generation from remaining VS (especially in case of dry fermentation) will not be sufficient to evaporate the corresponding amount of water. Additional energy input will be required. The water separated during mechanical dewatering can only partially be reused in the fermentation process for creating optimal process conditions. The inevitable water surplus needs to be treated in a waste water treatment plant.

DRY VERSUS WET FERMENTATION

For the comparison of aerobic and anaerobic processing of biowaste in the Netherlands the dry fermentation process (20 to 30% TS in the reactor) is taken into consideration. The wet fermentation process (approximately 10% TS in the reactor) has not yet been applied for full scale treatment of biowaste in the Netherlands. At industrial scale one biowaste fermentation plant is in operation at the moment. The investment costs for this plant are known. This does not include a waste water treatment plant. Its surplus water is treated in a nearby communal waste water treatment plant. Aerobic posttreatment takes place in a closed hangar without forced aeration. Such a treatment is believed to be insufficient. Therefore forced aeration has to be taken into account.

Application of the wet fermentation process for Dutch biowaste has some consequences. At 10% TS sand will settle at the bottom of the reactor and will cause obstruction of the process. Therefore sand has to be removed. This aspect is or can be incorporated in most wet fermentation systems. By means of sand separation TS is removed that contains relatively little moisture. Mechanical dewatering is applied to achieve 40% TS. Since the digestate contains less TS than with dry fermentation, a corresponding amount of water cannot be contained in the dewatered digestate. This will result in a larger amount of waste water and a smaller amount of compost than in case of dry fermentation. The composition of the compost is different as well. The compost from aerobic composting and dry fermentation contains more inert material (sand) whereas the compost from wet fermentation has a higher VS content.

FINANCIAL ANALYSIS

Table 1 shows the estimated capital investments. In particular, the costs for aerobic composting match fairly well with real prices obtained in practice. The treatment costs are based on plants provided with environmental protection facilities according to EC directives, including wastewater treatment and biofiltration for waste air treatment.

Table 1: Capital investment for indoor composting and fermentation of separately collected biowaste (Dfl million)

Capacity (tonne/yr)	Aerobic	Anaerobic
25,000	9.4 - 11.5	17.5 - 21.5
50,000	17.4 - 22.0	28.0 - 33.5
75,000	24.5 - 29.0	37.5 - 43.5
100,000	31.5 - 36.0	47.5 - 54.5
125,000	38.0 - 43.0	58.0 - 66.0

The estimated costs per tonne of biowaste for indoor aerobic composting plants and for anaerobic plants including aerobic post-composting, are summarised in Figure 2.

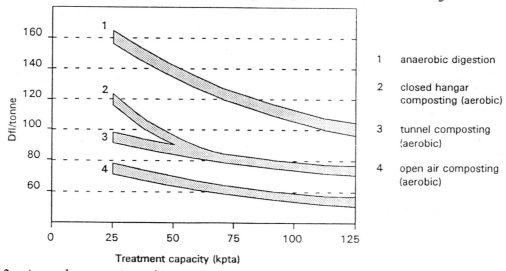

Fig. 2: *Annual costs of aerobic composting and anaerobic fermentation*

These cost calculations are based on 15-20 years amortisation, 8% interest, maintenance at 1-3% of capital investment, energy costs at Dfl 0.15/kWh, and utilisation of energy from biogas conversion at Dfl 0.08/kWh. Income from compost sales is not included, nor are costs related to the separate collection of biowaste, land acquisition, and VAT. The investment costs are related to the civil works and to the mechanical/electrical works. Generally speaking, the split between civil works and mechanical/electrical works for composting plants is approximately 55 : 45, and the ratio for anaerobic fermentation is about 35 : 65.

The evaluation of the treatment costs shows the following:

- About 60% to 70% of the treatment costs are taken up by amortisation costs, and hence by the amount of capital investment;
- the remaining 30% to 40% of the treatment costs are caused by:
    - personnel and maintenance :about 20%;
    - energy, disposal of residues etc. :10% to 20%.

The difference in annual costs between large scale operation of aerobic composting and anaerobic fermentation systems is considerable, approximately Dfl 25 per tonne of biowaste. Feasibility studies have shown that optimal use of the biogas produced, i.e. 100% recovery and reuse of electricity and heat, may reduce the annual costs of anaerobic fermentation by no more than Dfl 6.50 to Dfl 7 per tonne of biowaste. This is based on the possible current revenues from recovery of electricity and heat in the Netherlands.

REFERENCES

1. VAM; Research on composting of VFG-waste in the open air and under a roof (in Dutch); NOH 9226; 1992.
2. Heijo Scharff and Thijs Oorthuys; Operational experiences form indoor aerobic composting of separately collected household biowaste; European Conference on Sludge and Organic Waste; University of Leeds; April 1994.
3. Operational data of the SOW closed hangar composting plant; personal communication with the plant manager; 1990-1995.
4. Operational data of the OGAR tunnel composting plant; personal communication with the plant manager;1994-1995.
5. Operational data of the St. Oedenrode tunnel composting plant; personal communication with the plant manager; 1992-1995.
6. Operational data of the Bladel tunnel composting plant; personal communication with the plant manager;1994-1995.
7. Environmental Impact Statement; Fermentation plant for VFG-waste (in Dutch); Waste Authority Central Brabant; 1992.
8. Documentation of both wet and dry fermentation and personal communication with the suppliers; 1992-1995.
9. Biomass Technology Group; Conversionroutes for energy crops (in Dutch); Enschede, Netherlands; 1994.

Section 8

Product Quality, Marketing & End User Demands

Reduction of Pathogenic Bacteria and Viruses by Anaerobic Digestion

Bente Lund, H.J. Bendixen*, Per Have** and Birgitte Ahring, Danish Technological Institute, Department of Biotechnology, P.O.Box 141, DK-2630 Taastrup, Denmark, Danish Veterinary Service *, Rolighedsvej 25, DK-1958 Frederiksberg C, Denmark, Danish Veterinary Institute for Virus Research**, Ministry of Agriculture, Lindholm, DK-4771 Kalvehave, Denmark.

ABSTRACT

The pathogen reducing effect of anaerobic digestion was investigated in relation to the current recommended criteria for sanitation (time/temperature combinations). Faecal enterococci were used as indicator bacteria and laboratory experiments were done on the inactivation of viruses. Extensive studies in large scale centralised biogas plants have demonstrated that a treatment which results in a 4 $\log_{10}$ reduction of faecal enterococci will also cause elimination of most of the common pathogenic bacteria and viruses in manure and wastes. Certain virus types (e.g. parvovirus) and certain bacteria (i.e. the spores of bacteria) are thermoresistent and will not be eliminated but only reduced in numbers. Faecal enterococci are suggested as indicator bacteria in biogas plants at temperatures up to 60°C and porcine parvovirus as indicator in laboratory tests at temperatures between 50-80°C.

KEYWORDS

Anaerobic digestion, enteric bacteria, indicator microorganisms, manure, pathogen reduction, virus

INTRODUCTION

The large centralised biogas plants (CBP) in Denmark process daily about 300 to 400 tons of biomass such as manure, organic industrial waste and household waste. Manure is collected from a large number of herds and represents 80 to 85% of the total amount of biowaste. Manure from domestic animals contains bacteria, viruses and parasites corresponding to the health status in the herds of origin. The pathogens may be shed from sick animals and also from apparently healthy carriers.

Various industrial by-products and urban wastes constitute 15 to 20% of the organic matter. These products may be more or less heavily contaminated with various pathogens and, they are incorporated in the following categories according to the Danish regulations: A) sludge and waste from processing of vegetable raw material and from dairies, B) sludge from processing of raw material of animal origin, C) organic matters from household wastes and D) sewage sludge. Waste materials in categories C and D - and some raw products in category B - are regarded as highly contaminated.

After anaerobic digestion in the CBPs the digested materials are distributed as fertiliser on agricultural land. The risk of spreading pathogens of human and animal origin by this recycling of organic matter must be counteracted by the use of hygienic measures during collection, transport and handling of the untreated raw materials. Moreover anaerobic digestion in the CBPs must be carried out in order to achieve an adequate Pathogen Reducing Effect (PRE). This can be obtained by maintaining anaerobic digesters in the thermophilic range of temperature and allowing a Minimum Guaranteed Retention Time (MGRT) of the biomass in the reactors. In Denmark the following criteria are currently recommended: 52°C for 10 hours; 55°C for 6 hours; or other combinations of temperature and MGRT. The PRE which can be attained with anaerobic digestion operated at mesophilic temperature is less satisfactory. Biomass with heavy biological contamination should not be processed at this temperature level unless it is further treated in a special sanitation tank. The Danish regulation requires a 'controlled sanitation', i.e. at least 70°C for at least 1 hour, or equivalent temperature/time combinations. A survey concerning these aspects in highly contaminated waste products at 11 Danish CBPs has just been published [1].

The PRE in a digestion tank of a CBP can be measured by means of a microbiological test using faecal enterococci (FE) as indicator organisms. Extensive studies in Danish CBPs have demonstrated that a treatment which results in a 4 $\log_{10}$ reduction of FE will also cause elimination of most of the common pathogenic bacteria (e.g. salmonella, campylobacter, coli, clostridia, yersinia and listeria), viruses (e.g. classical swine fever virus, foot- and mouth disease virus and rota virus) and parasite eggs in manure and wastes [2]. It has been suggested that this criterion should be taken into account for the achievement of an adequate PRE. Certain virus types (e.g. parvovirus) and certain bacteria (i.e. the spores of bacteria) are thermoresistent and will not be eliminated but reduced in number. A standardised laboratory method - the FE method - has been tested and is proposed as official method for supervision and check of the PRE in the Danish CBPs at temperatures up to 60°C [3]. Inactivation of viruses has been evaluated by the use of enterovirus as indicator organism in model experiments [4]. Bovine enterovirus belongs to the relatively thermoresistent picornaviruses, a family consisting of pathogens like foot- and mouth disease virus, teschen-disease virus and swine vesicular disease virus as well as human viruses as echo-, coxsackie- and poliovirus. Since these viruses are not found in constant and high numbers in manure and organic waste, only laboratory experiments can be used.

Danish Technological Institute and Danish Veterinary Institute for Virus Research have performed laboratory experiments concerning reduction of virus in manure during anaerobic digestion in biogas reactors. The aim of the experiments was to study the sanitation effect of the anaerobic digestion process by means of the evaluation of the use of either faecal enterococci (FE) as indicators of virus reduction at temperatures under 60°C, or porcine parvovirus as indicator in laboratory tests at temperatures above 60°C. In the present study we used a laboratory set-up which reflects a modern biogas plant concerning biowaste, active anaerobic bacteria, temperature, hydraulic retention times. Further studies on inactivation of different pathogenic bacteria and viruses in the temperature interval 50-80°C are suggested.

METHODS AND EXPERIMENTAL PROCEDURES

Substrate: The biomass used was similar to the basic material used in CBPs; 95% homogenised manure (25% pig and 75% cow) with 5% bleaching clay added and adjusted to a VS content of 4%. All substrates used were prepared in one batch and stored at -18°C until use. **Cultures:** Two non-enveloped viruses with different thermal resistance were chosen for the experiments: *bovine enterovirus* (BEV), a field strain from faeces and *porcine parvovirus* (PPV) originally isolated from a porcine fetus. The viruses were cell-culture-propagated according to [4] with a final titre of $10^{7.8}$-$10^{8.2}$ TCID$_{50}$/200µl (BEV) and $10^{7.7}$ TCID$_{50}$/50µl (PPV). The *faecal enterococci* (FE) used was isolated from manure. **Microbiological analyses:** Analysis of faecal enterococci was according to Danish Standard Methodology (1993) with enumeration on Slanetz agar (Merch 5289). Plate Count Agar (Difco 0479-01-1) was used for propagation followed by wash off in sterilised physiological saline solution (8.5 g NaCl/l). **Chemical analyses**: Gas production, methane content, volatile fatty acids (VFA), ammonia and volatile solids (VS) were determined as described by Angelidaki [5].

Tests were performed in laboratory reactors (3 litre working volume) as previously described by [6]. The reactors were run until steady-state was reached with a mean hydraulic retention time of 15 days. Substrate was added 4 times a day giving a MGRT of 6 hours. A mesophilic (35°C) reactor and a thermophilic (55°C) one fed with manure and bleaching clay were used in the experiments. For experiments on virus inactivation, virae particles were added as a suspension and the feeding of the reactors was stopped until termination of the experiment. The initial theoretical virustitre were $10^{5.7}$ TCID$_{50}$/50µl for PPV and $10^{5.5}$-$10^{5.7}$ TCID$_{50}$/200 µl for BEV. For FE, a start of approx. 10^7 colony forming unit/ml was used. Counts of virus and FE in digester samples were done after incubation for 0.25, 0.5, 1, 1.5, 2, 4, 8, 16, 32 and 48 hours.

Parallel to the reactor experiments, survival of the same viruses and FE were followed in batch experiments using physiological saline solution incubated at 35°C and 55°C.

Furthermore, the survival of these microorganisms was tested in batch experiments placed at 70°C for 1 hour in either physiological saline solution or manure with bleaching clay. All experiments were done in duplicate and the experiments were replicated.

RESULTS AND DISCUSSION

The content of faecal enterococci (>6.1 log unit) and bovine enterovirus (>3.8 log unit) were effectively reduced by heat treatments in batch experiment at 70°C for 1 hour using physiological saline solution and manure with bleaching clay added. However, the content of the thermo-tolerant porcine parvovirus was only reduced 0.6 log unit.

Measurements of the concentration of viruses and FE during the experiments in continuous cultures and in batch cultures resulted in inactivation curves after linear regression of the results. In cases where it was possible to follow the inactivation of viruses over a longer period, a high initial reduction followed by a decreasing reduction rate was seen. Similar inactivation performances have been seen for different virus in manure and sludge by [7-9] probably due to particle association. As a biphasic inactivation was found, two death rates and corresponding MGRT have been calculated for PPV in the experiments at thermophilic temperatures during the first 4 hours, and a 'terminal MGRT' for the rest of the experiment (up to 48 hours). Both the initial and the 'terminal' reduction times for 4 $\log_{10}$ units reduction were calculated using extrapolation.

The time needed to obtain a 4 $\log_{10}$ unit reduction of the measured microorganisms was calculated for the reactors as well as for the batch experiments (table 1).

Table 1: *The time (hours) needed to obtain a 4 $\log_{10}$ reduction in microorganisms tested in continuously-fed reactors and in batch experiments, stated as the average of duplicates of replicates given with its standard deviation. substrate: manure with bleaching clay added or physiological saline solution. nd - not determined*

	Continuously-fed reactors. Substrate: manure with bleaching clay added		Batch tests. Substrate: physiological saline solution	
	35°C (h)	55°C (h)	35°C (h)	55°C (h)
Faecal enterococci	296 (124)	1.1 (0.08)	832 (76)	3.8 (0.2)
Bovine enterovirus	23 (1.6)	<0.5	219 (-)	2.3 (0.0)
Porcine parvovirus				
- 'initial'	ND	11 (4.1)		
- 'terminal'	ND	54 (10)	ND	148 (7)

As expected, the rate of inactivation of faecal enterococci was much higher at 55 °C than at 35°C, which resulted in a recommended MGRT (a 4 $\log_{10}$ units reduction) of 1-2 hours and 300 hours, respectively. The pathogen reducing effect of mesophilic digestion is thus negligible. Inactivation in physiological saline solution is considerably lower than in continuously fed biogas reactors based on manure as substrate. This indicates a strong inactivating effect of the environment in a biogas reactor with an active biomass. Correspondingly, a considerably higher rate of inactivation was seen for viruses under thermophilic digestion than under mesophilic digestion; for bovine enterovirus a MGRT of 23 hours at mesophilic temperatures was found compared to <0.5 hours at thermophilic conditions. For porcine parvovirus under thermophilic conditions, the calculated MGRT is 11 hours with manure. Using the later retarded inactivation, a MGRT of 54 hours (based on the period 4-48 hours after addition of virus) was calculated.

The rate of inactivation for the indicator bacteria, faecal enterococci, was generally higher (6-50 times) than the rate of inactivation for porcine parvovirus and lower (up to 13 times) than for bovine enterovirus during anaerobic digestion. Using FE as indicator for inactivation of enterovirus thus leaves a good safety margin.

Compared to batch tests in physiological saline solution, the inactivation in reactors with manure as substrate is considerably faster. The results shows that the reduction of pathogens in biogas reactors was not solely due to thermal treatment, and, therefore, the minimum guaranteed retention time (the length of treatment) before use of these materials on agricultural land can be shortened when materials are treated in anaerobic reactors. This reduction can be expressed as the ratio between the rate of inactivation in physiological saline solution and in the digesters (Table 2). Embedded viruses will be better protected than a virus culture added as free suspension. This is important as viruses is often embedded in tissue when present in organic waste. Experiments indicate that it is between 4 and 10 times lower than for added cell-culture-propagated virus (corresponding to a MGRT between 2 and 5 hours for BEV at 55°C) [1].

Table 2: Ratio between the rate of inactivation in physiological saline solution and in the reactors. At 70°c this relationship is about 1 for ppv. nd - not determined.

	35°C	55°C
Faecal enterococci	0.5	0.3
Bovine enterovirus	0.11	<0.15
Porcine parvovirus		
- initial		0.08
- 'terminal'	ND	0.40

The ratio between the initial and the terminal inactivation rate was approx. a factor 4 for porcine parvovirus. A factor 10 difference between the decimation in the initial phase and the later phase was found by [8] for a coxsackie virus (human enterovirus) in reactors based on wastewater sludge.

Parvovirus has been under consideration as an indicator virus in the temperature interval 50-80°C. Parvovirus is suitable for comparative studies in this temperature interval as they persist at high temperatures, making it possible to follow their inactivation over a longer period. Due to the biphasic inactivation curve and the heat resistance, care most be taken when extrapolating to other viruses. If a treatment corresponding to a heat treatment at 70°C for 1 hour is demanded, it is necessary to use a heat resistant indicator virus such as parvovirus for laboratory experiments.

The criteria are at present under re-evaluation to ensure a effective treatment in relation to the persistence of pathogens in different waste types.

CONCLUSION

The environment in an anaerobic digestion tank promotes inactivation of bacteria and viruses. The pathogen reducing effect of anaerobic digestion of is not only due to heat treatment of the manure and other organic waste. Faecal enterococci are suitable indicators for measuring the pathogen reducing effect of thermophilic anaerobic digestion concerning enteroviruses and more labile viruses. A 4 log reduction of FE and enterovirus can be obtained during thermophilic (55°C) anaerobic digestion with a MGRT of 2 hours if holding period and temperature are well controlled and regulated. The pathogen reducing effect of mesophilic digestion though is not adequate. Parvovirus is suitable for comparative investigations on inactivation in the temperature range 50-80°C due to the extreme heat resistance of this virus.

ACKNOWLEDGEMENT

This work was supported by a grant from the Danish Energy Agency.

REFERENCES

1. Bendixen, H.J. (1995): Smitstofreduktion i Biomasse. Det vetrinære forsøgsprogram i Biogasfællesanlæg. Bind I. Hovedrapport, Bind II. Delrapporter og bilag. Rapport fra Veterinærdirektoratet, Landbrugs- og Fiskeriministeriet. (Inactivation of pathogens in biomass. The veterinary research programme in Centralized Biogas Plants. I. Main Report. II. Part-Reports and Appendices. Edited by The Danish Veterinary Service)
2. Bendixen, H.J. (1994): Safeguards against pathogens in Danish Biogas Plants. Wat. Sci. Tech., vol.30, no.12, pp.171-180.
3. Bennetsen, O. & U.S. Mikkelsen (1993): FS- metodens anvendelighed som hygiejnisk kontrolparameter. (The usability of the FE-method as a control parameter of sanitation). Dansk Veterinær Tidsskrift 76, 14, pp.597-656.
4. Lund,B., V.F.Jensen, P.Have & B.Ahring (1995):'Inactivation of virus during anaerobic digestion of manure in laboratory scale biogas reactors. Proceedings from the International Meeting on Anaerobic Processes for Bioenergy and Environment. Copenhagen, 25.-27. jan. 1995. Section 6. The Nordic Energy Programme.
5. Angelidaki I (1992):'Anaerobic thermophilic biogasreactor process: The effect of lipids and ammonia'.Ph.D.Thesis, Dept. of Biotechnology, The technical University of Denmark.
6. Angelidaki I & Ahring BK (1993): Thermophilic anaerobic digestion of livestock waste: The effect of ammonia. Appl. Microbiol. Biotechnology, 38: pp. 560-564
7. Bøtner A (1990): Modelstudier vedr. overlevelse af virus i gylle under traditionel opbevaring og under udrådning i biogasanlæg. Delprojekt 1. Veterinær forskning og rådgivning i forbindelse med etablering og drift af biogasfællesanlæg (Modelstudies concerning survival of virus in manure under traditional storage and digestion in Biogas Plants. Part I). Statens Veterinære Institut for virusforskning, 64 pp.
8. Lund E, Lydholm B & Nielsen AL (1983): The Fate of virses during sludge stabilization especially during thermophilic digestion. In: Bruce, A.M., Havelaar, A.H. & L. Hermite, P. (Eds.): Disinfection of Sewage Sludge: Technical, Economic and Microbiological Aspects. Proceedings of a Workshop in Zürich, 11.-13. Maj, 1982.
9. McKain N & Hobins PN (1987): A Note on the Destruction of Porcine Enteroviruses in Anaerobic Digestions. Biol. Wastes, 22, 147-155.

Use of Sewage Sludge and Composted Household Waste as a Fertiliser Source

J. Petersen, Department of Soil Science The Danish Institute of Plant and Soil Science Research Centre Foulum, P.O. Box 23, DK-8830 Tjele, Denmark

ABSTRACT

The paper concerns the content of nutrients in municipal sewage sludge and compost in Denmark. The fertiliser value of these waste products correspond approximately to the content of mineral nitrogen. Leaching of nitrate and increased P-index in the soil are recorded at high application rates of sewage sludge. The effects on the environment are less at application rates respecting the thresholds. The content of heavy metals in the soil and the crop were increased by application of highly loaded sewage sludge, but today's quality generally respects the legislated thresholds and the risk of effects on the environment are reduced. In compost and manure the contents of heavy metals are less.

KEYWORDS

Nitrogen, phosphorus, potassium, heavy metals, soil content, yield, crop uptake, fertiliser value.

INTRODUCTION

Sewage sludge and composted household waste are not well defined products. The content of nutrients depends on the quality of the input to the treatment plants, and the type and operation of the treatment plants. There are 1800 wastewater treatment plants in Denmark, and 78% of the wastewater is treated by 210 plants [1]. The number of composting plants is small but will probably increase in the future. More than 69% (>144,000 tonnes DM) of the total amount of sewage sludge in Denmark is applied to agricultural land [2]. The amount of composted household waste applied to agricultural land was 2,000 tonnes DM in 1993 [2].

The aim of the paper is to present the Danish experiences on use of the waste products as fertiliser for agricultural land. To a great extent the paper is based on long-term experiments at Askov Experimental Station [3].

MATERIALS AND METHODS

The long-term sewage sludge field and lysimeter experiments [3] have to be described briefly due to the extensive amount of treatments and results. The field experiments were carried out at Askov (sandy loam), Lundgård (coarse sand) and Rønhave (sandy loam) Experimental Stations during 1974-79 and the nitrogen residual effect was investigated in 1980-84. Today the experimental field at Askov is maintained as permanent grass.

Two types of sewage sludge with different loads of heavy metals (I and II, Table 1) were both applied in two rates (7 and 21 tonne DM/ha/year) during 1974-79. The main crops were spring barley or oats, beet, grass, potatoes, carrots, and cabbage or kale. During 1980-84 the residual effects were investigated in two sections: with and without nitrogen. In this period the crop rotation was spring barley, grass, winter wheat/rye and beet.

Dry matter yield was measured, and the contents of N, P, Na, Ca, Mg, Cu, Mn, Zn, Ni, Cr, Co, Pb and Cd were analysed for the untreated plots and the plots receiving the high rate of sewage sludge. The soil was analysed prior to and following application for Mn, Cu, Zn, Ni, Cr, Co, Pb and Cd. In addition Cd in the soil was analysed 11 years after the last application.

The lysimeter experiments were carried out at Askov Experimental Station in lysimeters (0.8 m) filled with soil from Rønhave (sandy loam) and Lundgård (coarse sand). The two types of sewage sludge (I and II, Table 1) were applied in three rates (400, 800 and 1600 g DM/m/year) and the crop rotation was spring barley, beet, grass and oat. Only one crop was grown each year. The yields and uptakes were similar to the field experiment, but in the lysimeter experiment the leaching was also determined.

Other Danish experiments [4, 5] concern mainly the yield effect and have basically to be considered as single year experiments. By putting the results together some general conclusions appear about the nutrient value of waste products.

RESULTS AND DISCUSSION

Content of Nutrients

The fertiliser value of sewage sludge is related to the content of nitrogen and phosphorus. The mean content of total-N is 35 kg/tonne DM and of total-P 28 kg/tonne DM [2] (VI, Table 1). Only 10-20% of the total nitrogen is present as mineral nitrogen [3]. The content of phosphorus depend to a great extent of the quality of the waste water [6]. The urban waste water has the highest P-content (IV, Table 1) compared to industry and a town (III and V, Table 1).

The nutrient content in composted household waste depend very much on the origin (Table 2). The content of mineral nitrogen is less than 1 kg/tonne DM corresponding to less than 10% of total-N [7]. The potassium content in sewage sludge is low, 1-2 kg K/tonne DM [3], whereas the content in compost is significantly higher (Table 2).

Table 2: Total content of nutrients [kg/tonne DM] in typical composts [7]

	Household refuse	Garden/park waste
Nitrogen	17-30	3-8
Phosphorus	3-6	1-1,5
Potassium	8-13	3-6

Table 1: *Total-N and P content in waste products. Legislated thresholds for heavy metal and the heavy metal content in different types of waste products.*

Description of the origin of the organic matter [1]	Type		kg/tonne DM		ppm in DM						
			Total -N[3]	Total -P	Cad-mium	Mer-cury	Lead	Nic-kel	Chro-mium	Cop-per	Zinc
Legislation thresholds [2] [8]	L				.8	.8	120	30	100	1000	4000
'Low' in heavy metals, 1974-79 [3]	S	I	21	10	5		226	31	111	179	1033
'High' in heavy metals, 1974-79 [3]	S	II	27	23	23		1350	265	831	1120	2282
High industrial load, 1994 [6]	S	III	(79)	11	.6	5.1	39	22	36	149	207
City >0.5 mill. inhabitants, 1994 [6]	S	IV	(116)	38	1.5	1.3	46	24	117	155	414
Town without industrial load, 1994 [6]	S	V	(72)	21	1.5	1.9	48	46	93	255	433
DK-Survey 1993 [2]	S	VI	35 (21-51)	28 (13-34)	2.0 (1-5)	1.7 (1-4)	83 (24-192)	27 (10-61)	27 (17-53)	221 (101-410)	685 (520-1315)
Household refuse incl. soft garden waste [9]	C	VII		4.7	.2	.05	38	8			
Source sorted household refuse [10]	C	VIII	22	4	.3		74			84	224
Manure (FYM and slurry) [11, 12]	M	IX	30-33	7-12	.5 (0.3--0.7)		3 (3-4)	8 (5-15)	6 (2-12)	70-300	175-750

[1] L=Legislation, S=Sewage sludge, C=Compost, M=Manure.
[2] As an alternative the thresholds could be relative to phosphorus. For Cd, Hg, Pb and Ni these thresholds are 200, 200, 10000 and 2500 mg/kg P, respectively.
[3] The figures in brackets are calculated as a difference between input and output from the wastewater plants. Nitrogen is removed from the wastewater by nitrification-denitrification and lost from the sewage sludge.

Yield effect

The yield effect of applied mineral nitrogen in waste products corresponds in most cases to the effect of the same amount of nitrogen as mineral fertiliser (Table 3). Due to the low content of mineral nitrogen in sewage sludge often more than 1000 kg total-N/ha/year were applied in the experiments [3].

The fertiliser value is defined as the amount of mineral nitrogen fertiliser [kg/ha], which is able to replace 100 kg total-N/ha in the waste product for obtaining the same yield. In this way a fertiliser value of total nitrogen in sewage sludge of 20-30 was obtained in the first year after application [3,4]. The highest values were obtained by spring application [4]. The fertiliser value of total nitrogen in compost is 10-15 in the first year after application [5,10, 13].

Table 3: The effect on the yield [hkg DM/ha] of two types of sewage sludge compared to mineral fertiliser. Mean of 1974-79 [3]

	Untreated	Type I	Type II	Fertiliser	LSD.$_{95}$
Barley, grain	18	37	34	37	4
Barely, straw	10	31	23	32	3
Grass	12	80	49	76	11

The fertiliser value is defined as the amount of mineral nitrogen fertiliser [kg/ha], which is able to replace 100 kg total-N/ha in the waste product for obtaining the same yield. In this way a fertiliser value of total nitrogen in sewage sludge of 20-30 was obtained in the first year after application [3,4]. The highest values were obtained by spring application [4]. The fertiliser value of total nitrogen in compost is 10-15 in the first year after application [5,10, 13].

The obtained fertiliser values correspond approximately to the fraction of mineral nitrogen in the waste product. In the first year after application the nitrogen fertiliser value is therefore ascribed to the content of mineral nitrogen in the applied waste product. In the following years, the organic nitrogen is mineralised. The residual effect depends on applied mineral fertiliser (Table 4). In the unfertilised section there is a significant residual effect of applied sewage sludge, and for beets the yield nearly was the same as for fertiliser. Application of mineral fertiliser reduced the residual effect, but crops with a long growing season still benefit of the applied sewage sludge in the 1974-79 (Fertilised section in Table 4). The fertiliser value obtained in the second year after application of sewage sludge was 5-7 when the calculations are based on the first year application [4].

The effect of P in sewage sludge could not be calculated in short-term experiments due to the absent effect of fertiliser P in the unfertilised reference treatment [4]. In a three year experiment with growing cabbage the effect of 170 kg K/ha in compost correspond to the effect of the same amount of K in animal slurry [14].

Due to the low content of potassium in sewage sludge the content was attempted increased by addition of ash from straw used at central heating installations. The ash contained 120-130 kg K/tonne DM, whereas the phosphorus content was low. The effect of mixing these two waste products was investigated on soils poor in P and K status in a single year experiment with spring barley and investigating the residual effect in undersown grass [15]. The mixed products were fused together by heating and granulated to different extent. The more treatment during the mixing procedure reduced the nutrient effect of P and K compared to the mixture of untreated waste products. The yield effect of citrate extractable P in the untreated waste product corresponded to the effect of a equivalent amount of P in fertiliser. An yield effect of K was not observed even on the soil poor in potassium. Application of the mixture of untreated waste products increased the P and K content in the grass but not the yield.

Table 4: *Residual effect 1980-84 [hkg DM/ha] [3]*

	Crop	Treatment 1974-79				LSD.95
		Untreated	Type I	Type II	Fertiliser	
Without fertiliser 1980-84	Winter cereal	19	36	34	52 [1]	4
	Spring barley	13	31	27	34 [1]	3
	Beet	45	83	81	88 [1]	11
	Grass	19	56	48	85 [1]	8
Fertiliser applied 1980-84	Winter cereal	48	49	51	48	3
	Spring barley	32	33	30	32	4
	Beet	88	98	103	88	7
	Grass	72	95	94	80	5

[1] Fertiliser applied in 1980-84.

Application rate and leaching of nitrate

Current Danish legislation [8] limits the application of waste products to not exceed 250 kg total-N/ha/year, 40 kg total-P/ha/year as a mean of 3 years and 10 tonnes DM/ha/years as a mean of 10 years. Application of sewage sludge is mainly limited by the P-threshold, whereas application of composted waste is often limited by the DM-threshold.

An application rate of 200 kg total-N/ha/year (=1×SSR) increased the leaching compared to mineral fertiliser (Table 5) [3]. The difference was more noticeable at higher application rates. The sewage sludge was applied in November/December, four months before sowing spring cereals. The leaching was increased in this period for the application rates 2×SSR and 4×SSR compared to the base level of 0.8 g/m/year (Table 5). Still 2/3 of the leaching occurs after harvest and until next spring.

Table 5: *Leaching of nitrate-N [g/m/year]. Figures in brackets are the loss by leaching from application in November/December until sowing of spring barley or oat in April. Sewage Sludge Rate (SSR) = 400 g DM/m/year. [3]*

	1× SSR		2× SSR		4× SSR	
Type I	3.0	(1.0)	5.5	(1.7)	10.6	(2.9)
Type II	2.4	(1.0)	3.5	(1.4)	5.7	(2.0)
Fertiliser	1.7	(0.8)	2.0	(0.8)	3.3	(0.8)

The 1×SSR correspond to 100 kg P/ha/year, which is more than the crop requirement. Increasing the application rate 2 or 4 times results in accumulation of P in the soil (Table 6). The accumulation depends on the P-content of the sludge (Table 1 and 6). The risk for phosphorous accumulation is small when respecting the application threshold of 40 kg P/ha/year in both sewage sludge and compost.

Table 6: Content (P-index) of phosphorus in the top soil layer of the lysimeter experiment. Sewage Sludge Rate (SSR) = 400 g DM/m/year. [3].

Treatment	Year	Sandy loam			Coarse sand		
		Application rate, SSR					
		1×	2×	4×	1×	2×	4×
Before start	1973	8			5		
Type I	1981	11	19	29	9	15	24
	1987	10	15	26	8	14	26
Type II	1981	17	26	53	15	24	39
	1987	14	20	43	13	21	37
Fertiliser	1981	9	14	20	7	12	16
	1987	9	12	16	9	10	15

The legislated thresholds concerning N and P application rates seems reasonable when the nutrients in the waste product have to be utilised in an environmentally friendly way.

Heavy metals

The potential use of waste products as fertiliser is reduced by the content of compounds acting harmful to the environment, e.g. heavy metals. The content of heavy metals in waste products applied to agricultural land is limited by legislation [8] (Table 1). The waste products has to respect either the limits in Table 1 or the limits which relate the content of heavy metals to the phosphorus content (Table 1 note 2). 2/3 of the produced sewage sludge in Denmark respect these P-related limits. Otherwise the sludge is not allowed to be used on agricultural land.

High application rate of sewage sludge highly loaded with heavy metals (II, Table 1) resulted in increased plant uptake of Zn, Ni and Cd, whereas the uptake of Cr, Co and Pb was not affected (Table 7) [3]. The quality of today's sewage sludge is much better (III, IV, V and VI, Table 1). The contents of heavy metals in compost and manure are in more cases less, but particularly compost is applied in higher rates of DM/ha. Caused by the irreversible accumulation in the soil, heavy metals is still a problem, but the time until reaching an unacceptable and risky level in the soil is prolonged. The plant uptake of Cr, Co and Pb were not affected by soil type, whereas the Zn, Ni and Cd uptake was increased at Lundgård (Table 8). The availability of Cd is linked to pH [16], which normally is lower on coarse sandy soils.

Table 7: Content of heavy metals [ppm in DM] in grains of spring barley 1974-79 [3]

Metal	Untreated	Type I	Type II	Fertiliser	$LSD_{.95}$
Cu	4.0	5.5	5.5	3.4	0.8
Zn	28	58	39	31	10
Cd	0.06	0.12	0.14	0.08	0.03
Pb	0.4	0.5	0.6	0.5	-
Ni	0.5	0.5	0.7	0.5	0.1

Table 8: Zn, Ni and Cd content [ppm in DM] in crops 1974-79 in relation to soil type [3]

	Crop	Soil type			LSD$_{.95}$
		Lundgård, coarse sand	Askov, sandy loam	Rønhave, sandy loam	
Zn	Barley	48	36	29	4
	Oat	52	39	35	7
	Grass	51	39	32	9
Ni	Oat	5.6	1.9	2.3	1.5
	Grass	3.8	2.1	2.5	0.6
Cd	Barley	0.13	0.09	0.06	0.04
	Oat	0.20	0.12	0.09	0.07

The content of Zn and Cd in the straw of barley and oat were affected in the same way as the grain (Table 7). The Zn content in beet and curly kale were higher at Lundgård than at Askov and Rønhave [3].

In the lysimeter experiment the percolated water was analysed for heavy metals. None of the samples had a content above the detection limits (Cu, Mn, Ni < 10 ppm; Cr, Co < 5 ppm; Pb< 1 ppm; Cd < 0.1 ppm) [3]

The contents of Cu, Zn, Pb and Cd in the top soil layer of the lysimeter experiment were increased during application. By application of 2×SSR and 4×SSR the legislated limits of heavy metal content in the soil were surpassed [3]. The thresholds are for Cu, Zn, Pb and Cd 40, 100, 40 and 0.5 g/kg dry soil, respectively [8]. Six years after the last application the contents were still high. The soil at 20-40 cm depth was only slightly affected [3]. Similar results were obtained in the field experiments.

CONCLUSIONS

- The first year effect of mineral nitrogen and citrate extractable phosphorus corresponds to the effect of the nutrients in mineral fertiliser.

- Application of waste products at rates not exceeding the application thresholds does not supply the crop nitrogen requirement.

- Application of sewage sludge with a high content of heavy metals increased the contents in soils and crops.

- The content of heavy metals in sewage sludge has decreased during the last 30 years and seems not to be a great problem today.

- Satisfactory yield can not be obtained by application of waste products solely. Therefore supplementary mineral nitrogen has to be applied. Fertilisation by combined application of waste product and mineral fertiliser has be investigated.

REFERENCES

1. Miljøstyrelsen 1994 Vandmiljø-94. Redegørelse fra Miljøstyrelsen, nr. 2, 1994. 150 pp. (In Danish only).
2. Miljøstyrelsen 1995 Notat om Jordbrugsmæssig anvendelse af affaldsprodukter i 1993. Miljøstyrelsen, Ferskvand og landbrugskontoret, marts 1995. (In Danish only).

3. Larsen, K.E. & Petersen, J. 1993 Long-term field- and lysimeter experiments with large annual dressing of heavy metal loaded sewage sludge, SP-report no. 3, 1993, 69 pp. (In Danish with English summary).
4. Knudsen, L. 1995 Gødskning of kalkning. I C.Å. Pedersen (red.) Oversigt over Landsforsøgene 1994. (In Danish only).
5. Knudsen, L. 1993 Gødskning of kalkning. I C.Å. Pedersen (red.) Oversigt over Landsforsøgene 1992. (In Danish only).
6. Miljøstyrelsen 1994 Miljøfremmede stoffer i renseanlæg. Miljøprojekt nr. 278. Miljøministeriet. (In Danish with English summary).
7. Miljøstyrelsen 1994 Dyrkningsforsøg med kompost 1989-1993. Miljøprojekt nr. 258. Miljøministeriet. (In Danish with English summary).
8. Miljøministeriet 1995 Bekendtgørselse om anvendelse af affaldsprodukter til jordbrugsformål. Nr. 730 af 5. september 1995. (In Danish only).
9. Miljøstyrelsen 1993 Komposteringsanlæg i Århus Nord. Miljøprojekt nr. 238. Miljøministeriet. (In Danish with English summary).
10. Kjellerup, V. 1993 Evaluation of composted source-graded household refuse: Nitrogen effect. Tidsskr. Planteavl Report S2254, 38 pp. (In Danish with English summary).
11. Dam Kofoed, A. Kjellerup, V. 1984 The content of heavy metals i animal manure. Tidsskr. Planteavl 88, 349-352 (In Danish with English summary).
12. Christensen, B.T. (Eds.) 1989 Husdyrgødning og dens anvendelse, 2. reviderede udgave. Tidsskr. Planteavl Beretning S1809.
13. Arenfalk, O. & Hagelskjær, L. 1995 The use of different type of manure in organic vegetable growing. SP-report no. 6, 1995. 27 pp. (In Danish with English summary).
14. Mikkelsen, G. 1994 Stategier for anvendelse af komposteret husholdningsaffald ved økologisk grønsagsdyrkning. I Henriksen, K. (red.) Forskningsdag om grøntsager. SP-report no. 2, 1994. (In Danish only).
15. Hansen, J.F. & Kjellerup, V. 1994 The nutrition effect of phosphorus and sodium in sewage sludge and straw ash - Micro plot experiment. SP-report no. 14, 1994. 44 pp. (In Danish with English summary).
16. Tjell, J.C. & Christensen, T.H. 1992 Sustainable management of Cadmium in Danish agriculture. In Vernet, J.-P. (Eds.) Impact of Heavy Metals on the Environment. Trace Metals in the Environment 2. Elsevier.

Biowaste Composting - Constraints and Advantages

Yann Jomier (1), Hélène Maille (2) and Isabelle Paris (3); (1) ORVAL: BP 2125, 44203 Nantes Cedex, France, (2) CGEA: Parc des fontaines, 169 av. G. Clémenceau 92735 Nanterre cedex, France, (3) CREED: Zone portuaire de Limay, route du Hazay, 78520 Limay, France

ABSTRACT

As waste treatment becomes more and more sophisticated, waste streams tend to be collected and treated in separate fractions, each fraction being treated according to its specific characteristics. In such schemes, the organic fraction of Municipal Solid Waste (MSW) is separately collected in order to be composted. While raw MSW composting is a well-known process, very few studies have been carried out on the composting of the organic fraction from source separated household waste in France. Existing studies show that such waste is best treated mixed with green waste as bulking agent. However, no data are available on seasonal variations of the mixing ratio, waste composition or biowaste compost quality. The purpose of the pilot composting platform established by the CREED is double:
- To define the constraints to be applied to organic waste from source separated MSW in order to obtain a specific kind of compost: requirements for collection quality (fractions to be composted, maximum disturbing substances) and for tonnage;
- to compare the obtained compost to strict quality standards and marketing requirements.

The experiment puts the emphasis on characterisation and sampling: samples must be as close as possible to the collection in totality. The methodology used in this experiment has been established in order to make characterisation as reproducible as possible for incoming waste, waste in fermentation or final compost. The experimental programme is organised in the following steps:
- incoming waste characterisation (composition, nutrients and heavy metal contents)
- monitoring of the composting (turn-over, water content, chemical analyses)
- compost characterisation (sieving residue, nutrients and heavy metal contents, biotests).

The results will allow the evaluations of biowaste evolution over the four seasons, comparison of the final product with compost standards and finally, the study of the existing market for such a compost.

KEYWORDS

Biowaste, source separated MSW, waste characterisation, composting, compost market, European legislation, heavy metals.

INTRODUCTION

The first results of a French experimentation which has been carried out since July 1994 are presented in this paper. This programme is involved in the Eureka project INTEC (INtegrated TEChnology for a global waste management) and followed by the French Ministries of Industry, Environment and Research. The CREED, the Research Centre for waste management of the Générale des Eaux Group is the head of this programme which has been established for and in collaboration with all the Group subsidiaries involved in composting at each stage of the process [1].

An experimental composting platform has been established in Paris area, near the CREED office. Biowaste is characterised and blended with yard waste as bulking agent. In this paper composting is understood as an aerobic biological windrows process.

As an appropriate characterisation of the different types of waste is necessary, our programme included a comparison of all characterisation methodologies elaborated by our partners for the INTEC project, (Krüger, Yorkshire Water, Obrem, Bezner).

The experimentation protocol established by the CREED defines:

- Sampling methodology of each type of waste (biowaste and green waste).
- Analysis:
    - characterisation,
    - monitoring of the composting process,
    - evaluation of the final product.
- Sampling frequency.

The parameters studied before composting are:
- The quantity of the collection of biowaste (from 3 to 18 tons per week),
- The composition of the collection
    - distribution in size (large, medium and small parts, fines),
    - distribution in various constituents (11 classes),
    - the time variations of these distributions .
- The waste pre-treatment and its consequences on final compost
    - biowaste is shredded or not,
    - green waste is dilacerated in all cases.

The final objective of our study is to correlate the quality of the obtained compost from this experiment to the characteristics of the incoming waste.

In all this presentation, the word 'biowaste' represent the organic fraction from source separated municipal solid waste: It includes garden waste, kitchen remains and disturbing substances like plastics or metal.

DESCRIPTION OF THE COMPOSTING PLATFORM

The area of the test unit is about 500 m², it is located in Limay, 50 km west from Paris.
The program has been carried out over 4 seasons, series of 3 windrows are built successively and composting process is monitored daily.

The site is under the regulation of the French law on industrial emissions in the environment. The following plan describes the site design.

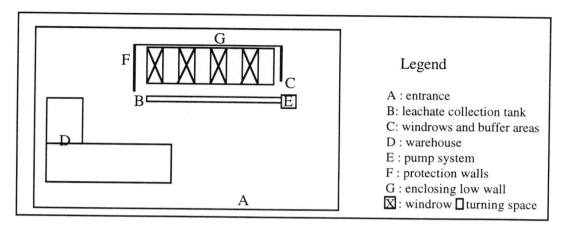

Fig. 1: Composting platform scheme: windrows in parallel

Incoming waste

For our experiment, the waste collected from AURORE, a global waste management facility near Paris, has been used. Biowaste is collected in a specific green bin. Yard waste is collected from a recycling centre. The incoming biowaste treatment (with or without shredding) and the biowaste / green waste ratios (80/20 or 60/40) are studied in various windrows, final composts will be analysed according to these differences. In all cases, green waste is dilacerated before being blended with biowaste. The following table sums up these parameters.

Table 1: Windrows parameters

	Windrows	Ratio (Bio/GW)	Biowaste shredding
Summer (July-August 1994)	A1	80/20	No
	A2	60/40	No
	A3	80/20	Yes
Fall (October-November 1994)	A4	60/40	No
	A5	80/20	Yes
	A6	60/40	Yes
Winter (January-February 1995)	A7	60/40	Yes
	A8	80/20	Yes
	A9	80/20	No

Waste treatment

The composting process consists in regular overturning of windrows with a monitoring of the moisture content and temperature. Leachates are collected, then analysed and sprayed, if necessary, on the windrows. There is no more overturning during the maturation phase. After about four months of composting, windrows are screened and the final product is obtained.

CHARACTERISATION OF WASTE TO BE COMPOSTED

Characterisation consists in determining the composition of a batch, according to chosen objectives. Sampling is the first step of characterisation. It has to face two main problems: make a sample as representative as possible of the original waste stream, and minimise the size of this sample, which is difficult in regard to the heterogeneous characteristics of such waste.

The second step is the analysis of several parameters: chemical characteristics (dry matter content, organic matter, pH, nutrient and heavy metal concentration) and physical characteristics (volume weight / bulk density, texture).

CREED protocol

In Europe, there is no specific characterisation methodology for incoming waste (like biowaste), but there are a lot of sampling and analyse methodologies for final compost.

Different methods are available in France for the characterisation of waste and each of them has its own sampling practice. The methodology used in this experiment has been established from these different methods (especially with the MODECOM method of the ADEME agency) and also from our tests [2-3].

The different steps of the characterisation are as follows: - Sampling - Fraction size determination - Sorting - Moisture measurements - Sample preparation for laboratory - Chemical analysis.

For the incoming characterisation (*cf.* fig 2), 11 fractions are separated:
- yard waste   - papers & cardboard   - glass   - other combustible waste (wood...)
- food waste   - plastics   - textiles   - other non combustible waste (stone...)
                - metals   - composite   - hazardous waste (paint, drugs...)

Table 2 sums up the frequency of analysis done for the incoming waste stream, during the composting process, and for the characterisation of the final product.
Analyses done during the composting process allow to evaluate the organic matter degradation by biological activity.
For incoming waste and final compost, the same chemical analyses are carried out. These chemicals parameters have been chosen according to the French requirements on organic soil improvement [4]. Biotests in final compost are carried out in order to evaluate maturation of the pile.

ANALYSES AND RESULTS

The first results are about the waste stream characterisation.

Table 2: Type of analysis done before, during and after composting

Product	Analysis	Description	Frequency
Dilacerated and added green waste	Characterisation	Visual estimation	1
	Physical	Moisture & pH	1
	Chemical	Heavy metals	1
		OM; C; N; P2O5, K2O, CaO, MgO	1
Incoming biowaste	Characterisation	Granulometry & 11 fractions	1
	Physical	Moisture & pH	1
	Chemical	Heavy metals	1+4 (per fraction)
		OM; C; N; P2O5, K2O, CaO, MgO	1
Windrow	Characterisation	Granulometry and Inert	1
	Physical	Moisture & pH	1 per week
	Chemical	OM; C; N	1 per week
Final compost	Characterisation	Granulometry & 11 fractions	1
	Physical	Moisture & pH	1
	Chemical	Heavy metals	1+4 (per fraction)
		OM; C; N; P2O5, K2O, CaO, MgO	1

Biowaste distribution

Biowaste tonnage varies with the month of collection (figure 3), it decreases in August (holidays, less gardening works) and increases in October-November (gardening works). It also fluctuates with the quantity of collected garden waste.

Taken as a whole, there is no hazardous waste in biowaste collected for this programme. Disturbing substances represent less than 10% of the global collection. The part of fines (smaller than 20 mm) is very fluctuant (from 13% to 43% in raw weight). Their composition varies with the week of collection: mineral substances like ground and sand + clipping grass in October or kitchen remains in February. On Figure 4, various observations can be done:

- There is a predominance of green waste in summer and fall with the higher percentage in October;
- there are less than 5% of food waste and paper-cardboard excepted in winter: 29% and 16%.

These organic fractions are collected in the other bin which must only contains dry constituents like packaging. People put first green waste in the bin then, if the bin is not full, other organic waste.

On the nine experimental windrows, the 2 theoretic biowaste / green waste ratios tested (80/20 and 60/40) didn't represent a reality. In fact, 3 different ratios [(food waste + cardboard) / (garden waste) were experimented. The analyse of these batches shows that, for most part of the year, biowaste composting can be done only (it's not necessary to add and mix green waste from other origin in the windrows).

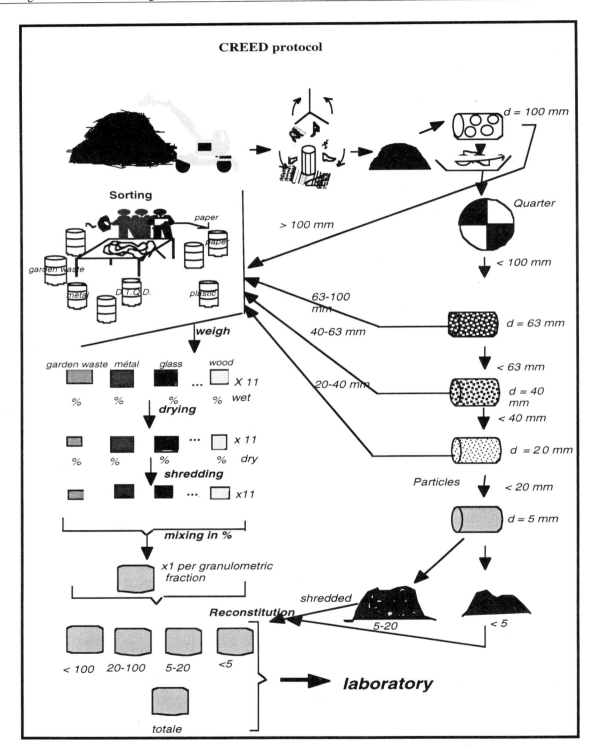

Fig. 2: CREED protocol for characterisation of the incoming waste

Compost quality

A study of analyses of compost quality in Europe shows that the same criteria are used to describe this quality. Only the thresholds of those criteria change according to the countries [5-7]. The results of analyses are expressed in various units: % in dry matter content, % in fresh matter content, in volume or weight units. On Table 3, analytic composition of 3 batches of compost are done.

Table 3: Analytic composition (before maturation phase)

Batch Criteria	A1	A4	A9
Organic matter (%FM)	28.7	21	29.8
Moisture content (%)	28	42	30
OM / N	21.5	19.6	27.2
N (%DM)	1.85	1.85	1.56

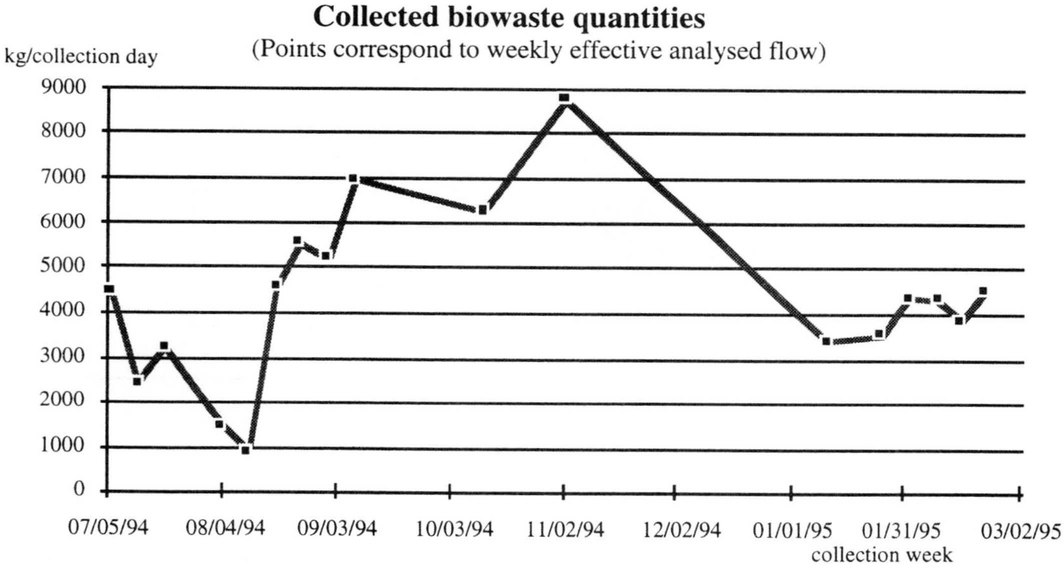

Collected biowaste quantities
(Points correspond to weekly effective analysed flow)

Fig. 3: Quantitative biowaste distribution

More severe tests have to be done for heavy metals contents (Cd, Hg, Pb, Ni, Zn, Mo, Cr, Cu, As, Se). Variation of admitted thresholds from one country to another can be quite important. Heavy metal content in CREED composts are compared in the following table to the European ecolabel (94/923/CE) requirements.

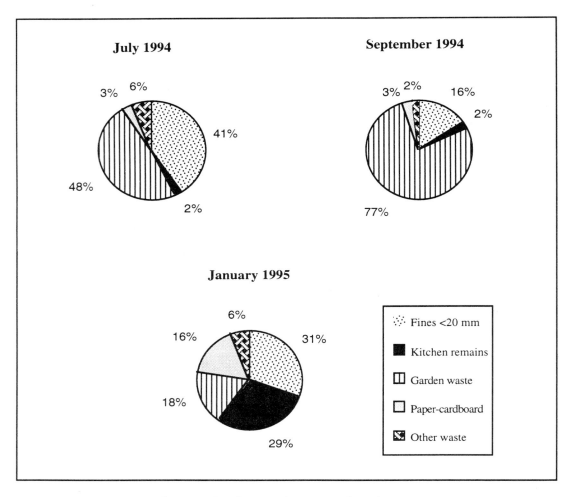

Fig. 4: Qualitative biowaste distribution

Table 4: Heavy metal contents in mg/kg dm

Element	European label	Compost A2 <10 mm	Compost A2 10-25 mm	Compost A3 <10 mm	Compost A3 10-25 mm
Cd	1.5	1.14	0.89	1.33	0.62
Cr	140	44.7	33.9	62.4	99.1
Cu	75	71.6	57.3	77.6	75.7
Hg	1	0.27	0.13	0.24	0.13
Ni	50	21.1	22.3	27.0	20.7
Pb	140	56.5	108	56.4	51.8
Se	1.5	0.34	0.16	0.34	0.27
Zn	300	189	170	196	170
As	7	15.1	16.7	12.9	3.98
Mo	2	2.09	2.18	2.32	2.71

A 10% divergence is acceptable on an analysis point of view. As & Mo contents are only required for compost derived from MSW or sludge. These two European thresholds are very low and are controversial.

CONCLUSION AND PERSPECTIVES

The characterisation of biowaste distribution in the organic household waste from separated collection in France has been done in order to follow the evolution of the biowaste composition overtime.

It has been checked that a greater part of green waste is collected at the end of summer and during the fall and a greater part of food waste is collected during winter. Nevertheless, green waste represent the majority of the collected organic waste during one year.

The different batches of compost from these collections present different properties: nutrients and organic matter content, heavy metal (Table 3). The link between biowaste and compost qualities has been established.

This programme gives already some indications concerning the management of an industrial platform. In order to insure the same quantity and quality of compost all along the year, the 2 following techniques must be processed:

- biowaste will be dilacered after checking its disturbing substances content (according to specifications on collection);
- ratios must be adapted to the biowaste composition: no added green waste is required except in August (to increase the tonnage) and in February (to insure the same quality of the batch).

The analyses of the compost show a very low content of heavy metals, well below the European standards except for lead and molybdenum.

Our programme will now put the emphasis on:
- the detailed specifications that must be applied to the collection of the organic fraction of MSW in order to obtain a marketable compost, (maximum percentage of glass and plastics. minimum percentage of compostable waste. optimal percentage of fines. ...);
- the distributions of heavy metals in incoming fractions and final compost in order to see if any element is concentrated in any particular fraction and if there is a direct link with compost quality. (*i.e.* could the sorting of any particular incoming fraction be a solution to decrease heavy metal content in compost ?);
- the marketability of the compost: evaluation of the compost quality versus European recommendations and with end-users' needs and uses.

ACKNOWLEDGEMENTS

This work has been done in collaboration with:

Collection / Valorisation	Valorisation	Valorisation / Marketing
CGEA - CFSP - SARM	CGC - OTV.D - SOGEA	ORVAL

REFERENCES

1. I. A. Paris, S. Opsomer. (1994). Experimental composting platform. CREED internal reports (in French).
2. S. Opsomer. (1994). Waste characterisation: CREED Protocol. (in French).
3. ADEME. (1993). MODECOM^{TM}: méthode de caractérisation des ordures ménagères. Angers. France.
4. AFNOR. (1976). Experimental standard U 44-101: organic products. organic soil conditionners. organic material for soil improvement sampling. (in French).
5. (1994) Ecological label for soil conditionners (94/923/CE). Journal officiel des communautés europeénnes 31/12/94 n° L 364/21.
6. (1993) Compost products: declaration and control of environmental and quality parameters. Nordiske Seminar- og Arbejds- Rapporter.
7. E.I. Stentiford, S. Kelling, J.L. Adams. (1986). Composting of refuse derived fines from mechanical sorting. *Yorkshire Water*.

Application of Compost in Amenity Areas

M. Carlsbæk & U. Reeh, Danish Forest and Landscape Research Institute, Hoersholm Kongevej 11, 2970 Hoersholm, Denmark

ABSTRACT

In Denmark there is a need for extended guidance concerning the application and use of compost. Practical experiences with compost use have been collected from people working with park and landscape construction and management, mainly through telephone interviews. The main reason to use compost is stated to be the soil improving properties and the expectations were in most examples fulfilled. Generally, a 10 cm layer composted garden and park waste is mixed with the top soil to a depth of 25 cm, or used as a mulch. Base fertiliser is often added. The result is an over-fertilisation as a 1.5 cm layer or less of compost equals base fertilisation as regards P, K and Mg. If the top soil or the growth medium is mixed with more than 20-30 vol% compost, it should be tested before application to 'new' plant species. Impaired plant growth has been recorded when top soil is substituted by garden and park waste compost or when it constitutes 40 vol% of the top soil at establishment of lawns. Before the compost is used for transplanting bedding plants and small bare-rooted plants, it should be tested for weed seeds and of viable plant parts. The texture of a peat-compost-mixture will often not allow proper drainage for outdoor use and will decompose and settle substantially within a few years. A N-effect of compost has been observed 2_ months after application of compost-amended topdressing to football fields. The incorporation of a 15 mm layer of topdressing lasted 3 weeks, longer during droughts.

KEYWORDS

Garden and park waste, yard waste, source separated biodegradable household waste, domestic waste, organic municipal solid waste, landscaping, soil improving, mulching, topdressing, football field, fertiliser value, liming effect, trees, bedding plants, perennials, grass.

INTRODUCTION

The amount of waste composted in Denmark increased rapidly from 1990 to 1993 by nearly 500%. 308,000 metric tons of waste were composted in 1993 and approx. 154,000 metric tons, equal to 205,000 m^3 compost, were produced. Some of the reasons for this increase might be found in the growing environmental consciousness in society and the political will to support recycling practices through the introduction of waste charges, 160 DKK = 29 US$ per metric ton if incinerated and 195 DKK = 35 US$ if landfilled. Most municipalities are now composting their park waste and substantial parts of the garden waste from the inhabitants who do not carry out home composting. Compost made solely from garden and park waste constitutes 73% of the produced compost in Denmark [1,2].

This paper summarises results from two ongoing projects, a survey concerned with the use of compost in amenity areas, and growth trials with compost as an ingredient in topdressing at football fields. The amenity area in general, is an important market for compost in Denmark and 13% of the compost produced in Denmark is used in these areas [2]. People working with park and landscape construction and management are expressing a need for extended guidance concerning compost application and use. The survey and the growth trials have been conducted to meet this demand.

Compost can be used for soil improvement because of its content of organic material and biological activity and its liming effect. When incorporated into clay soils, compost often improves drainage and air exchange and counteracts compaction and puddling. In sandy soils, compost usually increases soil water retention and the content of plant-available nutrients (CEC), and the risk of wind and water erosion is often reduced [3,4,5].

Compost may be used as a mulch. Mulching conserves moisture in the surface soil, moderates soil temperature variation, reduces soil compaction caused by surface traffic, reduces the growth of weeds and protects roots from mechanical injury. Mulching should not be used on wet or poorly drained soil because the evaporation of excess water is restricted by the mulch. If the mulch contains an abundance of fine particles the gaseous exchange between the atmosphere and the soil is reduced and this might be fatal for plant growth, particularly in compact clay soils [6].

In addition to soil improving properties, compost has a substantial fertiliser value. A slow release of nitrogen in compost is characteristic for organic fertilisers and this could be a positive quality when fertilising perennials. Prevalence of perennial plants with a substantial number of plant species is typical for amenity areas. Different plant species have different needs concerning the nutrient level and may also react differently towards organic fertilisers like compost [7]. It is not possible to predict the response of various plant species to application of compost, which is why it is necessary to continue growth experiments with more plant species. Precaution is thus important when developing general recommendations for application of compost in amenity areas.

Preliminary results from an experiment with compost as part of topdressing for football fields will be presented in this paper. Among the areas maintained by the municipalities in Denmark, the largest amounts of mineral fertilisers and pesticides are used at ball fields. Topdressing with compost has a fertilising value compared to traditional topdressing made of sand only.

In addition, many types of compost have demonstrated plant disease suppressive qualities and research has shown this effect of compost-amended topdressings towards several soil-born pathogens on golf greens. The predictability of the disease suppressive effects needs further research [8,9]. The disease suppressive qualities are not measured in the experiment.

METHODS

The survey, which was initiated in June 1994, is based on two sources of information concerning the locations of previous compost use and the opinions of the people who used the compost. One source of information was the references from a survey conducted in 1990, which also dealt with the use of compost in amenity areas. The places mentioned in this survey were visited and the persons in charge of the maintenance were interviewed concerning progress of plant growth since 1990 [10]. The second source of information was a national statistical record mentioning all major producers of compost in Denmark [2]. These producers provided the names of the professional users like gardeners, landscape architects, contractors, and road foremen who had used their compost in amenity areas.

In total, 40 persons were interviewed about purposes and ways of compost application, type of compost used, soil type, plant species, maintenance and progress of the plant growth. 120 different examples of compost use were described and 20 locations were visited.

The use of compost as part of topdressing and its supplementary fertilisation effect on ball fields are tested in experiments on three fields in the municipality of Vejle: one football field on sandy soil (plot size 247 m^2), one football field on clay soil (plot size 549 m^2), and one field partly used as a turf cutting nursery but with no ball playing activities (19 treatments, plot size 6 m^2). The experiments were initiated in the spring of 1993 and run until winter 1996. The clippings were left in situ the first year and removed the second year. Here only the main features of the different treatments will be described. 25-50 vol% composted source separated biodegradable household waste or 33-100 vol% composted garden and park waste were mixed with 0-75 vol% washed, 0-4 mm gravel. A 10, 15 or 20 mm layer of topdressing was applied yearly in May-June. Some of the plots with topdressing applications had supplementary mineral fertiliser nitrogen added. Reference plots with gravel only, or with mineral fertilisers only, or without any treatment at all were established for comparison. In the small plots, compost was applied by hand while a topdressing machine was used for large plots.

In general, 4-6 mm is the recommended thickness of topdressing layers for ball fields [11]. In these experiments thicker layers were applied to achieve an increased fertilisation effect without increasing the number of topdressing applications. Consequently, an increase in the economical expenses could be avoided.

The following parameters were measured for the various treatments: grass colour, grass height, grass incl. weed biomass production, ground area covered with grass/weed/bare soil, and the period where the sward was covered with topdressing. Soil samples were analysed for P-soluble in 0,5M $NaHCO_3$, K-exchangeable, and pH in 0,01M $CaCl_2$. Playing qualities of the swards with different treatments will be measured in 1996.

RESULTS AND DISCUSSION

General aspects from the survey

The compost used most often in amenity areas is made from garden and park waste, which is one year old and screened to < 19-22 mm. Often compost is produced and used by the municipality itself. Dry compost products are easier to handle than ordinary top soil. The price must be competitive with the price of good top soil, peat, wood chips or bark chips respectively, depending on the purpose of the compost use. 44 DKK (8 US$) per m^3 or 59 DKK (11 US$) per metric ton of fresh weight compost are average prices [2]. Compost is generally used in two ways: mixed with the top soil at establishment and as mulching for maintenance. Compost is mainly used in areas where gardeners expect the soil to benefit from the soil improving qualities. The prevalent ways of using compost in amenity areas in Denmark are outlined in table 1. The result from the survey will be presented with the same division as in table 1, i. e. divided according to the task in question.

Compost mixed with top soil at establishment

The Danish practical experiences show that thick layers of compost are incorporated into the top soil, and compost often constitutes 40 vol% of the top soil. According to many of the practical examples, normal fertilisation practice with mineral fertiliser is often carried out even though large amounts of compost have been applied. This practice must be regarded as over-fertilisation and may cause damage to new plants. Furthermore it constitutes a potential risk of leaching. However, no conclusive examples of scorching or impaired establishment of bushes or trees have been recorded, when garden and park waste compost were mixed into the top soil. Some examples of impaired plant growth have been recorded, when 3-10 cm layers of composted source separated biodegradable household waste or composted sewage sludge were mixed into the top soil.

Analyses of the composts used showed that the nutrient contents in a 1.4 cm layer of composted garden and park waste or in a 0.7 cm layer of composted source separated biodegradable household waste were equal to the amount of nutrients usually applied with base fertiliser in parks and gardens, except for N (referring to values in table 2). In addition, preservation liming can be omitted because of the liming effect of the compost. The nutrients in the compost, except for the large proportion of organically bound N, will be fully available for plant growth when the compost is worked into the soil. The nitrogen requirement of woody plants may be met with the mentioned layer thickness while most herbaceous plants will need additional nitrogen. The Danish legislation limits the use of compost to a maximum application of 2.5 kg N-total/100 m^2 * year. The only cases where there are no regulations of maximum application rates of nutrients are when compost is made solely from garden and park waste or if any type of compost is used in private gardens.

Examples of impaired establishment and reduced growth of lawns have been recorded when 10 cm garden and park waste compost were incorporated into the top soil to a total depth of 25 cm. In one example on a sandy clay soil, the germination ratio of the grass was very low.

In another example on sandy soil, the grass areas exposed to direct sun dried out and died during a sunny period without rain. In general, firmness of the grounds were lowered. Thus it must be advised against to use as much as 40 vol% compost in the top soil when establishing lawns or ball fields. An example of establishing large lawns in a growth medium consisting of 20 vol% 1 year old garden and park waste compost and 80 vol% top soil did not cause any

Table 1. Prevalent ways of using compost in amenity areas in Denmark, 1994-95 (not necessarily to be recommended).

	Registered way of using compost (GPW or SBHW)[*)]	**Comments**
General		
Establishment	Mixed with top soil to a depth of 25 cm: 10 cm (5-15 cm) GPW or 4 (3-10) cm SBHW.	Mostly for soil improvement of sandy soil, heavy clay soil, or poor urban soil. Base fertiliser is often added too, i. e. over-fertilisation of P, K and Mg. Watch out for increased conductivity when using SBHW. The compost should be free of weeds.
Maintenance	Mulching: 5 - 25 cm GPW or 2 - 5 cm SBHW.	Mostly the amounts of weeds are reduced in the first year. Weeding is facilitated. Root propagating weeds must be removed beforehand. Growth of older bosquets and trees are improved. Lower N-effect of GPW than expected. SBHW often substitutes fertiliser. Young compost can be used.
Specific		
Planting holes	25-50 vol% GPW mixed with sand or soil from the planting hole. Incorporated up to 50 cm depth.	Textures of mixtures with peat might be disadvantageous: drainage often hampered, breaking down too quickly. No compost deeper than 30 cm. If possible, use less compost. Consider mulching instead.
Pots/columns	10-100 vol% GPW, 90-0 vol% peat or top soil, 0-20 vol% gravel.	Watch out for increased conductivity and texture, incl. water retention capacity. Compost must not contain weeds!
Bedding plants	1-3 cm SBHW mixed with top soil or as for pots/columns.	See comments on pots/columns.
Lawns and football fields	*Establishment:* 10 cm GPW mixed with top soil. *Topdressing:* 25 vol%, screened 0-10 mm compost mixed with washed 0-4 mm gravel. 5-8 mm layer applied in May.	*Establishment:* poor germination, grass dying during droughts, reduced firmness of ground. Incorporate only 2-5 cm layer GPW! *Topdressing:* Substitutes part of mineral fertiliser and lime.

*) GPW: composted Garden and Park Waste. SBHW: composted Source separated Biodegradable Household Waste (or other nutrient rich compost)

Table 2. Nutrients supplied with base fertiliser or with 1 cm layer of compost (kg nutrient/100 m^2)

	N water soluble	N 25% of total	P citrate soluble	K water soluble	Mg total	CaCO$_3$ Scheibler method
NPK 16-4-12 + Mg, 7 kg/100 m^2	1.1	-	0.3	0.8	0.1	5 [1)]
Garden and park waste compost [2)]	0.1	0.8	0.5	0.8	0.5	5
SBHW compost [3)]	0.6	1.5	0.8	2.2	0.7	5

1) Preservation liming. 2) Average for 10 composting plants. Finished compost: 0.7 metric ton/m^3, 20% organic matter in dry matter. 3) Minimum 85 weight % Source separated Biodegradable Household Waste in starting materials. Average for 3 composting plants. Finished compost: 0.4 metric ton/m^3, 65% organic matter in dry matter. The Danish legislation allows 2,5 kg total-N/100 m^2 * year, use in private gardens is not regulated as regards fertilisation.

problems. The use of 20 vol% garden and park waste compost is in good accordance with German recommendations when a very thoroughly mixing is used [12].

In Germany, comprehensive recommendations have been developed for use of different types of compost in parks and landscaping. For soil improving, incorporation of 2-4 cm thick layer of compost is recommended. If mixing with top soil is optimal, compost may constitute 15-30 vol% depending on the nutrient contents of the compost. Compost should not be incorporated any deeper than 20 cm on heavy soils and 30 cm on sandy soils. The recommendations are also valid for planting holes. Liming and fertilisation with P, K, Mg and micronutrients are not necessary after addition of compost. Mineral fertiliser nitrogen is not necessary the first year as large amounts of compost, i. e. a 4 cm layer, is added. If a yearly layer of compost is applied in bosquets and flower beds, a 0.2-1.5 cm layer is recommended [12,13]. These recommendations correspond well with the nutrient contents of Danish compost types as well as with the need of most plants. However, the actual amounts used in Denmark are often much higher.

Compost substituting top soil

Some examples have been recorded where compost constitutes the entire top soil layer. Different bushes, *Spiraea cinerea, Ribes alpinum, Potentilla fructicosa,* and bouquet roses, were planted in 20 - 40 cm thick layers of 1-2 year old garden and park waste compost. The plants were irrigated normally but no fertilisers were added. The establishment and growth of the bushes were reported to be equal to or more vigorous than usual.

In another example, where a 25 cm layer of 1 year old garden and park waste compost was used as top soil in a number of large gardens, unusually low degree of establishment and wintering were recorded. During the first spring a yellow discolouring of the foliage occurred. In these examples plant species were not specified, but a range of bushes and trees were affected. The problems were solved by replanting the same season after mixing 15 cm of ordinary top soil into the former applied compost top layer. An immature compost might explain the impaired plant growth, exposing the plants to phytotoxic small fatty acids and anaerobic conditions during winter and nitrogen immobilisation during spring.

Examples of very low degree of germination have been recorded when establishing lawns on 25 cm layer of 1 year old garden and park waste compost.

Generally, it must be advised against to use compost as a substitute for top soil unless the response of the individual plant species to the specific type of compost used is already known. Thus some plant species like *Potentilla fructicosa* benefit from large compost additions according to this survey and the results of other researchers [7]. No matter how the plants react, this practice must be regarded as over-fertilisation and in addition, a substantial decomposition and settlement of the growth medium must be expected.

Compost used as a mulch

In general, the effects on plant growth were recorded to be positive when compost was used as a mulch. The fertilising effect of mulching with garden and park waste compost was in some examples lower than expected. The gardeners who used nutrient rich compost warned against application of too thick layers, since 5-10 cm layers had resulted in impaired plant growth in established bosquets in several examples.

In the County of Copenhagen, compost of garden and park waste was used as a mulch for road trees which had ceased growing and to a large extent were dying off. A 12 cm thick layer of compost was applied twice with 1 year interval to 260 oaks *Quercus robur*. The compost was applied about 14 years after planting the trees which were planted with a 12-14 cm stem circumference. The survival and growth of the trees were substantially increased in the following years. Three years after compost application, fine feeding roots were observed in the zone between the top soil and the settled compost. If a superficial root system is prevalent it constitutes a potential risk of drought damages, of mechanical injury from mowing and weeding and of damages from salt applied to roads for traffic safety in winter.

Mulching is often used to prevent and control weeds. The amount of weed seeds and viable plant parts in the compost itself depends on the operation of the actual composting plant from where the compost originates. Some composts contain no weeds, others contain low amount of weeds with only few species present. However, examples are known where the weed level in the compost seems to be equal to what is normally found in the top soil. An important statement which is repeated in most examples is that weeding is facilitated because of the loose structure of the compost layer. In some examples it was stated that the compost layer must be minimum 10 cm thick to give a substantial weed reduction and that a 10 cm layer of wood or bark chips was a better way of preventing weeds.

Root propagating weeds cannot be controlled by using compost as a mulch and must be removed before application. An example where compost was used as a mulch to control the root propagating weed grass *Elymus repens* without prior removal had the opposite effect as the growth of the weed grass increased tremendously.

Research has shown that a 3 cm layer of composted source separated biodegradable household waste almost completely prevents the emergence of smaller-seeded weeds when no viable plant parts are present. This is mainly caused by the covering effect of a layer of this thickness and to a lesser extent by the chemical effect of the compost [14].

Compost used for bedding plants

Many positive and some negative examples have been recorded concerning growing of bedding plants in compost-amended soil.

Using compost for planting of bedding plants demands a compost free of weed seeds. Very high expenses for weeding have been recorded in a few examples where nutrient enriched peat was substituted with compost as a soil improver. The used compost contained much higher amounts of weeds than the peat normally used.

In another example, bedding plants were scorched when 3 cm of a 1 year old nutrient rich compost was mixed 5 cm into the top soil before transplanting.

Propagation of bedding plants and bushes in a 1:1 volume mixture of peat and garden and park waste compost failed when the potted plants were moved outdoor for hardening. The growth medium did not drain properly during a period of rain, and subsequently many plants died. Eventually planting out was difficult because of the muddy structure of the growth medium.

The bedding plant *Impatiens spp* grows well in pure, 0-10 mm screened, 1 year old garden and park waste compost used in flower columns. The columns were irrigated with a 1 0/00 solution of NPK 23-3-7 from the top and the sides in the same way as when peat is used.

Growth trials with nursery plants in containers where 25 or 50 vol% of compost was mixed with peat showed opposite responses depending on the plant species. Among things to observe were mentioned: increased conductivity, increased pH, possible nitrogen immobilisation, and reduced waterholding capacity compared with peat [7]. In composts pH above 7 are common and this might induce phosphorous, iron or manganese deficiencies [15,16].

A general conclusion of these examples would be to use guaranteed weed free compost and to check possible increases in conductivity depending on type of compost and mixing rate. The texture of the growth medium must be considered, trying to optimise drainage ability as well as waterholding capacity. Too fine a texture of the growth medium will result in waterlogged conditions and hampered handling properties.

Compost in planting holes

Compost is used to improve the soil in planting holes when planting larger trees and bushes, bare root or with root ball. Normally, composted garden and park waste is mixed thoroughly with sand or with the soil from the planting holes and constitutes a maximum of 50 vol%. Thereby the growth during the second year was observed to be more vigorous than usual, or to be equal to growth obtained when using unamended top soil. There were no examples of the growth being inferior to growth in unamended top soil.

A number of 1 m Magnolia spp planted in a 1:1 volume mixture of garden and park waste compost and peat in 50 cm deep planting holes did all get pale green foliage, had no growth and started dying 2-3 years after planting. Refilling of the planting holes with additional growth medium was necessary after 2 years. The growth medium was greyish and sticky in some deeper parts when examined 3 years after planting. The reason to the bad condition of the plants might be the occurrence of anaerobic conditions caused by: settlement of the growth medium, too high water retention capacity and too few air filled pores, higher mineralization rates of both peat and compost compared to the organic matter of top soil. Thus, it is not advisable to use materials which consist of mixtures of peat and compost to refill planting holes without knowing the texture of the growth medium and the water retention. Secondly, the compost should be old and sufficiently decomposed if it is to be used as a constituent in refilling materials for planting holes.

The benefits of soil improvement for planting holes are often questioned. Alternatively, some researchers recommend to loosen the present soil in an enlarged planting hole with extended, flared shoulders and to choose plant species suitable for the actual soil type whenever possible. The organic material, here the compost, should then be used as a mulch [6,17,18].

Compost as part of topdressing for football fields

Results from 1993 and 1994 from the experiments in the municipality of Vejle show that topdressing layers with a thickness of 15 and 20 mm reduce the grass coverage permanently on fields without ball playing activities. This was true for all treatments. On fields in use, the length of the incorporation period for a layer of 15 mm topdressing with 67 vol% garden and park compost and 33 vol% gravel was 3 weeks. In periods without rain, there is almost no incorporation of the topdressing layer until grass growth is resumed. The percentage of grass that is not covered with topdressing is inversely proportional with the increase in thickness of the topdressing layer and without any differences between the different types of topdressing (an example in fig. 1).

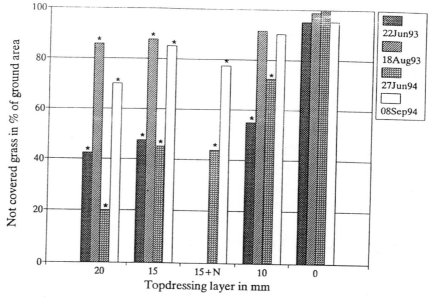

Fig. 1. Percentage of the sward that is not covered by the topdressing compared to the thickness of the topdressing layers and as a function of time. Topdressing with 33 vol% composted source separated biodegradable household waste and 67 vol% washed 0-4 mm gravel applied on the 6th May 1993 and on the 20th June 1994 to a ball field without playing activities. (*: 5% significant difference to 0-treatment on same day. +N: supplementary ammonium nitrate limestone added).

A nitrogen effect was observed 2 months after application of topdressing with garden and park waste compost with supplementary mineral fertiliser nitrogen and after application of topdressing with composted source separated biodegradable household wasted with or without supplementary fertiliser nitrogen. The N-effect was measured as increased height and increased fresh weight production compared with the plots without any treatments. No nitrogen effect was observed 2 months after sole application of ammonium nitrate limestone.

The growth season of 1993 indicated a 18-32% utilisation of N-total in compost the first year after application, highest for the composted source separated biodegradable household waste, lowest for composted garden and park waste. The value is based on measurements of grass colour and height. The same differences between the utilisation of N in the two types of compost were measured in 1994. Apparently, 4-8% of the total-N in the composts were utilised within the first week after application, the highest values were again obtained for composted source separated biodegradable household waste. If an average value of 25% utilisation of N-total is used, a 10 mm layer of topdressing consisting of sand and 67 vol% garden and park waste compost or 33 vol% composted source separated biodegradable household waste will provide 0.5 kg plant available N per 100 m^2 in the first year.

When composted source separated biodegradable household waste is applied to agricultural land in the spring, the utilisation of compost nitrogen in the first year is usually 10-20% [19,20]. A higher utilisation ratio may be expected in an already established and permanent grass crop as it has a very long growth period. However, the fact that compost is used as a mulch instead of being incorporated into the topsoil may partly counteract the increase in the utilisation ratio of compost-N. A 25% utilisation ratio for compost-N applied to ball fields seems reasonable.

No differences were observed regarding P and K levels in the soil in the spring of 1995 no matter whether the nutrients were supplied as citrate soluble phosphorous and water soluble potassium in compost or as a PK mineral fertiliser. In many cases P-total and K-total in compost are regarded as fertilisers fully comparable to PK in mineral fertilisers [12,19]. All treatments with topdressing had a liming effect, 0.7 increase in soil pH, compared with plots solely given ammonium nitrate limestone and PK mineral fertiliser.

CONCLUSION AND RECOMMENDATIONS

Compost is mainly used for its soil improving effects and mainly by incorporation of the compost into the top soil. Compost should not be incorporated deeper than 20 cm in heavy clay soils and 30 cm in sandy soils. Garden and park waste compost may constitute maximum 30 vol% of the top soil, however only 20 vol% when establishing lawns and ball fields. Both percentages imply a thorough mixing of the materials. When using larger amounts of compost for soil improvement, only old and sufficiently decomposed compost with a low nutrient contents should be used. Positive results from experiences where more than 20-30 vol%

compost was used in the growth medium cannot be applied to other plant species without prior testing in each case.

The compost should be free of weed seeds and viable plant parts before it is used. This is especially important when planting bedding plants, small bare-rooted plants and the like, where no competition from other plants can be tolerated in the beginning.

It would be advantageous to use younger and coarsely screened compost when compost is used for mulching. Root propagating weeds cannot be controlled with a thick mulch of compost.

Special attention should be paid to the physical properties of mixtures of compost and peat when they are used as refilling material for outdoor planting holes or as a growth medium for potted outdoor plants. The texture of the medium can be very fine not allowing proper drainage and the growth medium decompose within few years enhancing risks of anaerobic conditions in the root zone.

For specific purposes, the fertilising effect of P, K and Mg in compost and part of the N in compost has been demonstrated and thereby proved to reduce the need for mineral fertiliser. In general, the application of mineral fertiliser should be reduced accordingly to the amounts of nutrients supplied by the compost.

When applying topdressing on football fields, the layers should be thinner than 15 mm to ensure rapid incorporation into the thatch and ground. Compost-amended topdressings can substitute part of the mineral fertiliser and lime.

ACKNOWLEDGEMENTS

The survey is funded by Council of Recycling and Cleaner Technology affiliated to the Danish Environmental Protection Agency. All persons who provided information in the survey are greatly acknowledged. The topdressing experiments are funded by Municipality of Vejle, National Olympic Committee and Sports Confederation of Denmark, and Danish Gymnastics and Sports Associations. Finn Jørgensen, Gitte Rasmussen and Carsten Damgaard carried out the practical tasks. The results from the first growing season were processed by Finn Jørgensen, M.Sc.

REFERENCES

1. Domela, I., A. Nielsen, U. Reeh, & J. Martinus (1993): Kompoststatistik 1991. Materialstrømsovervågning. (Compost Statistics 1991. Material-Flow-Monitoring). 37 pp. Rendan A/S & Danish Forest and Landscape Research Institute, Denmark.
2. Domela, I. & B. Mortensen (1994): Kompoststatistik 1993. Materialstrømsovervågning. (Compost Statistics 1993. Material-Flow-Monitoring). 62 pp. Rendan A/S, Denmark.
3. Shiralipour, A., D.B. McConnell & W.H. Smith (1992): Physical and Chemical Properties of Soils as Affected by Municipal Solid Waste Compost Application. Biomass and Bioenergy: 3:3-4:261-266.
4. Martin, O. & R. Kowald (1988): Auswirkung des langjährigen Einsatzes von Müllkompost auf einen mittelschweren Ackerboden. Zeitschrift for Kulturtechnik und Flurbereinigung 29:4:234-244.
5. Vaughan, D. & B.D. Ord (1984): Soil Organic Matter - A Perspective on its Nature, Extraction, Turnover and Role in Soil Fertility. In: Vaughan, D. & R.E. Malcolm (eds): Soil Organic Matter and Biological Activity. Developments in Plant and Soil Sciences, vol. 16:1-35. Martinus Nijhoff.
6. Craul, P.J. (1992): Urban Soil in Landscape Design: 231-239, 314-320, 364-369. 396pp. John Wiley & Sons, Inc., New York.
7. Jauch, M. & P. Fischer (1993): Kompost im Substrat. Deutscher Gartenbau 46:2888-2891.

8. Nelson, E.B. & C.M. Craft (1992): Suppression of Dollar Spot on Creeping Bentgrass and Annual Bluegrass Turf with Compost-Amended Topdressings. Plant Disease 76:9:954-958.
9. Nelson, E.B., L.L. Burpee & M.B. Lawton (1994): Biological Control of Turfgrass Diseases. In: Leslie, A. (ed.): Integrated Pest Management for Turf and Ornamentals, rev. ed.: 409-427. CRC Press, Inc., Florida, USA.
10. Petersen, A.V. (1990): Anvendelse af kompost i grønne områder. (Application of Compost in Amenity Areas). 41pp. Arbejdsrapport nr. 40, Miljøstyrelsen, Denmark.
11. Petersen, M. (1981): Græsplæner. Principper og funktioner. (Lawns. Principles and Functions). p246. 362pp. A/S L. Dæhnfeldt, Odense, Denmark.
12. Bundesgütegemeinschaft Kompost e.V. (1994): Kompost mit Gütezeichen für den Garten- und Landschaftsbau mit Öffentlichem Grün und Rekultivierung. 24pp. Köln, Germany.
13. Jauch, M. & P. Fischer (1994): Kompost richtig dosiert. Garten+Landschaft 9:14-17.
14. Ligneau, L.A.M. & T.A. Watt (1995): The Effects of Domestic Compost upon the Germination and Emergence of Barley and six Arable Weeds. Annals of Applied Biology 126:153-162.
15. Fischer, P. & M. Jauch (1991): Grüngutkompost als Substratbestandteil bei Container kulturen. Baumschulpraxis 21:2:60-62.
16. Jauch, M. (1993): Versuche zum Kompost in Container Substraten. Qualitätskriterien für Kompost in Baumschulsubstraten. Taspo:127:13:8.
17. Craul, P.J. (1988): Assessing Soil for Urban Tree Survival. Grounds Maintenance Sep:6pp.
18. Corley, W.L. (1984): Soil Amendments at Planting. Journal of Environmental Horticulture 2:1:27-30.
19. Bundgaard, S., M. Carlsbæk, U. Juul & C.E. Jørgensen (1993): Jordbrugsmæssig værdi af produkter fra organisk dagrenovation. (Agricultural Value of Products from Biodegradable Household Waste). 90pp. Arbejdsrapport nr. 64, Miljøstyrelsen, Denmark.
20. Nielsen, L.K. (1994): Dyrkningsforsøg med kompost 1989-1993. (Growth Trials with Compost 1989-1993). 35pp. With English summary. Miljøprojekt nr. 258, Miljøstyrelsen, Denmark.

Organic Waste Recycling: A New Market with the Farmers

I. Coulomb, SITA, 94 rue de Provence, 75 009 Paris, France

ABSTRACT

Biowaste represents a potential of 52 to 63 MT/year of biomass, an equivalent of 50% of the total household waste, at the European scale. However biological treatment today is applied to less than 6% household waste on average. This situation seems to be a result of psychological perceptions by farmers with regards to use of recycled end-products from biological treatment. A reuse of organic waste: biowaste, green waste, sewage sludge, waste from supermarkets, from restaurants, from food industry, and/or agriculture still remains.

The group SITA and five European partners (Norsk Hydro, Degrémont, MAT (Müll und Abfalltechnik Gmbh), Lyonnaise des eaux, Agro-développement) have started a EUREKA program called VALORBIO to address research potential development of organic waste recycling markets. The program aims to determine requirements of agriculture and identify new trends, for example the need to integrate compost into fertilisation plans. Adaptation of processes to the characteristics of the waste and the definition of quality criteria should enable us to ensure development of the market share in agriculture.

This program is divided into three phases. The first phase aims at characterising the organic wastes (quantities and qualities) as defined by parameters that will be important for recycling into agriculture markets. In order to adapt the organic recycling process to specifically defined product quality standards, mixing of various bio-wastes has been tested. The second phase of the program will involve industrial trials of the ratios resulting from the initial phase of the program with an objective of developing an industrial plant in the third phase.

It is hoped that the results of the study will develop the partners knowledge, know-how and a common understanding of the needs of the farmers as this is the necessary preconditions to implement a successful program of recycling of organic waste in agriculture.

KEYWORDS

Compost, biowaste, organic waste characterisation, biological treatment, markets.

INTRODUCTION: THE LACK OF BIOWASTE TREATMENT IN EUROPE

Biowaste represents a potential of 52 to 63 MT/year of biomass at the European scale for a base of about 105 MT of household waste. Nevertheless biological treatment comprise less than 6% of the total household waste in the European union and up to 10% in Denmark [1]. The differences are due to the development of selective collection in northern Europe where citizens, by various regulations and incentives, are required to separate organic waste.

In other countries, like France, the recovery rate of organic waste remains low because composting methods vary greatly and the outlets for the produced compost are uncertain. In France, 600 000 tons of compost are produced from non-sorted household waste. But this amount is becoming more and more difficult to recycle the composts into agriculture markets. The standards for compost quality are very low, for example elevated heavy metals and plastic content limits which resulted in negative social perceptions. In effect, the market for compost produced from biological waste has virtually disappeared and compost is currently being stored until other outlets can be found.

This situation could be rectified by a change in attitude by the waste companies that have previously considered biological treatment of waste only as a way of treatment instead of a mean to recycle. Therefore the quality of the end-products (i.e. plastic content, heavy metals, fertilisers contents) and the agricultural requirements were not considered as necessary conditions for waste treatment standards. Agriculture's demand was neither identified nor satisfied.

COMPOST IN THE CHAMPAGNE VINEYARD: A HISTORICAL EXAMPLE

The history of composting in the Champagne Vineyard is an interesting example of how the market is affected by the quality of recycled products. The use of raw compost in this vineyard has decreased consistently over the last ten years and was completely stopped three years ago. Compost had not been used for fifteen years in planting because the region feared potential disease factors. Main Champagne companies were considered as leaders in agricultural operations and their decision to ban waste compost from their field was then followed by small vineyard farmers. The decision to ban compost stemmed from several reasons.

First, the compost looked bad because undesirable particles could be found in it, in particular plastic. This resulted in bad public relations and a marketing problem for this internationally well-known vineyard.

Secondly, mineralization of compost nitrogen was not controlled which led, in some cases, to an excess of plant vigour. Some consequences for example were to increase the risk of some diseases, like Botrytis, or a decrease in the quality of the wines produced.

According to CIVC ('Comité Interprofessionel des Vins de Champagne'), a professional association:

- compost material was not characterised thoroughly ;
- only minimal analyses was performed and end-users were not aware of them ;
- odour management was insufficient.

A media campaign 'Champagne is grown on a dustbin!' resulted in negative attitudes from potential compost users.

Further, since agriculture demand was linked to a struggle against erosion, a maximum soil cover was requested. As compost in thick layers presented toxicological disadvantages, farmers turned to a new product: barks. Fresh or composted barks coming from eastern and northern France were used in place of the compost produced from biowaste.

Bark, although convenient is very expensive. Therefore, the agricultural market is interested in a more cost-effective replacement. Given the negative story of waste composts in this region, a condition for a new development of the market will be to fit the quality requirements of the farmers while producing a cost-effective quality compost product.

VALORBIO: A PROJECT TO ADDRESS POTENTIAL DEVELOPMENT OF ORGANIC WASTE RECYCLING MARKETS

A situation that must be changed through market understanding and development

The historical reduction of compost market share does not preclude future development. Possibilities of recycling organic waste into agriculture markets still exist. Potential interest in sorting and treating organic wastes still remains: including biowaste, green waste, sewage sludge, waste from supermarket, from restaurants, from food industry or agriculture.

In order to reopen the market it is necessary to start from the bottom up and include the requirements of agriculture in the definition of the quality criteria for the compost end-products. These criteria should respond to environmental and market demands. Depending on local conditions several products will have to be designed. This new perspective implies that a closer relationships must be developed with the farmers.

VALORBIO a program which aims at developing closer relationships with agriculture

Improvement of compost quality is the primary condition necessary to reopen the market. This is why the group SITA and five European partners (Norsk Hydro, Degrémont, MAT, Lyonnaise des Eaux, Agro-Développement) have started a EUREKA program called VALORBIO. This project initiated, December 2nd, 1993 aims at developing biological treatment of organic waste in Europe and developing markets for the end-products resulting from treatment.

The partners represent the different parties involved in the recycling chain:

- Lyonnaise des Eaux, a water supply company who must find outlets for sewage sludge produced in its sewage water treatment plants ;
- SITA, a waste service company that must provide an organic program for its clients ;
- Degrémont and MAT (Müll und Abfalltechnik GmbH) who are involved in different types of treatment plant engineering and construction ;
- Agro-Développement who has implemented the recycling of sludge in agriculture and has a strong network in this field ;
- Norsk Hydro who produces fertilisers and who brings a down stream point of view through its working knowledge of agriculture fertilisation.

The program aims to characterise agriculture requirements and identify new trends. A base of waste composition is also hoped to be acquired. Processes will be adapted jointly based on compost requirements and waste streams to be treated. Control of waste stream input and end-product quality should then enable us to ensure development of the share of the market for fertilisers and soil improvers in agriculture.

VALORBIO planning

The program is divided into three phases.

The first phase aims at characterising organic wastes as defined by parameters that will be important for recycling the end-product in agriculture markets and thus aims at defining agronomic requirements.

Waste production and composition quality have already been determined for a range of waste streams.

Variations in waste streams have been analysed and compared to seasonal changes, regional conditions, waste producers (citizens, industry, landscape gardeners), collection systems, etc..

Composting trials have then been carried out to test the blending of these wastes and the effect on product quality.

Industrial trials and engineering of a multi waste treatment plant are planned for the second phase to verify the importance of input control in compost production.

It is hoped that optimisation and validation of the plant will be performed in the third phase of the project.

PROVISIONAL RESULTS: WASTE CHARACTERISATION

Selected Parameters

The composition of incoming waste influences the composting process and therefore the end-product composition. The parameters for input wastes were selected in reference to the process and required end-product characteristics.

Waste preparation is used to sort undesirable materials, crush and mix components in order to get the required water content, and aerate and limit the potential risks for agriculture use of end-products.

Fermentation and composting abilities depend on waste composition. Micro-organism are sensitive to their growth conditions. Their activity level varies depending on aeration, water content, nutrients, etc... Waste biodegradability can be determined by organic matter content, and ligno-cellulosic fractions content. Carbon, nitrogen, size of particle, water content, nutrient content (P, K, Ca, Mg..) all must be recorded because they affect both the process and the final quality of the compost produced [2-3].

The material then must be refined in a final step, composed of crushing and/or sorting by particle size to reach the required levels of quality.

Therefore, identification of waste streams, in particular undesirable materials and chemical composition are useful guides for selection of the types of wastes to be processed. A data base has been established to record the various parameters of wastes.

Product Quality, Marketing & End User Demands

Examples of useful results

Through trials, it has been demonstrated that collection of grey waste (residues after recyclables and biowaste collection) is not well adapted to the production of a high quality compost because as much as 55% of the mass input is contaminated with undesirable materials (i.e. plastic...). Such levels of contamination pose the same types of problems as one experiences when composting non-sorted household waste (Cf. fig. 1), namely poor compost quality with visible contaminants and high level of heavy metals.

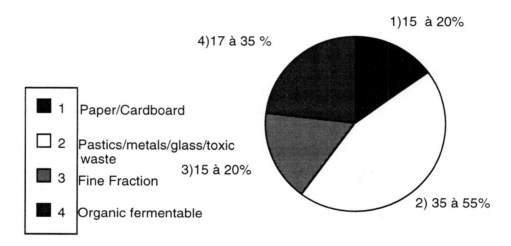

Figure 1: Grey waste composition

The content of kitchen waste (ref. in fig 1, organic fermentable) ranges between 17 and 35% of the grey waste which enables fermentation. In this case, biological treatment could be interesting as a stabilising process before final disposal but not as a recycling process.

Comparison of a collection where residents could mix kitchen waste and yard waste together and a collection where residents where told to put only kitchen waste in their waste bin, demonstrated that banning of yard waste from kitchen waste is not useful. In fact, preliminary results demonstrated that yard waste is already being mixed in with kitchen waste: in analysing the organic fermentable fraction of total kitchen waste, it was found that up to 50% of this material consisted of yard waste.

Chemical data are useful to predict final composition through simulation as well. In order to adapt the process to a required quality standard, mixing of waste was tested. Simulations and trials have been combined in order to produce well characterised compost. Risk for the environment could then be evaluated by comparing the compost with standards and regulations. For a compost based on 90% kitchen and 10% wood heavy metal content does not seem to be a problem (cf. Table 1)

Table 1: Simulation of heavy metal content in a biowaste compost based on 90% kitchen waste and 10% wood

mg/kg	Cd	Cr	Cu	Hg	Ni	Pb	Zn
Compost produced from 90% organic ferment-able, 10% wood	<9*	9	8	<1.5	9	21	106
French standard compost - Class A	<8			<8	<200	<800	
Netherlands MSW (compost proposition)	1,5	100	50	1,5	50	150	250

*Precision of analysis: 4 ppm

CONCLUSION

Today a minor part of the potential biowaste is treated biologically and recycled. One can expect, however, an increase in this treatment method if the process and products are improved. A 15%-20% recovery rate of organic waste is expected to be reached within the next ten years in France because organic waste will be banned from landfills and few other treatment method exist. Working with the farmers will be a key to success.

By combining a better knowledge of waste streams, processes, and more particularly mixing methods, market demand and compost applications should open the door to securing biological treatment and agricultural recycling market opportunities.

REFERENCES

1. ISWA (1995). Status and trends for biological treatment of organic waste in Europe, Ed. W. Rogalski and J. Charlton, Vienna. Austria.
2. Mustin (1987). Le compost, gestion de la matière organique. Ed. Dubusc.
3. Diaz et al. (1993). Composting and recycling municipal solid waste. Lewis publisher.

Consumer Demands

Jens Bo Holm-Nielsen, Institute of Biomass Utilization and Biorefinery, South Jutland University Centre, Niels Bohrs Vej 9, 6700 Esbjerg, Denmark

ABSTRACT

Consumers are farmers, land users in general, private as well as public, and citizens overall. The concept of centralized biogas plants has been developed in Denmark. At present 16 plants are in operation with capacities ranging from 50-500 tons of biomass per day. The biomass consists of approximately 75-80% animal manure co-digested with 20-25% organic waste from abattoirs and various food industries. For some plants there are programs to introduce sewage sludge and municipal solid waste. The digested slurry is returned as a sanitary and nutritionally declared fertilizer product, partly to farms that supply fresh manure to the digestion process, partly to crop farms in need of organic fertilizers. Environmental and agricultural benefits include savings for farmers as a consequence of improved fertilizer efficiency and thereby reduction of needs of chemical fertilizers. A side effect is less greenhouse gas emission and cheap, environmentally sound waste recycling.

KEYWORDS

Recycling, biological waste, consumers demand, anaerobic digestion, environmental chemicals, controlled sanitation, heavy metals.

INTRODUCTION

New methods for handling and distributing manure and slurry have been introduced in Denmark as a result of using modern technology in livestock production. A Danish intergovernmental biogas demonstration program started in 1987 [1,2].

Correspondingly, plans for recycling organic matter by distributing urban and industrial by-products and wastes after composting or treatment in biogas plants, are implemented in order to replace traditional dumping or incineration. This is a part of the Danish schemes to protect the environment and to promote better use of bioenergy resources.[3]

The objective here is to look into what controls the recycling of biowaste in Denmark in the middle of 90'ties and what concerns are expressed by consumers and farmers.

WHAT CONTROLS THE RECYCLING OF BIOWASTE RIGHT NOW?

There are many concerns among farmers and citizens as to how far to go in terms of recycling of as much as possible of biowaste from the entire society. Besides this, there is much public awareness on the same issues. Biowaste, in its different categories, needs safe guidelines to secure human health and environmentally sound recycling.[4]

The Danish Ministry of the Environment have made regulations concerning application of animal manure, sewage sludge and compost.[5,6,7] Some orders regulating the areas are to be reviewed. The Danish Veterinary Service, under the Ministry of Agriculture, sets standards and runs monitoring programs for controlled sanitation of digested manure from Centralized Biogas Plants.[8]

Advantages and possible disadvantages of recycling biological waste. Consumer demands and concerns at present time are mainly linked to pathogen risks, heavy metal pollution and persistent organic compounds in the environment.

- Digested manure is more homogeneous than normal animal manure and therefore easy to pump and handle.
- Nutrient contents are more well declared and precise to integrate in farming fertilizer planning systems than normal animal manure. Improved use of digested manure results in the possibility of replacing chemical fertilizer[9]
- .Less odour nuisance of digested products. Digested manure does smell the first few hours. It does not smell as traditional animal slurry, but of ammonia. Specila injection and drag hoses equipment used in growing crops minimize the problems, and the applied effluent infiltrates more easily the top soil.
- Weed seeds are killed through the thermophilic process.
- Anaerobic digestion processes in Biogas Plants can be controlled so that an adequate reduction of pathogens is achieved (table 1 and 2).
    - The Danish Veterinary Service recommends that all centralized biogas plants treat slurry at thermophilic temperatures, at least 53-550C for a minimum of several hours (table 2).
    - Regarding pathogens, sewage sludge and MSW (organic household waste) are considered to be higher risk waste compared with manure and most of organic industrial waste. When digesting sewage or MSW, the Danish Biogas Plants must meet the present standard of treatment at 70oC for at least one hour.
- Heavy metal contents have to be lower than environmental standards in recommendations from the Ministry of Environment of 1989, and new lower standards apply from July 1, 1995. Analyses of biowaste from industries are controlled by Regional County authorities. Farmers do not want contamination of their land by heavy metals (table III).
- The consumers do not want plastics or chemicals that are not completely degradable. Persistent compounds and risk of leaching are increasing concerns. This area is still very open, but artificial oestrogenic and related compounds create increasing awareness. Consumers want safe products to be spread. It is a matter of soil and soil-water safety, and production of food and feedstuffs with no risk.

'The growing number of reports on environmental contaminants that possess estrogenic activity, presents the working hypothesis that the adverse trends in human male reproductive health may be, at least in part, associated with exposure to estrogenic chemicals, during fetal and childhood development. The reproductive health trends in men are consistent with this hypothesis. While exposure levels to estrogenic chemicals are not at all well known for humans, the large number of chemicals in numerous environmental categories suggests adequate availability'.[10]

For example, chemicals reported to be estrogenic include, but are not limited to:

- Organochlorine pesticides.
- Polychlorinated biphenyls (PCBs).
- Dioxins and Furans.
- Alkyl phenol polyethoxylates.
- Phytoestrogens.
- Other Xenoestrogens.

Taken into account the end user's concern regarding this high risk group of organogenic chlorinated compounds, it is evident that biowaste has to be well declared, before use for recycling, to protect both human health, and the environment.

A detailed total analysis system for all potential chemicals and compounds will be illusory, but consumers can make demands for certain ecotoxicological tests and screening of biowaste.

New standards for controlled sanitation

Controlled sanitation is defined as follows in the newest Danish recommendation [6] from 1995:

A) Treatment in reactors, which secure a temperature at minimum 70^-C, for minimum 1 hour. The treatment has to be documented by registration of time and temperature. The biomass product has to fulfill the following demands before delivery to the end user:
   - Salmonella may not be present.
   - Streptococcus faecalis to be lower than 100/g.

B) Treatment in an anaerobic digestion reactor at thermophilic digestion temperature by normal retention time (10-15 days), and with a guaranteed minimum retention time of the following combinations:
   $58°C$, minimum 4 hours
   $57°C$,   -   6 hours
   $56°C$,   -   8 hours
   $55°C$,   -   10 hours
   $54°C$,   -   12 hours
   $53°C$,   -   14 hours
   $52°C$,   -   16 hours

Table 1: Hygienic conditions for the use of biowaste belonging to various categories.[7,8]

Waste Category	Not Treated	Stabilized	Controlled Composting	Controlled Sanitation
CATEGORY A Sludge etc. Vegetable Production	+	+	+	+
CATEGORY B Sludge etc. from Fresh Water Fisheries	May not be used in horticulture	+	+	+
CATEGORY C Sludge from Animal Production	May not be used for agricultural purposes 3)	To be injected directly before 12 hours. May not be used in horticulture.	+	+
CATEGORY D Segregated Household Waste etc.	May not be used for agricultural purposes.	Not for consumable crops or horticulture. To be injected before 12 hours.1) 3)	1)	+
CATEGORY E Effluent Sludge	May not be used for agricultural purposes 3)	Not for consumable crops or horticulture. To be injected before 12 hours.2) 3)	Not for consumable crops or horticulture 2)	+

+) May be used without any hygiene motivated restrictions.
1) At farms with cloven-hoofed animals compost may be spread on bare soil if subsequently covered.
2) At areas where sewage sludge has been applied, up to one year after the last supply, only grain or seed crops until ripe may be grown, and grass etc. only for industrial production of dry fodder. Further, non-edible crops may be grown. It is forbidden to grow potatoes, grass and maize for forage production, fodder- and sugar beets. For at least six months after spreading sludge in a forest, the latter must be kept closed to the public. Picking of berries, mushrooms, or similar should be forbidden by signposting. Storage, transport, spreading and injection equipment shall be cleansed properly immediately after use.
3) Until first of July 2000 not-treated sludge from animal production can be used, if it is incorporated immediately after spreading. Until first of July, year 2000 non-treated effluent sludge can bee used as well, when it is injected 10 cm below ground surface, and the rules in no 2) have to be fulfilled at the same time.
The demand of injection of stabilized segregated household waste and effluent sludge starts the first of July, year 2000. The products must not be used in horticulture.

The biomass product has to fulfill the following demand before delivery to the end user:
- Salmonella not to be present.
- Streptococcus faecalis to be lower than 100/g.

C) Treatment by adding lime, so that all the material reaches pH 12 for minimum 3 months. The treatment has to be documented by registration of time and pH measurements, representative of all the treated material. The product has to fulfill the following demands before delivery to the end user:

- Salmonella may not to be present.
- Streptococcus faecalis to be lower than 100/g.

Table 2: Analysis parametesr and limit values on heavy metals [6]

	Mg pr kg dry matter	Mg pr kg phosphorus total
Cadmium [1]	0.8	200
Mercury	0.8	200
Lead [2]	120	10,000
Nickel	30	2,500
Chromium	100	
Zinc	4,000	
Copper	1,000	

Analysis parameters: cadmium, mercury, lead, nickel, chromium, zinc, copper, dry matter, phosphorus total, nitrogen total.

[1] Ashes containing up to maximum 3 mg cadmium pr kg dry matter, can be used with agricultural purpose, with County council permission. Concerning ashes containing between 0,8 and 3 mg cadmium pr kg dry matter, the use of this matter is limited in a pursuance of section 15, pkt. 15 of Government order.

[2] Detected quantity for lead is 60 mg pr kg dry matter for horticulture. For application in horticulture is the detection limit of 25 mg lead pr kg dry matter further more available.

CONCLUSION

Implementation of already available knowledge on biowaste recycling can prevent further deterioration of the environment. Local/regional socioeconomic pressure together with lack of information and concern frequently prevent the society from obtaining the ideal balance between human activity and a healthy environment. The ideal situation will be to recycle as much as possible of biowaste to farmland based on safe guidelines.

REFERENCES

1. Update on Centralized Biogas Plants, Danish Energy Agency, October 1992.
2. Progress Report on the Economy of Centralized Biogas Plants, Danish Energy Agency, February 1995.
3. Centralized Biogas Plants Combine Agricultural and Environmental Benefits with Energy Production. S. Tafdrup, Danish Energy Agency, 7th International Symposium on Anaerobic Digestion, 23-27 January 1994, Cape Town, South Africa, pp 460-468.
4. Safeguards against Pathogens in Danish Biogas Plants. H.J. Bendixen, Danish Veterinary Service, 7th International Symposium on Anaerobic Digestion, 23-27 January 1994, Cape Town, South Africa, pp 629-638.
5. Statutory order no. 736 of October 26, 1989, on application of Sludge, Sewage and Compost etc. for Agricultural Purposes. Ministry of the Environment, Denmark, National Agency of Environmental Protection, Translation LK, November 1990.
6. Statutory order no. 730 of September 5, 1995, on application of Sludge, Sewage and Compost etc. for Agricultural Purposes. Ministry of the Environment, Denmark, National Agency of Environmental Protection.
7. Statutory order from the Ministry of the Environment no. 1121 of December 15, 1992 on Professional Livestock, Livestock Manure, Silage etc. Ministry of the Environment Danish Environmental Protection Agency, Translation LK, September 1993.

8. Safeguards against Pathogens in Biogas Plants. Practical Measures to prevent Dissemination of Pathogens and Requirements for Sanitation. H.J. Bendixen and S. Ammendrup, The Danish Veterinary Service, Veterinary Research, Monitoring and Consulting on Establishment and Operation of Joint Biogas Plant, 1992.
9. Joint Biogas Plant Agricultural Advantages - circulation of N, P and K. J. B Holm-Nielsen, N. Halberg and S. Huntingford for the Danish Energy Agency, March 1993.
10. Male Reproductive Health and Environmental Chemicals with Estrogenic Effects. Environment-project no. 290. Ministry of Environment and Energy, Danish Environmental Protection Agency, 1995.

Section 9

Status & Opinion

Urban Biodegradable Wastes - Status and Opinion

Jens Aage Hansen, Environmental Engineering Laboratory, Aalborg University, Sohngaardsholmsvej 57, DK-9000 Aalborg, Denmark

ABSTRACT

Biological treatment of separately collected wastes is on the increase in Europe while landfilling of bulk wastes is being reduced. Anaerobic treatment and methane production is interesting where gas yields and energy prices are high. Aerobic treatment could be limited by still more restrictive air pollution and odour control as well as a non-established market for the end product. In both cases, end product quality and producer responsibility are issues to be resolved to obtain a better basis for operation, control and decision-making regarding new plants. Ecological farming seems to possess a great potential for end use of good quality urban biodegradables.

KEYWORDS

Biological processes, biodegradables, collection, occupational health, ecological farming, producer responsibility, end product quality.

INTRODUCTION

Farmers and villagers in India have been composting for centuries. All over the world house owners are practising composting in their backyard. They appreciate the disappearance of their organic wastes and the appearance of compost for use as a soil conditioner and having a certain fertiliser value. Farmers and villagers in China have for many years produced biogas from excreta and hog manure in order to provide fuel for light, cooking and heating; as well as basic fertilisation and conditioning of the arable soil by application of the degasified slurry from the collection tank. Biological conversion of organic wastes, even in reactors made for that purpose, is therefore a human activity of ancient design and knowledge.

So why was biological treatment in the 1970'ies at a record low activity level where less than 1% of the European household wastes generated were treated biologically? At least one reason could be the then cheap alternative of direct landfilling of the wastes collected from consumers. Why bother to operate a relatively costly biological process and end up with a product, e.g. compost, the quality of which was questionable due to high levels of heavy metals and risks of communicable diseases due to pathogens?

However, a change has occurred. Wannholt [1] shows that in 1994 in Europe the installed capacity for biological waste treatment is about 4% of household waste generated. Further, most of this capacity has been installed during the last five years. So clearly, a new and significant trend is observed where the relative increase of installed capacity is higher for biological than for other treatment, e.g. incineration. This means that there is a net rerouting of urban organic wastes away from direct landfilling and into biological processing.

It is not so obvious how the product from biological processing will finally be used or disposed of. The increase in biological processing is due rather to a need for an alternative to landfilling than to a demand for the product. Nevertheless, in many reports on new projects and plants for biological treatment it is clear that attempts are being made to produce an end product with an acceptable quality for a specified usage, e.g. soil conditioning in vini- or agriculture.

Collection, processing and routing of urban organics to land areas for use as a soil conditioner or fertiliser is the topic of this status and opinion paper. Organics from households, catering and other service industries as well as park and garden wastes are included. In order to limit the scope, sludge from wastewater treatment, hazardous wastes and agricultural (farm) wastes are mentioned only where relevant for comparison or comment.

The term *organic* is often used to describe *waste meant for biological degradation*, i.e. biodegradable waste. However, a lot of organics are non biodegradable or not biodegradable within realistic periods of time and the term is therefore not relevant, where biodegradable wastes is the topic matter. The use of the word organic should be limited and use of the more specific term *biodegradable* favoured. The reader should accept that organic is used incorrectly on a number of occasions, even in this *status and opinion* contribution.

The primary background for this status and opinion is the information provided by the authors of the other chapters of this book. Where directly quoted these authors are identified by name in parenthesis, e.g. (Kristensen).

LEGISLATION AND REGULATION

Less waste production, more recycling and less landfilling seems to be the guiding principles for decision making on waste management in several countries, e.g. Germany, Sweden, Austria and Denmark. Use of the residual products from biological treatment is normally considered recycling, which means that biological treatment of organic wastes is relatively high on the political priority list.

The limited use of biologically processed organic wastes indicates that problems exist, some of which are of legal and administrative rather than technical nature.

- *Inconsistent quality criteria* exist for different products applied to land; e.g. in the same country different acceptance criteria are applied for sewage sludge, compost from urban organics and farm slurry or manure.
- *Criteria are different* from one country to another for the same product and receiving soil. This could apply to product quality as well as dosage rate. At best this can be used as inspiration to improve criteria where they are inadequate; at worst it creates insecurity as to the validity of the criteria. In the latter case the end users will prefer tested and known alternatives.

- *Producer responsibility* seems only vaguely or not defined in most legislation and regulation. This means that the end user runs a non specified risk when accepting a product for use on his land. This makes the whole marketing of end products at jeopardy, because a mutual basis for agreements between producers and users is missing.

There is a need for administrators and legislators to consider these issues, if the use of processed organics is to be increased according to the political goals to increase recycling of substances, including urban organics.

At the same time it is important to consider new developments that will influence the recycling and reuse of organics. As an example, ecological farming is on the increase and implies strict regulation regarding quality of the products that can be used and bans on those that cannot. As seen (Knudsen), the farmers will not accept one set of strict rules for ecological farming and a less strict set for others. Such acceptance would imply a stamp of inferior quality to conventional farming.

Another example is the German concept of Closed Substance Cycle Waste Management (Schenkel). It means that balanced input/output should be established at all levels to the extent possible. But difficulties are easy to identify; e.g. regarding imports/exports over the national or regional borders; or related to a city; or related to an area of land where organic end products are to be used. Legislation and regulation is not yet geared to manage this challenge.

There is a need to cope with these legal and regulatory issues. Much could be done by international bodies such as ISWA, the International Solid Wastes Association, or regional bodies such as EU, the European Union, to secure good dissemination and analysis of existing information as well as trends and developments in the field of organic waste management. At the bottom of this evaluation lies the question of wastes that cannot pass borders or products that can. Given sufficiently strict and mutually agreed product quality criteria it would seem illogical to allow free trade with primary products and no trade with secondary (recycle) products. That would often work against the "closed substance cycle concept".

COLLECTION

In order to limit landfilling and increase reuse and recycling, source separation of wastes becomes an interesting challenge to citizens and managers. Based on end-product and process quality criteria, the wastes are categorised in different fractions that are kept separate or only intentionally mixed during collection and processing. E.g. food waste is wrapped and kept separated from other wastes from the kitchen to the processing site.

In the past mixing of wastes during collection took place to minimise collection costs and increase convenience to the system users. But increasing treatment costs and reduced capacity for disposal makes it necessary to reconsider what is optimal in terms of the total system economy. Also, the system users may be ready to sacrifice some of the traditional convenience and contribute some of the labour necessary, if they feel convinced that such behaviour is to the benefit of the environment.

Food waste separation is important with regard to compost or biogas processes in order to secure a well defined end product with acceptable quality criteria for land use. The separation secures against heavy metal or contamination from other waste categories; and these are in turn easier to handle and process in the absence of the putrescible organic matter. Food waste separation is practised increasingly (Goldschmid, Herbst, Domela, Nilsson, Kristensen), but there is a gap between potential and actual amounts; e.g. only 30-50% of the potential amount are reported as results from several full-scale, long-term collections in European cities. It seems that behaviour is to be further studied and changed in order to improve participation and efficiency of collection. Separate collection is reported (Nilsson) to be more expensive, but a higher degree of factual participation by the users would improve economy and overall performance of the waste management system.

Source separation has induced a new development in collection system and receptacle design (Domela, Kristensen, Herbst, Goldschmid and others). Some of these designs include improved services to the users, e.g. house owners, as well as improved ergonomics and health protection to workers. The functional and visual design of some of the equipment is of very high and commendable quality.

Home or local composting of vegetable organics from kitchens and gardens combines the issues of reduced collection and increased recycling. It contains a significant element of local user participation that would be in line with some of the "green city" and "city ecology" projects that are locally established in many cities; e.g. involving composting of organic wastes, reduced water consumption, local reuse of waste waters, local use of rain water, reduced energy consumption, use of solar energy, etc. For that reason home or local composting will stay on the urban waste management agenda. It will, however, require ad-hoc consideration in terms of public health, e.g. control of communicable diseases, pests and allergies in densely populated cities.

ANAEROBIC PROCESSING

Energy recovery is a primary goal in anaerobic processing of organic wastes. Some of the organic carbon is transformed into carbondioxide and methane (CH_4), the latter being a useful and valuable gas, e.g. 35 MJ or 10 kwh per m3 of dry CH_4.

It seems that the efficiency in producing biogas varies widely, cf. the range 50-700 m3 CH_4/t VS (volatile solids) reported in papers in this book. The 700 refers to carefully controlled laboratory scale experiments that are not realistic for full scale operation. The results reported for full scale are in the range 50-150 m3 CH_4/t VS, but the numbers are poorly documented. Based on stochiometric calculation the potential biogas production would be app. 400 m3 CH_4 per ton of cellulose and 1100 m3 CH_4 per ton of rape seed oil. Generally the yields reported in this book are low compared to these potentials when transforming organics into biogas. Lack of feed characteristics and inefficient process control may explain some of this.

The feed is variable in volatile solids. E.g. slurry from ruminants will be relatively low compared to wasted food organics from a kitchen. So far the biogas plants receive what comes as comes and little is done to characterise the feed, categorise and mix it intentionally, and then use it as controlled input to the reactor. Often the yields are given in m3 gas per ton of feed, not specifying neither dry matter contents nor volatile solids; not to speak of fractions of carbohydrates, lipids or proteins or species thereof. Better feed characterisation is necessary to improve biogas plant performance.

Process control is often limited and relies more on remedial action when problems occur than preventive measures. This is related to the absence of relevant information on the feed characteristics. But it is also due to the non-established experience in operation of digesters under variable feed conditions. Models of the processes and the systems in which the processes take place are necessary to improve this situation. Some initial steps have been taken (Angelidaki and Pavan) to identify key process parameters and use them in models. But there is a long way before the plant operators will possess a tool to improve the performance of their plants.

Revenues from energy production differ from country to country, e.g. in the range 0.2-0.8 DKK/kwh. This may - together with the widely differing gas yields - explain why anaerobic digestion is on the increase in areas where both gas yields and energy prices are high. The span in revenue per ton of feed may be from a low 5 to a high 280 DKK per ton of dry matter fed into the reactor, given the above assumptions on yields and prices. Such variation in revenues could easily change the budgets from negative to positive balances and hence the decisions regarding anaerobic processing of organic wastes.

Biogas processing takes place in closed reactor systems in order to secure anaerobic conditions. This implies relatively high investment costs but also a working environment with good possibilities of control of odours, aerosols, dust and direct contact between operator and waste. The capitalisation of these reduced risks should be part of the economic assessment of biogas plants.

When treating bulk wastes, e.g. agricultural slurries or separately collected urban organics the effluent after the anaerobic digestion is not much reduced in weight or volume; only gaseous compounds have been removed. A slurry remains and land application is so far the major option. However, access to land becomes more and more restricted, i.e. acceptable only during shorter periods and requiring soil injection as well as adjusted dosages of nitrogen, phosphorus, potassium and limited amounts of pathogens and toxic compounds. All this necessitates increased product storage capacity, e.g. equivalent to 6-9 months of production and advanced equipment for soil application.

There seems to be a need for new technologies to separate and dewater these slurries to increase flexibility in production, storage and applications. Efficient dewatering may open for use of the dewatered residuals as fuels, directly or indirectly, but it may also facilitate a more flexible land application, if liquid and solid phases can be applied independently. In any case, a thorough characterisation of the end products with regard to fate in the soil after application is necessary. Future land use of digested slurries will be subject to better characterisation of the product as well as specific dosages at specified times during growing seasons and regulated according to crop, soil type and state of fertilisation.

Ecological farming may be a particularly interesting niche or area of future use of treated urban organics, because only organic conditioning and nutrient supply is permitted while chemical fertilisation is banned. Recycling of urban organics is in line with the ecological farming concepts, but it requires a clean and well characterised product in terms of organics, nutrients, pathogens and potentially toxic compounds.

The work in recent years on anaerobic digestion of organic wastes has focused too much on energy recovery and too little on end product quality. Cleaner technologies in industry and preventive measures in cities and agriculture may reduce the contamination of the organic wastes and the residuals after treatment. But the bulk flow of organics and nutrients through animals and humans cannot be changed drastically. Future management of these residuals requires an improved understanding of soil processes with emphasis on organic matter/nitrogen interactions, and balanced input/output accounts for other substances for local as well as regional areas.

AEROBIC PROCESSING

Aerobic processing or composting turns putrescible organics, including residues from anaerobic processing, into a biologically relatively stable end product called compost. Its physical characteristics are determined by the type of technology and operation; the chemical characteristics are a result of the feed composition and the process operation. It is produced for use on land, e.g. in gardens, vineyards or agriculture.

Successful composting implies a significant volume and mass reduction of the feed; typical for the putrescible organics from household wastes would be a volume reduction of 50% and a mass reduction of 60%. The process is extremely robust in terms of accepting a wide range of the feed C/N and moisture ratios and still - given time and good management - delivering the desired end product.

Composting is energy consuming, e.g. 30-50 kwh/t of feed material. Heat is developed during the processing and can when properly managed lead to time and temperature conditions that will render the end product free of pathogens. Heat recovery - e.g. using heat pumps - is feasible but does not seem economically justified.

The energy consumed is used to secure the oxygen supply throughout the composting body of organics. The mass reduction is due to the release to the atmosphere of mainly water, carbon dioxide and - depending on the process control - ammonia.

Scientifically composting is poorly reported and described. But it is well documented as an art and a practice in which both highly professional managers and simple garden amateurs can participate. It is feasible in industrialised as well as developing countries and in hot as well as cold climates.

So what is the problem? Given a non-contaminated (e.g. low heavy metals and toxic organics) feed material there need not be problems in producing a compost that is free of pathogens, can be stored and eventually used on land. However, when considering composting as an option for biological treatment of organic wastes a number of concerns seem relevant.

- Nitrogen may be lost to the atmosphere as ammonia during processing and storage of compost. Limitation of this loss requires sophisticated process control and technology and is one reason for in-house rather than open-air composting.
- As most commonly practised, composting involves a lot of turning of the material which in turn may create occupational health (respiratory tract or allergy) problems. Their prevention and control, in-door and outdoor, may involve more costly installations, technology and control in the future. Cf. the section in this book on occupational health. The increased concern for occupational health is relatively new and it remains to be seen to which extent the risks are real and what consequences it will have on organic waste management, including composting.
- Odour control may be important, depending on site location and local regulations. Odour and occupational health control are likely to be combined issues in most waste management cases, including composting as a vulnerable case if not carried out in closed systems.
- Cost of composting could easily increase as a result of increased demands on nitrogen, health and odour control. And only revenues from sale of compost are available to compensate for these costs. Therefore, better assessment and documentation of benefits are needed for compost used as a soil conditioner and fertiliser for given crops, soils, seasons and compost characteristics. Without such documented benefits the odds for more composting are poor.
- If a "clean" compost product can be guaranteed and beneficial soil and fertiliser effects proved there seems to be a new niche for use of urban compost in ecological farming. In fact, ecological farming is based on the concept of closed cycling of organic matter and nutrients, and the urban organics originating from farmed crops and manufactured food belong to this recycling system. Because ecological farming will not accept chemical fertilisers they are sure to be short of nutrients and would therefore - subject to clean end products - be an existing and non saturated market for compost, or otherwise processed urban organics.
- Home and local composting may be relevant and locally important activities. They will not, however, significantly reduce the amounts of compost potentially available to farmland from urban areas.

PRODUCT QUALITY

End user demands must be the starting point for establishing quality criteria for products from biological processing of organic wastes. This has obviously been forgotten for some time where feasibility and cost of collection and processing were major criteria in choosing waste management systems. Product quality criteria with regard to end use possibilities is now trendsetting for better management of urban organic waste. This concept relates to the source as well as the sink of the product.

Source separation and pre-considered, intentional mixing of specific waste fractions during processing now seems an accepted and effectively working procedure to achieve a better and more controlled end product quality. It applies to all types of processing of organic wastes, including anaerobic as well as aerobic processes. An alternative or supplementary procedure seems to be pre-treatment, e.g. through wet separation prior to biological processing (Scharff & Oorthuys).

Land is the natural sink for the biologically processed organic waste products. Their fertiliser value is limited according to product analyses as well as measured crop yields in growth experiments. In traditional farming compost and digester residuals are normally considered low value fertilisers compared to chemical trade products. Although there is an inherent slow release of nutrients from organic products, the nutrient value is limited and deemed insignificant in traditional farming, in part because of the much higher cost of spreading a unit of N, P or K in compost as compared to high strength mineral fertilisers. These judgements are, however, based on short term plot experiments where the alternatives are commercially available chemical fertilisers versus compost or digester residues. On this background and considering the recurring discussion on erosion and disappearance of soil organics it seems relevant to reconsider existing information on the application of waste organics to soil. It may be necessary to supplement the existing information on soil conditioning and fertilising effect of organic waste products. And long term experiments may need be established for an improved and more valid assessment. Additionally, the difference in short-term and long term effects of different products should be investigated and assessed, including various origins of the waste organics (e.g. food or garden waste) and various treatment processes (e.g. composting or anaerobic digestion).

In general, where organic farming is the targeted activity, urban organics may find a niche for utilisation where capacity exceeds potential production. This presupposes fulfilled quality criteria but applies equally to compost and digester residuals.

MARKETING, MARKETS AND PRODUCER RESPONSIBILITY

'You cannot sell biowaste, but certainly organic products of high quality. What is technically feasible at your treatment plant is secondary to the issue of specific end user demands'.

What the marketing and media people tell in this message is that in the past the collection and treatment was too often seen as the challenge in terms of optimising waste management. The planners, engineers and managers obviously missed the point of asking the end users about their needs and demands.

Possibly this point should be readily accepted. But additionally, the perception of the human health and environmental quality issues has changed dramatically over the last two decades.

What used to be good practice in production, transportation and consumption is now questioned by the accumulated effects of e.g. heavy metals in human and animal organs. There is a need to change practice and improve technology to solve these inherited problems. But also new chemically or microbiologically produced compounds and products need careful pre-assessment of life cycles and impact on health and environment. And it seems that even with careful pre-assessment there will be a need for corrective and remedial action when things develop differently from the planned scenarios. Therefore, risk assessment and producer responsibility will be new and challenging issues. This is not particular to marketing of urban organics as a high quality product, but it is essential also in this context where the organic wastes of the society are returned to the land where crops are grown for food, feed or other industrial purposes. Risk and responsibility are becoming still more important issues with increasing economic activity and at both local and global scales.

To accomplish marketing of high quality fertiliser and soil conditioning products it may be necessary to be more strict about the specific routing of substances and products. Source separation, in the households, catering business etc., has been repeatedly mentioned as an example of sensible consumer participation and potentially improved and-product quality. By source separation cross contamination is avoided during collection and treatment of organic wastes. However, source separation should not be taken for an effective mantra or panacea. Reports exist to tell that consumers, by incident or default, misuse their bags or bins. In such cases the end products could be contaminated and the recycling concept made obsolete.

Marketing people also tell that end users are willing to pay the price of a good product. Examples of such practice are few, so far, because marketing of high quality urban organic products is not yet commonplace. Therefore, for a long time to come, systems for collection, treatment and final use of urban organics must be based on relatively high fees to the waste producers and not on revenues from sale of products.

Generally for reuse and recycling of urban organic wastes there is much to be done in studies on consumer behaviour, motivation and willingness to pay. Similarly, studies are needed on fee structures, e.g. penalties, charges or rewards. Effectively operated systems will depend on such socio-economic tools to be developed in parallel with new cleaner technologies and better use of the existing processes and infrastructures.

Producer responsibility is to be thoroughly analysed as concept and management tool. As pointed out (Knudsen) this is the core issue in reaching an agreement between Danish farmers and municipalities regarding land use of sludge. But the issue is far more general and is sure to enter future management schemes where residuals from one party, e.g. a city, is to enter the premises of another party, e.g. the farmer, as a collective or a single judicial person.

Most certainly, researchers in both technical and social sciences need get together and prepare the ground for new organic waste management.

RECOMMENDATIONS

The authors of the other contributions in this book have made specific and concrete reports on management of urban biodegradable wastes. Their results and proposals for practical application or supplementary studies are to be considered in their own context, but they also serve as a basis for the following observations and recommendations.

Separate collection

Separate collection of organics is now an established practice in several cities in Europe and is planned in more to follow. This development is relatively new and results are limited. Product quality, end user acceptance, soil and crop effects, costs and revenues, and economies of scale are all interesting aspects for which only scattered and inconclusive results exist. There is a need for both medium and large scale testing over several years to assess successes as well as failures in source separation, separate collection and treatment of biodegradables. But separate collection seems to be an important step towards more and cleaner recycling as well as reduced landfilling.

Occupational health

This needs extra consideration if the number of manual operations with waste is increasing, e.g. as a result of source separation and more recycling. Risks of exposure to pathogens, spores, dust or smell need new assessment when new systems are taken into operation. Waste producers, collectors and end-product users need be taken into account. In a few countries, e.g. Denmark and Norway, occupational health is considered an important issue and intensive monitoring and research is carried out. Occupational health problems are most obvious around plants where the wastes are unloaded and treated. They could serve as an interesting topic for comparative studies in different countries and for different collection and treatment systems. The result could be a more efficient, more representative and cost effective assessment of occupational health measures relevant to management of biodegradable and other wastes.

Processing of organics

The scientific background for choosing or operating a process or a technology for treatment of biodegradables seems poor, according to the reports in previous chapters of this book. The number of new plants in Europe for compost as well as biogas is nevertheless on the increase and would permit studies to obtain much better rationales for selecting processes and operating plants. Based on combinations of laboratory and full scale studies this could lead to improved decision making; taking into account occupational health, process control, product quality and economy.

Product quality

There is - at least in Europe - reason to harmonise the quality criteria for end-products arising from biological treatment of organic wastes. If there is to be free trade with goods, e.g. food products grown in agriculture, there is also a need to apply commonly agreed quality criteria regarding what is added to the soil as fertilisers or soil conditioners. If the same quality criteria are agreed and applied, the processed waste end products should flow over borders as do the goods and primary products.

Producer responsibility

This must be more clearly addressed in legislation and regulation, both nationally and internationally. Only then can producers end users of processed waste products establish the necessary agreements and share liabilities in a legally acceptable manner.

Ecological farming

This farming concept (no pesticides, no chemical fertilisers) is of interest to those who produce, collect and treat biodegradable wastes. It identifies an interesting market for the end-product, in particular in a time where the demand for ecologically grown products is on the increase. Recycling is enhanced when food organics are returned to agricultural land, which is in accordance with the ecology concept. And on the overall, the demand for the organic end-products is going to exceed the production, because losses in the cycling of nutrients and organics are inevitable.

ACKNOWLEDGEMENT

Much of the information behind this contribution comes from the participants of the BIOWASTE '95 conference in Aalborg, May 21-24, 1995, including the many authors whose papers were accepted for presentation. The inspiring and lively discussions during each session contribute, as do the reports from each session, carefully prepared by individual reporters. I am grateful for inspiration and help and hope that some added value is found in this status and opinion.

REFERENCES

1. Wannholt, L. (1994) Biological Waste Treatment in Europe, RVF Report no 94:5, Swedish Association of Solid Waste Management, Malmö, Sweden.

Authors who are referenced by name in parantheses and without year of publication are found in other chapters of this book.

Author Index

Ahring, B.K	132, 142, 281	Lund, B.	281
Angelidaki, I.	132, 142	Maille, H.	295
Arndt, M	172	Malmros, P.	106, 115
Battistoni, P.	125	Mark, C.	91
Bendixen, H.J.	281	Mata-Alvarez, J.	125
Bjerre, A.B.	209	Mathrani, I.	132
Breum, N.O.	98	Messner, K.	91
Brinkman, J.	181	Mingarini, K.	65
Carlsbaek, M.	305	Mochty, F.	9
Cecchi, F.	125	Musacco, A.	125
Christiansen, J.U.	106	Nielsen, B.H.	98
Coulomb, I.	317	Nielsen, E.M.	98
Dalemo, M.	65	Nielsen, L.K.	237
Ellegaard, L.	142	Nilsson, P.	35
Fernqvist, T.	209	Nybrant, T.	65
Frostell, B.	65	Oorthuys, T.	153, 262, 272
Fuchs, L.	253	Paris, I.	295
Garcia, H.	132	Pavan, P.	125
Gautam, R.	201	Petersen, J.	287
Hack, P.J.F.M.	181	Plöger, A.	209
Hansen, J.AA.	331	Poulsen, O.M.	98
Hansen, J.C.	106	Reeh, U.	25, 305
Hauer, W.	42	Rilling, N.	172
Have, P.	281	Rogalski, W.	82
Herbst, W.	54	Scharff, H.	153, 262, 272
Hochrein, P.	162	Schenkel, W.	1
Hofman, A.	193	Schmidt, A.S.	209
Holm-Nielsen, J.B.	323	Schwinning, H.-G.	253
Hoppenheidt, K.	228	Shin, H.-S.	218
Hwang, E.-J.	218	Sigsgaard, T.	106
Jeong, Y.-K.	218	Six, W.	193
Jomier, Y.	295	Sonesson, U.	65
Jönsson, H.	65	Stegmann, R.	172
Karki, A.B.	201	Sundqvist, J.-O.	65
Knudsen, L.	17	Thyselius, L.	65
Kristensen, K.H.	76	Tränkler, J.	228
Lindberg, C.	245		

C628 /Han

until 5/12/02

This book is due for return on or before the last date shown below.

2 2 DEC 1999

-4 DEC 2002

1 3 JAN

WITHDRAWN

Don Gresswell Ltd., London, N.21 Cat. No. 1208

DG 02242/71